Zariskian Filtrations

K-Monographs in Mathematics

VOLUME 2

This book series is devoted to developments in the mathematical sciences which have links to *K*-theory. Like the journal *K-theory*, it is open to all mathematical disciplines.

K-Monographs in Mathematics provides material for advanced undergraduate and graduate programmes, seminars and workshops, as well as for research activities and libraries.

The series' wide scope includes such topics as quantum theory, Kac-Moody theory, operator algebras, noncommutative algebraic and differential geometry, cyclic and related (co)homology theories, algebraic homotopy theory and homotopical algebra, controlled topology, Novikov theory, transformation groups, surgery theory, Hermitian and quadratic forms, arithmetic algebraic geometry, and higher number theory.

Researchers whose work fits this framework are encouraged to submit book proposals to the Series Editor or the Publisher.

Series Editor:
A. Bak, *Dept. of Mathematics, University of Bielefeld, Postfach 8640, 33501 Bielefeld, Germany*

Editorial Board:
A. Connes, *Collège de France, Paris, France*
A. Ranicki, *University of Edinburgh, Edinburgh, Scotland, UK*

The titles published in this series are listed at the end of this volume.

Zariskian Filtrations

by

Li Huishi

Shaanxi Normal University,
Xian, People's Republic of China

and

Freddy van Oystaeyen

University of Antwerp, UIA
Antwerp, Belgium

KLUWER ACADEMIC PUBLISHERS
DORDRECHT / BOSTON / LONDON

Library of Congress Cataloging-in-Publication Data

Li, Huishi.
    Zariskian filtrations / by Li Huishi and Freddy van Oystaeyen.
        p.   cm. -- (K-monographs in mathematics ; v. 2)
    Includes bibliographical references and index.

    1. Filtered rings. 2. Filtered modules.   I. Oystaeyen, F. van,
    1947-   . II. Title. III. Series.
    QA251.4.L5   1996
    512'.4--dc20   96-32738

ISBN 978-90-481-4738-0

Published by Kluwer Academic Publishers,
P.O. Box 17, 3300 AA Dordrecht, The Netherlands.

Kluwer Academic Publishers incorporates
the publishing programmes of
D. Reidel, Martinus Nijhoff, Dr W. Junk and MTP Press.

Sold and distributed in the U.S.A. and Canada
by Kluwer Academic Publishers,
101 Philip Drive, Norwell, MA 02061, U.S.A.

In all other countries, sold and distributed
by Kluwer Academic Publishers Group,
P.O. Box 322, 3300 AH Dordrecht, The Netherlands.

Printed on acid-free paper

CONTENTS

Chapter III
Auslander Regular Filtered (Graded) Rings

Chapter IV
Microlocalization of Filtered Rings and Modules, Quantum Sections and Gauge Algebras

0. Introduction

In Commutative Algebra certain I-adic filtrations of Noetherian rings, i.e. the so-called Zariski rings, are at the basis of singularity theory. Apart from that it is mainly in the context of Homological Algebra that filtered rings and the associated graded rings are being studied not in the least because of the importance of double complexes and their spectral sequences. Where non-commutative algebra is concerned, applications of the theory of filtrations were mainly restricted to the study of enveloping algebras of Lie algebras and, more extensively even, to the study of rings of differential operators. It is clear that the operation of completion at a filtration has an algebraic genotype but a topological fenotype and it is exactly the symbiosis of Algebra and Topology that works so well in the commutative case, e.g. idèles and adèles in number theory or the theory of local fields, Puisseux series etc, ... In Non-commutative algebra the bridge between Algebra and Analysis is much more narrow and it seems that many analytic techniques of the non-commutative kind are still to be developed. Nevertheless there is the magnificent example of the analytic theory of rings of differential operators and $\mathcal{D}$-modules à la Kashiwara-Shapira.

In this book we develop an algebraic theory of filtered rings using as main tools :

1. The category equivalence between the category of filtered R-modules and the category of X-torsionfree graded $\widetilde{R}$-modules where $\widetilde{R}$ is the Rees ring of the filtration on R and $X \in \widetilde{R}_1$ is its canonical homogeneous element of degree 1,

2. The dehomogenization from certain graded objects to filtered objects (cf. Proposition 7, Ch. I, § 4.3.)

The Rees ring has also been used extensively in commutative algebra where it is sometimes called the "blow-up ring" or the "form-ring", however the simple facts referred to above seem to have been missed. The category equivalence mentioned reduces "filtered problems" to graded problems" over the Rees ring up to keeping track of some "torsionfree" condition. A good example of the simplification this leads to is given in Chapter IV, § 1, where algebraic microlocalization is obtained in a very elegant way from constructions on the Rees ring level.

In Chapter I we provide the general theory of filtered rings and modules together with their associated graded objects. The so-called "good filtrations" introduced in § 5 are one of the main ingredients in the theory developed in further chapters.

In Chapter II we define (non-commutative) Zariskian filtrations and provide several characterizations of these. It turns out that several notions appearing in the literature (earlier treated as different) are in fact equivalent to our notion of Zariskian filtration. The Zariskian property allows to lift information from the associated graded ring to the filtered ring; for example in Ch. II § 3.2. we lift the property of being a maximal order. In § 4 we study Zariskian filtrations on simple Artinian rings and this yields a link to non-commutative valuation theory (cf. also [Vo 1] or [VG 2]). At the end of the chapter we include an application to the calculation of K_0 for a ring with Zariskian filtration, slightly extending results of Quillen (cited in [M-R]).

Chapter III is devoted to the theory of Auslander regular filtered rings, in particular to the lifting of Auslander regularity from the associated graded ring to the Rees ring.

We have been attracted to the problems dealt with in Chapter III by the work of J.E. Björk, cf. [Bj. 1], [Bj 2], but it was the use of the categorical relation between filtered modules and X-torsionfree graded Rees-modules that allowed us to obtain a unified treatment and elegant solutions to those problems. In this chapter the interplay between filtered and graded properties is most clear. After the study of Auslander regularity of Zariskian filtered rings we also establish the invariance of the grade number in § 2.5., i.e. $j(M) = j_{G(R)}(G(M)) = j_{\widetilde{R}}(\widetilde{M}) = j_{\widehat{R}}(\widehat{M})$ for any M with a good filtration and under the condition that $G(R)$ is Auslander regular (here $\sim$ denotes the Rees objects, $\wedge$ denotes completion). Section 3.4. deals with holonomic and pure modules over Zariski rings with Auslander regular associated graded ring. It is known that not every simple module needs be holonomic, however, if R is Auslander regular it is true that every simple R-module is pure; so the class of pure modules maybe more interesting that the class of holonomic modules (but for other reasons one may also state the converse !) One of the main results here is Theorem 13 where a.o. it is proved that purity of M follows from purity of $G(M)$ and that on any pure R-module there exists a good filtration such that its associated graded module is pure over $G(R)$. Further, we also include the Gabber-Kashiwara theorem and a useful characterization of holonomic modules over a Zariski ring R with commutative Auslander regular $G(R)$. In the final section of Chapter III we consider codimension calculations of characteristic varieties and their geometric purity. For a Zariski ring R with commutative regular $G(R)$ which is equidimensional the purity of M is equivalent to the geometric purity of the characteristic variety plus the fact that this variety consists of the associated primes of $G(R)$, cf. 4.3.6. This shows that, even if most simple modules over Weyl algebras and enveloping algebras of finite dimensional Lie algebras are not holonomic they are still well-behaved becaused their characteristic varieties are geometrically pure.

In Ch. IV we connect to a trendy topic : quantum algebras. In fact we do not focus on the "real" quantum groups given in terms of some (quasi-triangular) Hopf algebra with bijective antipode but rather on the algebras stemming from "quantum spaces". Via algebraic microlocalization and rings of quantum sections we obtain methods for putting a "scheme"-structure on a big class of quantum algebras. The generalized Gauge algebras we introduce do contain the Gauge algebras appearing in Witten's work on Chern-Simmons Gauge theory. The filtered techniques, in particular the lifting results from the associated graded ring to the Rees ring, provide an iterative construction of nice gauge algebras (cf. Theorem 3.1. in Ch. IV, § 3). The approach via quantum sections fits nicely in the categorical approach à la M. Artin, where Proj of a suitable graded ring is defined to be the quotient category of the category of finitely generated graded modules modulo those if finite length. This leads to the definition of "schematic algebras"; this class includes iterated Ore extensions, gauge algebras iterated from the quantum plane, colour Lie superalgebras, quantum Weyl algebras etc...

In the final section we provide concretre calculations of rings of quantum sections for enveloping algebras of finite dimensional Lie algebras. We point out how to use this in a calculation of the point variety of homogenized enveloping algebras.

Surprizingly quick we arrive at the point where the reader will convinced that "there are more questions than answers" ... just to mention a few : describe rings of quantum sections of rings of differential operators on algebraic varieties ... determine the class of schematic algebras more narrowly ... use quantum sections of pure modules to obtain a "local" theory of pure modules ... calculate the point (also the line -) variety of Rees rings of Zariskian filtered rings R with suitably nice $G(R)$ ... etc...

So, as the song goes : "And the more I find out, the less I know..."

CHAPTER I
Filtered Rings and Modules

§1. Topological Prerequisites

It is assumed that the reader has some familiarity with the basic concepts of general topology. However, for his convenience, we briefly recall those that are relevant here.

A set X becomes endowed with a **topology** as soon as there is given a collection Ω of subsets of X having the following properties:

- (a). both X and the empty subset belong to Ω;
- (b). if $\{U_i\}_{i \in I}$ is an arbitrary family of subsets of X such that each U_i belongs to Ω, then $\cup_{i \in I} U_i$ also belongs to Ω;
- (c). if U_1, U_2, $\cdots$, U_n (n is an integer) belong to Ω, then $U_1 \cap U_2 \cap \cdots \cap U_n$ belongs to Ω as well.

The members of Ω are then referred to as the **open sets** of the topology and (a), (b), (c) are known as the **open sets axioms**. They can be summarized by saying that the class of open sets is closed under arbitrary unions and finite intersections.

Let X be a topological space. (This means that X is a set and we are given a collection of subsets which obeys the axioms for open sets.) A subset F, of X, is called a **closed subset** if its complement with respect to X is open. The union of a finite number of closed sets is necessarily a closed set and so is the intersection of a completely arbitrary family of closed sets. Of course it can happen that a set is both open and closed.

Suppose that K is a subset of X. The intersection of all the closed sets containing K is itself a closed set which will be denoted by $\overline{K}$. This is called the **closure** of K in X. If $x \in X$, then it is easy to show that x belongs to $\overline{K}$ if and only if every open set containing x contains at least one element of K. If it happens that $\overline{K} = X$, then one says that K is **everywhere dense** in X or that K is an **everywhere dense subset** of X.

Now assume that V is a subset of a topological space X. It is then possible to define a topology on V whose open sets are precisely the intersections of V with the open subsets of X. This is known as the **induced topology**. When V is endowed with the induced topology, we say that V is a **subspace** of X.

The elements of a topological space are often referred to as **points**. By a **neighbourhood** of a point x, of X, is meant any set which contains an open set to which x belongs. Suppose that X has the following property: **whenever x and x' are distinct points there exists a neighbourhood N of x and a neighbourhood N' of x' such that $N \cap N'$ is empty.** We then say that X is a **Hausdorff space**.

Once again, let X be a general topological space. A collection Σ of subsets of X is called a **base for the topology of X** or a **base for the open sets of X** if

(i). every set in Σ is an open set of X, and

(ii). every open set of X is the union of a (possibly empty) family of sets belonging to Σ.

Of course, one can obtain a base by letting Σ consist of all the open sets, but, in practice, interesting topologies tend to be those which have a base whose members are (in some sense) of a rather special kind. The notion of a base is useful in defining the **product** of two topological spaces. To this end let X and X' be topological spaces and suppose that Σ and Σ' are bases for their respective topologies. The Cartesian product $X \times X'$ consists of all ordered pairs (x, x'), where $x \in X$ and $x' \in X'$. If now $B \in \Sigma$ and $B' \in \Sigma'$, then $B \times B'$ is a subset of $X \times X'$. It is easy to see that there is a unique topology on $X \times X'$ which has the subsets $B \times B'$ as a base. Furthermore, this topology is unaltered if we replace Σ and Σ' by other bases for the open sets of X and X' respectively. The topological space $X \times X'$, so obtained, is called the **product** of X and X'. The topology on the product space is referred to as the **product topology**.

We conclude by recalling certain terms and definitions connected with the idea of a **continuous mapping**. Let X and X' be topological spaces and $f: X \to X'$ a mapping. Then f is said to be **continuous at the point** x, of X, if for each neighbourhood N' of $f(x)$ there exists a neighbourhood N of x such that $f(N) \subseteq N'$. Should it happen that we are given bases Σ and Σ' for the topologies of X and X' respectively, then the above definition is equivalent to the following: given B' in Σ' such that $f(x) \in B'$, it is always possible to find B in Σ such that $x \in B$ and $f(B) \subseteq B'$. When f is continuous at every point of X we say simply that f is **continuous**. It is well known that the following three statements are equivalent:

(a). the mapping $f: X \to X'$ is continuous;

(b). for every open set U' in X', $f^{-1}(U')$ is open in X;

(c). for every closed set F' in X', $f^{-1}(F')$ is closed in X.

Particularly important is the situation in which f is a one-one mapping of X onto X' and both f and f^{-1} are continuous. Such a mapping f is called a **homeomorphism** of X onto X' and f^{-1} is known as the **inverse homeomorphism**. We then say that the spaces X and X' **are homeomorphic**. If a one-one correspondence between X and X' is a homeomorphism then **a subset of one space is open if and only if the corresponding subset of the other space is open.**

§2. Filtrations on Rings and Modules

All rings considered in this book are **associative rings with identity element**. All ring homomorphisms **respect the identity element**. Let R be a ring. The category of left R-modules will be denoted by R-mod. Unless otherwise stated the term "**module**" will refer to "**left module**", even if we sometimes stress the one-sided character of some statement by adding the prefix "left" anyway. As usual, we use the characters $\mathbb{Z}$, $\mathbb{N}$, $\mathbb{Q}$, $\mathbb{R}$ and $\mathbb{C}$ to denote the sets of integers, natural numbers, rational numbers, real numbers and complex numbers respectively.

2.1. Definition. A ring R is said to be a **filtered ring** if there is an ascending chain of additive subgroups of R, say $FR = \{F_nR, n \in \mathbb{Z}\}$, satisfying $1 \in F_0R$, $F_nR \subseteq F_{n+1}R$ and $F_nRF_mR \subseteq F_{n+m}R$ for all $m, n \in \mathbb{Z}$.

From the definition it is clear that if R is a filtered ring with filtration FR then F_0R is a subring of R.

2.2. Definition. Let R be a filtered ring with filtration FR. An R-module M is said to be a **filtered module** if there exists an ascending chain, say $FM = \{F_nM, n \in \mathbb{Z}\}$, of additive subgroups of M satisfying: $F_nM \subseteq F_{n+1}M$ and $F_nRF_mM \subseteq F_{n+m}M$ for all $m, n \in \mathbb{Z}$. If R and S are filtered rings and M is an R-S-bimodule, denoted $M \in R$-mod-S, then M is said to be a **filtered R-S-bimodule** if there exists an ascending chain of additive subgroups of M, say $FM = \{F_nM, n \in \mathbb{Z}\}$, satisfying: $F_nM \subseteq F_{n+1}M$, $F_nRF_mM \subseteq F_{n+m}M$, $F_mMF_nS \subseteq F_{n+m}M$ for all $n, m \in \mathbb{Z}$.

Clearly, any filtered ring R may be viewed as a filtered left (right) R-module; Also R may be viewed as a filtered R-R-bimodule.

2.3. Examples

A. An arbitrary ring R may be viewed as a filtered ring if we define the **trivial filtration** FR of R by putting $F_nR = 0$ for $n < 0$ and $F_nR = R$ for $n \geq 0$. If M is an arbitrary R-module then any ascending chain of submodules of M defines a filtration and the structure of a filtered R-module on M. The **trivial filtration** on M is defined by $F_{-n}M = 0$, $F_nM = M$ for $n > 0$ and $F_0M = M$.

B. Let R be a ring and I an ideal of R. The **I-adic filtration** of R is obtained by putting $F_nR = R$ for $n \geq 0$, and $F_nR = I^{-n}$ for $n < 0$. On an R-module M we may then define the **I-adic filtration** of M by putting $F_nM = M$ for $n \geq 0$ and $F_nM = I^{-n}M$ for $n < 0$.

C. If R is a filtered ring with filtration FR and M is any R-module then we may define a filtration on M by putting $F_nM = F_nRM$, where F_nRM is the additive group generated by elements of the form rm with $r \in F_n R$ and $m \in M$. This filtration on M is called the **deduced filtration** (of FR); in this sense the I-adic filtration on an R-module M is the deduced filtration of the I-adic filtration of R.

D. Let C be a commutative ring and consider a C-module V. The C-module $V \otimes_C \cdots \otimes_C V$ will be denoted by $V^{\otimes m}$ if there are m-copies of V in the tensorproduct. The C-module $T(V) = \oplus_{m \in N} V^{\otimes m}$ may be made into a C-algebra, called the **tensor-algebra** of V over C, if one defines the product by C-linear extension of $(v_1 \otimes \cdots \otimes v_m) \cdot (w_1 \otimes \cdots \otimes w_n) = v_1 \otimes \cdots \otimes v_m \otimes w_1 \otimes \cdots \otimes w_n$, with v_i, $w_j \in V$. Put $F_n T(V) = \oplus_{p \leq n} V^{\otimes p}$, $n \in N$. Then $T(V)$ becomes a filtered ring; note that we put $F_0 T(V) = C$.

E. Let R be a ring and $R[t]$ the polynomial ring over R in the commuting variable t. Then $R[t]$ is a filtered ring with the **natural filtration** defined as follows: $F_n R[t] = 0$ if $n < 0$, $F_n R[t] = \sum_{p \leq n} R t^p$ if $n \geq 0$.

More generally, let $\varphi \colon R \to R$ be an injective ring homomorphism and $\delta \colon R \to R$ a φ-derivation, that is an additive morphism satisfying $\delta(xy) = \delta(x)y + x^\varphi \delta(y)$ for all x, $y \in R$. The ring of skew polynomials $R[t, \varphi, \delta]$ is obtained by adjoining a variable t to R and defining multiplication according to $tr = r^\varphi t + \delta(r)$ for $r \in R$. Obviously, R becomes a subring of $R[t, \varphi, \delta]$. The fact that φ is injective allows to define the degree function, denoted deg, in the obvious way and so we may define the **degree-filtration** by putting: $F_n R[t, \varphi, \delta] = \{f(t) \in R[t, \varphi, \delta] \,, \deg f(t) \leq n\}$, $F_0 R[t, \varphi, \delta] = R$.

F. Suppose that S is a ring with a subring R, and that S is generated, as a ring, by R together with elements $\{x_i, \ i \in I\}$. Here S is called a **ring extension** of R, and a **finitely generated ring extension** if I is finite. An element of S of the form

$$r_0 x_{i_1} r_1 x_{i_2} \cdots x_{i_n} r_n,$$

with $r_i \in R$, is called a **word** of length n. Define $F_j S$ to be the additive subgroup of S generated by all words of length j or less, with $F_0 S = R$. This yields the **standard filtration** of S with the respect to those generators. At this point, Example E. becomes a special case of standard filtrations.

G. Recall that a Lie algebra $\mathbf{g}$ over a field k, or k-Lie algebra for short, is a k-vector space equiped with a Lie product, i.e., a k-bilinear map $\mathbf{g} \times \mathbf{g} \to \mathbf{g}$, $(x, y) \mapsto [x, y]$ such that $[x, y] = -[y, x]$, $[x, x] = 0$, and satisfying the Jacobi identity

$$[x, [y, z]] + [y, [z, x]] + [z, [x, y]] = 0.$$

Note that if A is an associative k-algebra one can define a k-Lie algebra structure on A by setting $[a, b] = ab - ba$, this product being called the **bracket product**. A **representation** of a k-Lie algebra $\mathbf{g}$ is defined to be a Lie algebra homomorphism from $\mathbf{g}$ to any such A. The

universal enveloping algebra of a k-Lie algebra $\mathbf{g}$ is an (associative) k-algebra $U = U(\mathbf{g})$ together with a representation $f\colon \mathbf{g} \to U$ which is universal; i.e., given any (aasociative) k-algebra A and representation $\varphi\colon \mathbf{g} \to A$, there exists a unique algebra homomorphism $\psi\colon U \to A$ such that $\psi \circ f = \varphi$. If $\{x_i,\ i \in I\}$ is a k-basis for $\mathbf{g}$, then U may be described as the associative k-algebra generated by elements $\{x_i,\ i \in I\}$ with the relations $x_i x_j - x_j x_i = [x_i, x_j]$. In other words, $U(\mathbf{g})$ is the factor of the free k-algebra F on the set $\{x_i,\ i \in I\}$ by the ideal generated by the elements $x_i x_j - x_j x_i = [x_i, x_j]$. Thus $U(\mathbf{g})$ is an extension of the subring k with generators $\{x_i,\ i \in I\}$, and one can make the standard filtration on $U(\mathbf{g})$ as defined in Example **F**.

H. Let k be a commutative ring and A a commutative k-algebra. Recall that a k-module homomorphism $\delta\colon A \to A$ is called a k-derivation of A to A if for all a, $b \in A$ we have $\delta(ab) = a\delta(b) + \delta(a)b$. Let $\mathrm{Der}_k A$ denote the set of all k-derivations of A to A, then it is an A-module via $(a\delta)(b) = a(\delta(b))$. Since the commutator $[\delta, \gamma]$ of two derivations is again a derivation, this product gives $\mathrm{Der}_k A$ the structure of a k-Lie algebra provided k is a field. Now let $\mathbf{d}$ be any A-submodule of $\mathrm{Der}_k A$ closed under the Lie product, and $A[\mathbf{d}]$ be the k-algebra generated by A and $\mathbf{d}$. Then $A[\mathbf{d}]$ has a standard filtration over A based on $\mathbf{d}$ as a generating set. If $\mathbf{d} = \mathrm{Der}_k A$, then we write $\Delta(A) = A[\mathrm{Der}_k A]$ and call it the **derivation ring** of A.

I. In Example **H.** if $A = k[y_1, \cdots, y_n]$ is the commutative polynomial k-algebra, then it is well known that $\Delta(A) = A_n(k)$, where $A_n(k)$ is the n-th Weyl algebra over k, i.e., $\Delta(A) = k[y_1, \cdots, y_n][\partial/\partial y_1, \cdots, \partial/\partial y_n]$. For the sake of convenience $A_n(k)$ is usually defined as the k-algebra with $2n$ generators $x_1, \cdots, x_n, y_1, \cdots, y_n$ and relations

$$x_i y_j - y_j x_i = \delta_{ij},\ \text{the Kronecker delta,}$$

and

$$x_i x_j - x_j x_i = y_i y_j - y_j y_i = 0.$$

Two different filtrations on $A_n(k)$ have been used very often in literatures:

(i). The standard filtration (or Bernstein filtration): $F_j A_n(k)$ is defined to be the k-subspace generated by all $x_1^{\alpha_1} \cdots x_n^{\alpha_n} y_1^{\beta_1} \cdots y_n^{\beta_n}$ with $\alpha_1 + \cdots + \alpha_n + \beta_1 + \cdots \beta_n \le j$, $j \ge 0$;

(ii). The Σ-filtration (cf. [Bj2] or [Bor]): $F_j A_n(k)$ is defined to be the $k[x_1, \cdots, x_n]$-submodule of $A_n(k)$ generated by $\{1, y_1^i, \cdots, y_{n+1}^i;\ 1 \le i \le j\}, j \ge 0$.

J. Let A be a commutative k-algebra, where k is a field, and let M and N be A-modules. Give $\mathrm{Hom}_k(M, N)$ the structure of an $A \otimes A$-module by defining $((a \otimes b)\vartheta)(m) = a\vartheta(bm)$ for a, $b \in A$, $\vartheta \in \mathrm{Hom}_k(M, N)$, and $m \in M$. Define $\mu\colon A \otimes A \to A$ to be the multiplication map, $\mu(a \otimes b) = ab$, write J for $\mathrm{Ker}\,\mu$. Recall that the space of k-**linear differential operators**

from M to N of order at most n is defined by

$$\mathcal{D}_A^n(M,N) = \{\vartheta \in \text{Hom}_k(M,N),\ J^{n+1}\vartheta = 0\},$$

where $J^0 = A \otimes A$. Set $\mathcal{D}_A(M,N) = \cup_{n=0}^\infty \mathcal{D}_A^n(M,N)$. Note that J is generated by $\{1 \otimes a - a \otimes 1,\ a \in A\}$, if we write $[\vartheta, a] = (1 \otimes a - a \otimes 1)\vartheta = \vartheta a - a\vartheta$ for $a \in A$, $\vartheta \in \text{Hom}_k(M,N)$, then $\mathcal{D}_A^0(M,N) = \text{Hom}_A(M,N)$, and $\mathcal{D}_A(M,N) = 0$ if and only if $\text{Hom}_A(M,N) = 0$. Given the above description of J, one may alternatively define $\mathcal{D}_A^n(M,N)$ inductively by $\mathcal{D}_A^{-1}(M,N) = 0$ and for $n \geq 0$,

$$\mathcal{D}_A^n(M,N) = \{\vartheta \in \text{Hom}_k(M,N),\ [\vartheta, a] \in \mathcal{D}_A^{n-1}(M,N) \text{ for all } a \in A\}$$

It follows that $\mathcal{D}_A^n(M,N) \subseteq \mathcal{D}_A^{n+1}(M,N)$ for $n \geq 0$. If f and g are elements of $\mathcal{D}(A) = \mathcal{D}_A(A,A)$, then fg belongs to $\mathcal{D}(A)$ also; indeed, if f has order p and g has order q then fg has order $\leq p + q$. Thus $\mathcal{D}_A^n(A) \cdot \mathcal{D}_A^m(A) \subseteq \mathcal{D}_A^{n+m}(A)$ and $\mathcal{D}(A)$ is a filtered k-algebra with filtration $F\mathcal{D}(A) = \{F_n\mathcal{D}(A) = \mathcal{D}_A^n(A),\ n \in I\!N\}$. $\mathcal{D}(A)$ is called the **ring of differential operators** of A.

We refer the reader to [Bj1], [M-R], [SST] and [Bor] for some detail concerning Weyl algebras and rings of differential operators with positive filtrations (see definition 2.4. below).

K. Let R be a ring and Q an overring of R. Recall that an R-sub-bimodule I of Q is said to be **invertible** (in Q) if there exists another R-sub-bimodule J in Q such that $IJ = JI = R$. If I is an ideal of R such that it is invertible then $R \subseteq I^{-1} = \{q \in Q,\ qI \subseteq R\} = \{q \in Q,\ Iq \in R\}$. Hence we have

$$\cdots \subseteq I^2 \subseteq I \subseteq R \subseteq I^{-1} \subseteq I^{-2} \subseteq \cdots$$

and $R' = \cup_{n \in \mathbb{Z}} I^n$ is a filtered ring with filtration FR':

$$F_n R' = \begin{cases} R, & \text{if } n = 0 \\ I^{-1}, & \text{if } n \neq 0 \end{cases}$$

L. Let $A = \oplus_{n \in \mathbb{Z}} A_n$ be an arbitrary $\mathbb{Z}$-graded ring (we refer to §4. for some preliminaries on graded rings and modules). There are two (natural) filtrations on A given by putting

$$F_n^{(1)}A = \oplus_{k \leq n} A_k,\ n \in \mathbb{Z}$$
$$F_n^{(2)}A = \oplus_{k \geq n} A_k,\ n \in \mathbb{Z}.$$

We call these two filtrations on A the **grading filtrations**. As an example let us look at the $n + 1$-st Weyl algebra $R = A_{n+1}(k) = k[x_1, \cdots, x_{n+1}, y_1, \cdots, y_{n+1}]$ over a field k

of characteristic zero. If one put x_{n+1}-degree -1 and y_{n+1}-degree $+1$ while all the other generators $x_1, \cdots, x_n, y_1, \cdots, y_n$ have degree equal to zero, then R becomes a $\mathbb{Z}$-graded ring. The grading filtration $F^{(1)}R$ on R is just the so-called **Fuchsian filtration** (cf. [Bj2]).

2.4. Definition. Let R be a filtered ring with filtration FR and M a filtered R-module with filtration FM.
(a). If $F_nM = 0$ for $n < 0$, then FM is called a **positive filtration** (in this case we also say that M is positively filtered); If there exists an $n_0 \in \mathbb{Z}$ such that $F_pM = 0$ for all $p < n_0$, then FM is called **discrete**.
(b). If $M = \cup_{n\in\mathbb{Z}} F_nM$ then FM is called **exhaustive**.
(c). If $\cap_{n\in\mathbb{Z}} F_nM = 0$, then FM is called **separated**.

It may be clear from the examples that many different filtrations may be defined on a ring R or on an R-module M: even if we fix the filtration FR on R then there are still many possible definitions of particular filtrations on M that make it into a filtered R-module. Certain filtrations however are more desirable than others. The main idea is that the filtrations should carry some information about the module considered. In general $M' = \cup_{n\in\mathbb{Z}} F_nM$ is an R-submodule of M if we start from an exhaustive filtration FR on R, so in a sense it is not restrictive if we restrict attention to exhaustively filtered modules if FR is itself exhaustive.

2.5. The category R-filt
If R and S are two filtered rings with filtrations FR and FS respectively, then a ring homomorphism $f\colon R \to S$ is said to be a **filtered ring homomorphism of degree** p, $p \in \mathbb{Z}$, in case $f(F_nR) \subseteq F_{n+p}S$ for all $n \in \mathbb{Z}$.
If R is a filtered ring with filtration FR and M, N are filtered R-modules with their respective filtrations FM and FN, then an R-linear map $f \in \mathrm{Hom}_R(M,N)$ is said to have **degree** p if $f(F_nM) \subseteq F_{n+p}N$ for all $n \in \mathbb{Z}$. Homomorphisms of finite degree form an additive subgroup of $\mathrm{Hom}_R(M,N)$, denoted $\mathrm{HOM}_R(M,N)$, and homomorphisms of degree p form a subgroup of $\mathrm{HOM}_R(M,N)$, denoted $F_p\mathrm{HOM}_R(M,N)$. It is easily checked that the following properties hold:

(1). If $p \le q$ then $F_p\mathrm{HOM}_R(M,N) \subseteq F_q\mathrm{HOM}_R(M,N)$;
(2). $\mathrm{HOM}_R(M,N) = \cup_{p\in\mathbb{Z}} F_p\mathrm{HOM}_R(M,N)$;
(3). For filtered R-modules M, N and P if $f\colon M \to N$ has degree p, $g\colon N \to P$ has degree q then $g \circ f\colon M \to P$ has degree $p + q$.

Therefore, if R is a filtered ring with filtration FR we may define a category R-**filt** where the objects are are filtered **left** R-**modules** and the category-morphisms are the R-linear maps of **degree zero** (we will then write $F_0\mathrm{HOM}_R(M,N) = \mathrm{Hom}_{FR}(M,N)$). An element

f of $\mathrm{Hom}_{FR}(M, N)$ is called a **filtered morphism.**

If $M \in R$-filt with filtration FM and M' is an R-submodule of M, we say that M' is a **filtered R-submodule** provided that M' has a filtration FM' such that $F_n M' \subseteq F_n M$ for all $n \in \mathbb{Z}$.

Any R-submodule M' of a filtered R-module M can be made into a filtered R-submodule of M. Indeed, define the filtration FM' on M' as follows: $F_n M' = M' \cap F_n M$, $n \in \mathbb{Z}$, then it is obvious that M is a filtered R-submodule of M with respect to FM'. The filtration FM' on M' obtained in such a way is called the **induced filtration.**

It is clear that R-filt is an additive category and if $f \in \mathrm{Hom}_{FR}(M, N)$ then $\mathrm{Ker}f$ and $\mathrm{Coker}f$ exists in R-filt. For $\mathrm{Ker}f \subseteq M$ we may consider $F\mathrm{Ker}f$ induced by FM. Furthermore, if H is a submodule of M then one can define the **quotient filtration** on $Q = M/H$ by putting $F_n Q = F_n M + H/H$ for $n \in \mathbb{Z}$. In particular, for $\mathrm{Coker}f = N/\mathrm{Im}f$ we may consider the quotient filtration induced by FN. It immediately follows that monomorphisms and epimorphisms in R-filt are just the injective resp. surjective morphisms.

One also easily checks that arbitrary direct sums, direct products as well as inductive limits and projective limits do exist in R-filt (note: $F_p(\varinjlim M_i) = \varinjlim F_p M_i$, $F_p(\varprojlim M_i) = \varprojlim F_p M_i$).

2.6. Remark.
(a). Let $\{M_i\ i \in J\}$ be a family of objects of R-filt such that each FM_i is exhaustive, then the filtration defined on $\oplus M_i$ by $F_n(\oplus M_i) = \oplus F_n M_i$ is again exhaustive. Note however that this property may fail for the direct product $\prod M_i$ (if J is infinite). On the positive side it is true that inductive limits of exhaustive filtered modules are again exhaustive.

(b). The category R-filt is pre-abelian but not abelian. Let $R = k[t]$ be the polynomial ring over a field k in variable t. Let $I = (t)$ be the ideal of R generated by t, and $J = (t^2)$ the ideal of R generated by t^2. Take the I-adic filtration FR and the J-adic filtration $F'R$ on R respectively, and write G resp. H, for the filtered modules from R by considering FR resp. $F'R$. The identity map of R is in $\mathrm{Hom}_{FR}(H, G)$, but not in $\mathrm{Hom}_{FR}(G, H)$, so the identity map of M is a bijectivity mapping failing to be an isomorphism.

2.7. Some functors on R-filt
Let R be a filtered ring with filtration FR.

The **forgetful functor**: R-filt $\rightarrow$ R-mod is defined by forgetting he filtration of a filtered module. We also have a functor D: R-mod $\rightarrow$ R-filt that associates an R-module M the filtered module obtained by considering M with the deduced filtration. The problem with the functor D is (among others) that the deduced filtration will not necessarily be separated even if FR is separated.

For each $n \in \mathbb{Z}$ we define the n-th **shift functor** $T(n)$: R-filt $\rightarrow$ R-filt associating the filtered R-module M with filtration FM the filtered R-module $T(n)M$ obtained by filtering the R-module M by $F_p T(n)M = F_{p+n}M$ for all $p \in \mathbb{Z}$. The shift functors satisfy: $T(0) = \mathrm{Id}$, the identity functor, $T(n) \circ T(m) = T(n+m)$. In particular, every $T(n)$ is an equivalence of categories that commutes with direct sums, products, inductive as well as projective limits. Moreover, for M, N in R-filt and for $n \in \mathbb{Z}$ we have $F_n \mathrm{HOM}_R(M, N) = \mathrm{Hom}_{FR}(T(-n)M, N) = \mathrm{Hom}_{FR}(M, T(n)N)$.

Let R be a filtered ring with filtration FR. Put $R' = \cup_{n \in \mathbb{Z}} F_n R$. Then R' is a subring of R with exhaustive filtration by putting $F_n R' = F_n R$ for all $n \in \mathbb{Z}$. Similarly, if $M \in R$-filt with filtration FM then $M' = \cup_{n \in \mathbb{Z}} F_n M$ is a filtered R'-module with exhaustive filtration defined by putting $F_n M' = F_n M$ for all $n \in \mathbb{Z}$. In this way we obtain a functor $\varepsilon\colon R$-filt $\to$ R'-filt carrying a filtered R-module M into an exhaustively filtered R'-module.

Convention.
In the sequel of this book we always assume that the filtrations considered are exhaustive filtrations.
This assumption will not create any problems but we will be very careful when products of filtered modules are being considered because of Remark 2.6.(a).

§3. Complete Filtered Modules and Completions

Let R be a filtered ring with filtration FR, and let M be a filtered R-module with filtration FM. Let x be an element of M and $K_1, \cdots, K_n$ be subgroups of M. If we consider the cosets $x + K_i$ of K_i in M then it is easy to see that

$$(x + K_1) \cap \cdots \cap (x + K_n) = x + (K_1 \cap \cdots \cap K_n).$$

Now, let us form a collection Ω of subsets of M in the following way:

A subset U of M is to belong to Ω provided that whenever $u \in U$ there exists an integer n such that $u + F_n M \subseteq U$.

Obviously M and its empty subset belong to Ω, and Ω is closed under arbitrary unions. By the remark given above it is not difficult to see that the Ω is also closed under finite intersections. Accordingly there is a topology on M whose open sets are precisely those subsets of M which belong to Ω. We shall refer to it either as the **topology derived from the filtration** or, more briefly, as the **filtration topology**.

3.1. Some properties of filtration topology

Let R be a filtered ring with filtration FR and $M \in R$-filt.
(a). The sets of the form $x + F_n M$, $x \in M$, $n \in \mathbb{Z}$, form a base for the filtration topology of M.
(b). Let N be a subgroup of M. Then N is open in M (with respect to the filtration topology of M) if and only if $F_n M \subseteq N$ for some $n \in \mathbb{Z}$. If a subgroup N is open then it is also closed; In particular, each $F_n M$, $n \in \mathbb{Z}$, is both open and closed in M.
(c). Let N be a subset of M. Then the closure $\overline{N}$ of N (with respect to the filtration topology of M) is given by

$$\overline{N} = \cap_{n \in \mathbb{Z}}(N + F_n M)$$

where $N + F_n M = \cup_{x \in N}(x + F_n M)$. In particular, the closure of the zero group of M is $\cap_{n \in \mathbb{Z}} F_n M$.

Proof. By the definition of filtration topology (a) is clear.
(b).If N is open, then $F_n M = 0 + F_n M \subseteq N$ for some $n \in \mathbb{Z}$. Conversely, if there is an $n \in \mathbb{Z}$, such that $F_n M \subseteq N$, then $N = \cup_{x \in N}(x + F_n M)$, hence N is open. Moreover, suppose that N is open, then the complement of N in M is the union of all sets $x + F_n M$, where $x \in M$, $x \notin N$. Therefore N is itself closed.
(c). Let $x \in \overline{N}$, then for any $n \in \mathbb{Z}$, $(x + F_n M) \cap N \neq \emptyset$. It follows that $x \in \cap_{n \in \mathbb{Z}}(N + F_n M)$, i.e., $\overline{N} \subseteq \cap_{n \in \mathbb{Z}}(N + F_n M)$. Conversely, if $x \in \cap_{n \in \mathbb{Z}}(N + F_n M)$, then for any $n \in \mathbb{Z}$ we have $x = y + x_n$, where $y \in N$, $x_n \in F_n M$. Hence $y \in N \cap (x + F_n M)$ and this shows that $x \in \overline{N}$. Therefore $\overline{N} = \cap_{n \in \mathbb{Z}}(N + F_n M)$.   □

Again let M be a filtered R-module with filtration FM, then one easily derives the following properties for the filtration topology of M:

(d). M is a Hausdorff space if and only if $\cap_{n\in\mathbb{Z}} F_n M = 0$, or in other words, if and only if the filtration FM on M is separated.

(e). With respect to its filtration topology, M becomes a **topological module**, i.e., the map $\nu: M \to M$ with $\nu(x) = -x$, is a homeomorphism of M to M; the additive operation on M is continuous; moreover, the module operation of R on M is also continuous.

(f). If $N \in R$-filt, then any filtered morphism $f \in \mathrm{Hom}_{FR}(M, N)$ is continuous.

3.2. Equivalence of filtration topologies

Let R be a filtered ring with filtration FR.

Let M be a filtered R-module with two filtrations, say FM and $F'M$. We say that FM and $F'M$ are **topologically equivalent** if for every n, $m \in \mathbb{Z}$, there are n_1 $m_1 \in \mathbb{Z}$ such that $F'_{n_1} M \subseteq F_n M$, $F_{m_1} M \subseteq F'_m M$; we say that FM and $F'M$ are **algebraically equivalent** if there is an integer $c \in \mathbb{N}$ such that for all $n \in \mathbb{Z}$

$$F_{n-c} M \subseteq F'_n M \subseteq F_{n+c} M$$

Throughout this book we will simplify terminology and speak about **equivalence** when algebraic equivalence has to be understood. For certain good filtrations on finitely generated R-modules this notion of equivalence will be a very natural one and topological equivalence will even reduce to equivalence. But in general topological equivalence of filtrations is too weak in order to derive very stringent algebraic properties of the module.

1. Definition. Let M, N be in R-filt and $f: M \to N$ a filtered morphism. f is said to have the **Artin-Rees property** if there exists an integer $c \in \mathbb{Z}$ such that for all $n \in \mathbb{Z}$

$$\mathrm{Im} f \cap F_n N \subseteq f(F_{n+c} M);$$

f is said to be **strict** if for every $n \in \mathbb{Z}$

$$f(F_n M) = \mathrm{Im} f \cap F_n N.$$

Obviously, every strict filtered morphism has the Artin-Rees Property. Moreover, let N be a submodule of a filtered R-module M, where M has filtration FM. If N is endowed with the filtration FN induced by FM, then the inclusion mapping $N \hookrightarrow M$ is a strict filtered morphism. However, this is not necessarily true if N is a filtered R-submodule with filtration FN which is not induced by FM.

Again let N be a submodule of the filtered R-module M. If M/N is endowed with the quotient filtration: $F_n(M/N) = F_n M + N/N$, $n \in \mathbb{Z}$, then the natural mapping $M \to M/N$ is a strict filtered morphism.

But on the other hand, it is a fundamental although elementary observation that filtered morphisms of degree zero, even if they are bijective, need not be strict! For example, let M be a filtered R-module with two filtrations FM and $F'M$. Suppose that there exists an integer $c > 0$ such that $F_{n-c}M \subseteq F'_n M \subseteq F_{n+c}M$, i.e., FM and $F'M$ are equivalent. We may write M for the filtered R-module M with filtration FM and M' if we use the filtration $F'M$. The identity map $1_M : M \to M'$ has degree c; but if we view the identity map as a filtered morphism $M \to T(c)M'$, where $T(c)$ is the c-th shift functor, then it represents an element of $\mathrm{Hom}_{FR}(M, T(c)M')$ and in fact we may view M as a filtered submodule of $T(c)M'$. Of course the filtration FM is not necessarily induced by $FT(c)M'$. Moreover, define g: $M \to M \oplus T(c)M'$, $m \mapsto m \oplus m$, and h: $M \oplus T(c)M' \to T(c)M'$, $m \oplus m' \mapsto m'$, where $m \in M$, $m' \in T(c)M'$. The filtration on $M \oplus T(c)M'$ is defined by $F_n(M \oplus T(c)M') = F_n M \oplus F'_{n+c}M$. It is easily checked that both g and h are strict morphisms of degree zero but the composition $h \circ g$: $M \to T(c)M'$ is not strict! Therefore it is impossible to define a category of filtered modules and strict morphisms for the category morphisms.

Let M and N be filtered R-modules with FM and FN as their respective filtrations. Further let f: $M \to N$ be a filtered morphism. Then since $f(F_n M) \subseteq F_n N$, f induces the homomorphism f_n: $M/F_n M \to N/F_n N$ which is such that the diagram

$$
\begin{array}{ccc}
M & \xrightarrow{\;f\;} & N \\
\varphi_M^n \downarrow & & \downarrow \varphi_N^n \\
M/F_n M & \xrightarrow{\;f_n\;} & N/F_n N
\end{array}
$$

is commutative, where φ_M^n resp. φ_N^n is the natural mapping of M onto $M/F_n M$ resp. N onto $N/F_n N$. Hence for each $n \in \mathbb{Z}$, φ_M^n gives rise to a homomorphism $\mathrm{Ker} f \to \mathrm{Ker} f_n$.

2. Lemma. With notations as above, f is a strict filtered morphism if and only if for each $n \in \mathbb{Z}$ the associated mapping $\mathrm{Ker} f \to \mathrm{Ker} f_n$ is an epimorphism.

Proof. Suppose that f is strict. if $x \in \mathrm{Ker} f_n$, then $x = \varphi_M^n(e)$ with $e \in M$ and $f_n \varphi_M^n(e) = 0$. It follows that $\varphi_N^n f(e) = 0$ and therefore $f(e) \in f(M) \cap F_n N = f(F_n M)$. Hence there exists $x_n \in F_n M$ such that $f(e) = f(x_n)$. Consequently $e - x_n \in \mathrm{Ker} f$ and $\varphi_M^n(e - x_n) = \varphi_M^n(e) = x$. Thus $\mathrm{Ker} f \to \mathrm{Ker} f_n$ is an epimorphism.

Conversely, assume that $\mathrm{Ker} f \to \mathrm{Ker} f_n$ is an epimorphism for all n and let x' be in $f(M) \cap F_n N$. Then $x' = f(x)$ for some $x \in M$. Now $f_n \varphi_M^n(x) = \varphi_N^n f(x) = \varphi_N^n(x') = 0$ because $x' \in F_n N$. Thus $\varphi_M^n(x) \in \mathrm{Ker} f_n$. But by the assumption $\varphi_M^n(x) = \varphi_M^n(y)$ with $y \in \mathrm{Ker} f$. It follows that $x - y \in \mathrm{Ker} \varphi_M^n = F_n M$ and $f(x) = f(x - y)$. Thus $x' = f(x) \in f(F_n M)$ and the proof is complete. $\square$

3. Corollary. Let M be in R-filt with filtration FM. If $f\colon M \to M$ is a filtered morphism such that $f^2 = f$, then f is strict.

3.3. Complete filtered modules

Let R be a filtered ring with filtration FR and $M \in R$-filt with filtration FM.

A sequence $(x_i)_{i>0}$ of elements of M is said to be a **Cauchy sequence** if for every integer $p \geq 0$ there is an integer $N(p) > 0$ such that $x_s - x_t \in F_{-p}M$ for all s, $t \geq N(p)$. Obviously, in this definition it is sufficient to have $x_s - x_{s+1} \in F_{-p}M$ for any $s \geq N(p)$. A sequence $(x_i)_{i>0}$ **converges** to $x \in M$ if for every integer $p \geq 0$ there exists an integer $N(p) > 0$ such that $x_s - x \in F_{-p}M$ whenever $s \geq N(p)$. In this case we often write $x_i \to x$.

If the filtration FM on M is separated, or in other words M is a Hausdroff space with respect to its filtration topology, then any convergent sequence converges to a unique limit.

1. Lemma. Let K be a subset of M and let x belong to M, where $M \in R$-filt with filtration FM. Then x belongs to the closure $\overline{K}$ of K in M if and only if there exists a sequence $(x_i)_{i>0}$ such that $x_i \in K$ for all i and $x_i \to x$.

Proof. If such a sequence exists, then every neighbourhood of x contains at least one x_i and therefore it contains a point of K. Hence x belongs to the closure of K.

Suppose now that x belongs to the closure of K. Then $x + F_n M$ meets K and therefore we can find $x_n \in K$ such that $x - x_n \in F_n M$. Evidently $x_i \to x$, so the proof is complete. $\square$

2. Definition. An $M \in R$-filt with filtration FM is said to be **complete** (with respect to its filtration topology) if the following two conditions hold:

(a). FM is separated, i.e., M is a Hausdorff space;

(b). every Cauchy sequence converges to some element of M.

Let $(x_i)_{i>0}$ be a seqquence of elements of a filtered R-module and suppose that it converges to x. Then x_2, x_3, x_4, $\cdots$ also converges to x and therefore $(x_{m+1} - x_m) \to 0$.

3. Proposition. Let $M \in R$-filt with filtration FM. Suppose that M is complete with respect to its filtration topology and let $(x_i)_{i>0}$ be a sequence of elements of M. Then $(x_i)_{i>0}$ converges if and only if $\lim(x_{i+1} - x_i) = 0$.

Proof. Suppose that $\lim(x_{i+1} - x_i) = 0$. Let $p \geq 0$ be a given integer, then there exists integer $N(p) > 0$ such that $x_{s+1} - x_s \in F_{-p}M$ whenever $s \geq N(p)$. Suppose now that $s > N(p)$ and $t > N(p)$. Then

$$x_s - x_{N(p)} = \left(x_{N(p)+1} - x_{N(p)}\right) + \left(x_{N(p)+2} - x_{N(p)+1}\right) + \cdots + (x_s - x_{s-1})$$

which belongs to $F_{-p}M$. Similarly $x_t - x_{N(p)}$ is in $F_{-p}M$. Consequently $x_s - x_t \in F_{-p}M$. This shows that $(x_i)_{i>0}$ is a Cauchy sequence and hence it converges.

The converse implication has already been established above. $\square$

It is sometimes convenient to make use of the notion of a **convergent series**. To explain what this means let $(u_i)_{i>0}$ be a sequence of elements of a filtered module M and put $S_i = u_1 + u_2 + \cdots + u_i$. We then say that the series

$$u_i + u_2 + \cdots + u_i + \cdots$$

converges to s if $S_i \to s$. If this is the case and if also M is a Hausdorff space, then we write $s = \sum_{i=1}^{\infty} u_i$.

4. Proposition. Let $M \in R$-filt which is complete with respect to its filtration topology. Then a series $u_1 + u_2 + \cdots + u_i + \cdots$, of elements of M, converges if and only if $\lim u_i = 0$.

Proof. This follows immediately by applying Proposition 3.3.3. to the partial sums $u_1 + u_2 + \cdots + u_i$ of the series. $\square$

5. Remark.

(1). If $M \in R$-filt with a discrete filtration FM, then M is complete.

(2). Any finite direct sum of complete filtered R-modules in R-filt is also a complete filtered R-module.

3.4. The completion of a filtered module

Let R be a filtered ring with filtration FR and $M \in R$-filt with filtration FM.

1. Definition. A **completion** of M with respect to FM (or to the filtration topology of M) is a composite object consisting of a filtered R-module $\widehat{M}$ with filtration $F\widehat{M}$, and a filtered R-morphism $\varphi_M: M \to \widehat{M}$. These are required to satisfy the following conditions:

(1). $\widehat{M}$ is complete with respect to its filtration topology;

(2). the homomorphism $\varphi_M: M \to \widehat{M}$ is a strict filtered morphism;

(3). $\varphi_M(M)$ is everywhere dense in $\widehat{M}$;

(4). $\mathrm{Ker}\varphi_M = \cap_{n \in \mathbf{Z}} F_n M$.

The morphism φ_M is usually called the **canonical morphism**.

Before showing the existence of a completion for the given filtered R-module M, let us first see some basic properties of a completion.

2. Theorem. Let $\widehat{M}$ be one of the completions of M. If $\varphi_M: M \to \widehat{M}$ is the canonical morphism, then $F_n\widehat{M}$ is the closure of $\varphi_M(F_nM)$ in $\widehat{M}$ and $F_nM = \varphi_M^{-1}(F_n\widehat{M})$.

Proof. Since φ_M is strict we have $\varphi_M(F_nM) = \varphi_M(M) \cap F_n\widehat{M}$ and therefore $\varphi_M(F_nM) \subseteq F_n\widehat{M}$. Next, by property 3.1.(b)., $F_n\widehat{M}$ is closed in $\widehat{M}$. consequently the closure of $\varphi_M(F_nM)$ in $\widehat{M}$ is contained in $F_n\widehat{M}$.

let $\xi \in F_n\widehat{M}$. By the definition of completion $\varphi_M(M)$ is everywhere dense in $\widehat{M}$. Hence for every integer $k \geq 0$ the intersection of $\xi + F_{n-k}\widehat{M}$ and $\varphi_M(M)$ is not empty. Choose $\eta_k \in F_{n-k}\widehat{M}$ so that $\xi + \eta_K \in \varphi_M(M)$. Then $\xi + \eta_k$ belongs to $\varphi_M(M) \cap F_n\widehat{M} = \varphi_M(F_nM)$. Now $\lim \eta_k = 0$ and therefore $\lim(\xi + \eta_k) = \xi$. Thus ξ is the limit of a sequence of elements each of which belongs to $\varphi_M(F_nM)$. Consequently, by Lemma 3.3.1., ξ belongs to the closure of $\varphi_M(F_nM)$ in $\widehat{M}$. It follows that the closure of $\varphi_M(F_nM)$ is $F_n\widehat{M}$ as was asserted. Finally,

$$\varphi_M^{-1}(F_n\widehat{M}) = \varphi_M^{-1}(\varphi_M(M) \cap F_n\widehat{M}) = \varphi_M^{-1}(\varphi_M(F_nM)) = F_nM + \mathrm{Ker}\varphi_M = F_nM$$

because $\mathrm{Ker}\varphi_M$ is contained in F_nM. This completes the proof. □

3. Theorem. Let $\widehat{M}$ be a completion of M. Then there is a one-one correspondence between the open subgroups U of M and the open subgroups V of $\widehat{M}$. This is such that if U and V correspond and $\varphi_M: M \to \widehat{M}$ is the canonical morphism, then $U = \varphi_M^{-1}(V)$ and V is the closure of $\varphi_M(U)$ in $\widehat{M}$. In this situation φ_M induces an isomorphism of the group M/U onto the group $\widehat{M}/V$.

Proof. Suppose that V is an open subgroup of $\widehat{M}$ and put $U = \varphi_M^{-1}(V)$. Then since φ_M is continuous, U is an open subgroup of M. Choose positive integer s large enough to ensure that $F_{-s}\widehat{M} \subseteq V$. By property 3.1.(b)., $\varphi_M(M) + F_{-s}\widehat{M}$ is an open and therefore also a closed subgroup of $\widehat{M}$. But $\varphi_M(M)$ is everywhere dense in $\widehat{M}$. Hence $\varphi_M(M) + F_{-s}\widehat{M} = \widehat{M}$. Similarly for any $k \geq s$ we have $\varphi_M(M) + F_{-k}\widehat{M} = \widehat{M}$. Now, $\varphi_M(U) = \varphi_M(M) \cap V$ and therefore if $k \geq s$ we have

$$\begin{aligned}
\varphi_M(U) + F_{-k}\widehat{M} &= (\varphi_M(M) \cap V) + F_{-k}\widehat{M} \\
&= (\varphi_M(M) + F_{-k}\widehat{M}) \cap V \\
&= \widehat{M} \cap V \\
&= V
\end{aligned}$$

i.e., $V = \cap_{k \geq s}(\varphi_M(U) + F_{-k}\widehat{M})$, it follows from property 3.1.(c). that the closure of $\varphi_M(U)$ in $\widehat{M}$ is V. Again φ_M induces a homomorphism $M/U \to \widehat{M}/V$. Also, if s is a large positive integer,$\varphi_M(M)+V$ contains $\varphi_M(M)+F_{-s}\widehat{M} = \widehat{M}$. Thus $\varphi_M(M)+V = \widehat{M}$ which shows that the homomorphism in question is an epimorphism. On the other hand, if $\varphi_M(e) \in V$, then $e \in U$. This shows that $M/U \to \widehat{M}/V$ is also a monomorphism and hence an isomorphism. To complete the proof we must now show that if U' is a given open subgroup of M, then there exists an open subgroup V' of $\widehat{M}$ such that $U' = \varphi_M^{-1}(V')$. By property 3.1.(b)., we can find an integer n so that $F_n M \subseteq U'$. Put $V' = \varphi_M(U') + F_n\widehat{M}$. Then V' is certainly an open subgroup of $\widehat{M}$ and $U' \subseteq \varphi_M^{-1}(V')$. Moreover, if $e \in \varphi_M^{-1}(V')$, then there exists $u' \in U'$ such that $\varphi_M(e - u') \in F_n\widehat{M}$. Consequently, by Theorem 3.4.2., $e - u' \in F_n M$ and hence $e \in U'$. It follows that $U' = \varphi_M^{-1}(V')$ and now the proof is complete. $\Box$

4. Corollary. With notations as before, φ_M induces an isomorphism $M/F_n M \to \widehat{M}/F_n\widehat{M}$ of groups, for each $n \in \mathbb{Z}$.

Proof. By Theorem 3.4.2., we may take $U = F_n M$ and $V = F_n\widehat{M}$ in Theorem 3.4.3.. $\Box$

5. Theorem. Let M and N be in R-filt with filtrations FM and FN respectively. Furthermore let $\widehat{M}$ and $\widehat{N}$ be completions of M and N respectively. If now $f\colon M \to N$ is a **continuous** R-homomorphism, then there exists a unique continuous mapping $\widehat{f}\colon \widehat{M} \to \widehat{N}$ such that the diagram

$$
\begin{array}{ccc}
M & \xrightarrow{f} & N \\
\varphi_M \downarrow & & \downarrow \varphi_N \\
\widehat{M} & \xrightarrow[\widehat{f}]{} & \widehat{N}
\end{array}
$$

is commutative. (Here φ_M and φ_N are the canonical morphisms of M and N into their respectively given completions.) The mapping $\widehat{f}$, so obtained, is necessarily a homomorphism of R-modules. Moreover, if f is a filtered morphism of degree d ($d \in \mathbb{Z}$), then $\widehat{f}$ is a filtered morphism of degree d, too.

Proof. Since f is continuous, for any $k \in \mathbb{Z}$, there is a ν_k such that $f(F_{\nu_k}M) \subseteq F_k N$. It follows that $f(\cap_{k\in\mathbb{Z}}F_k M) \subseteq \cap_{k\in\mathbb{Z}}F_k N$, that is to say $f(\mathrm{Ker}\varphi_M) \subseteq \mathrm{Ker}\varphi_N$. Hence f induces an R-homomorphism $f^*\colon \varphi_M(M) \to \varphi_N(N)$ with $f^*(\varphi_M(m)) = \varphi_N(f(m))$ such that the diagram

$$
\begin{array}{ccc}
M & \xrightarrow{f} & N \\
\downarrow & & \downarrow \\
\varphi_M(M) & \xrightarrow[f^*]{} & \varphi_N(N)
\end{array}
$$

is commutative. Further we have:

$$(*) \qquad f^*(\varphi_M(M) \cap F_{\nu_k}\widehat{M}) \subseteq \varphi_N(N) \cap F_k\widehat{N}$$

Let $\xi \in \widehat{M}$. Since $\varphi_M(M)$ is everywhere dense in $\widehat{M}$, there exists a sequence (u_n) of elements of $\varphi_M(M)$ such that $\lim u_n = \xi$. Thus (u_n) is a Cauchy sequence and now we see, from $(*)$, that $(f^*(u_n))$ is a Cauchy sequence in $\widehat{N}$. But $\widehat{N}$ is complete. Consequently $(f^*(u_n))$ has a limit in $\widehat{N}$ and, since $\widehat{N}$ is a Hausdorff space, this limit is uniquely determined by (u_n).

Suppose that we have a second sequence, (v_n) say, such that $v_n \in \varphi_M(M)$ for all n and $\lim v_n = \xi$. Then, of course, $\lim f^*(v_n)$ also exists. Now $\lim(u_n - v_n) = 0$. Hence by $(*)$, $\lim f^*(u_n - v_n) = 0$ and therefore $\lim f^*(u_n) = \lim f^*(v_n)$. This conclusion may be rephrased by saying that $\lim f^*(u_n)$ depends only on ξ and is otherwise independent of the choice of the sequence (u_n). We may therefore define a mapping $\widehat{f}: \widehat{M} \to \widehat{N}$ by means of the equation $\widehat{f}(\xi) = \lim f^*(u_n)$.

If $r \in R$, then since $\widehat{M}$ and $\widehat{N}$ are filtered R-modules and both f and φ_N are R-morphisms we have $\lim(ru_n) = r\xi$ and $\lim f^*(ru_n) = r \lim f^*(u_n)$. Thus $\widehat{f}(r\xi) = r\widehat{f}(\xi)$. Again, if $\xi^{(1)}$ and $\xi^{(2)}$ belong to $\widehat{M}$, then we can find sequences $(u_n^{(1)})$, $(u_n^{(2)})$ in $\varphi_M(M)$, which convergw to $\xi^{(1)}$ and $\xi^{(2)}$ respectively. But then $\lim(u_n^{(1)} + u_n^{(2)}) = \xi^{(1)} + \xi^{(2)}$ and so $\widehat{f}(\xi^{(1)} + \xi^{(2)}) = \lim f^*(u_n^{(1)} + u_n^{(2)}) = \lim f^*(u_n^{(1)}) + \lim f^*(u_n^{(2)})$. Hence, $\widehat{f}(\xi^{(1)} + \xi^{(2)}) = \widehat{f}(\xi^{(1)}) + \widehat{f}(\xi^{(2)})$. Thus we see that $\widehat{f}$ is a homomorphism of R-modules.

Next assume that $\xi \in F_{\nu_k}\widehat{M}$. By Theorem 3.4.2., $F_{\nu_k}\widehat{M}$ is the closure of $\varphi_M(F_{\nu_k}M)$ in $\widehat{M}$ and therefore we can arrange that the sequence (u_n) is composed of elements of $\varphi_M(F_{\nu_k}M) = \varphi_M(M) \cap F_{\nu_k}\widehat{M}$. But in that case $f^*(u_n) \in F_k\widehat{N}$ for all n by virtue of $(*)$. Now, by property 3.1.(b)., $F_k\widehat{N}$ is closed in $\widehat{N}$ and therefore $\widehat{f}(\xi) = \lim f^*(u_n)$ belongs to $F_k\widehat{N}$ as well. It follows that $\widehat{f}(F_{\nu_k}\widehat{M}) \subseteq F_k\widehat{N}$ for all k. This shows that the homomorphism $\widehat{f}$ is continuous at the zero element. Hence it is continuous everywhere.

Again, if $\xi \in \varphi_M(M)$, then we may take for (u_n) the sequence whose terms are all equal to ξ and from this we see that $\widehat{f}(\xi) = f^*(\xi)$. Thus $\widehat{f}$ and f^* agree on $\varphi_M(M)$ and therefore the diagram

$$
\begin{array}{ccc}
M & \xrightarrow{f} & N \\
\varphi_M \downarrow & & \downarrow \varphi_N \\
\widehat{M} & \xrightarrow{\widehat{f}} & \widehat{N}
\end{array}
$$

is commutative.

Finally, we observe that if two mappings, $\widehat{f}_1$ and $\widehat{f}_2$ say, have the properties described in the statement of the theorem, then they must agree with f^* on $\varphi_M(M)$. Thus $\widehat{f}_1$ and $\widehat{f}_2$

are continuous and agree on an everywheredense subset of $\widehat{M}$. They must therefore agree everywhere. This completes the proof. (The last assertion of the theorem is obvious from the proof.)   □

6. Corollary. In the situation of the statement of Theorem 3.4.5. and supposing that the filtered homomorphism $f\colon M \to N$ has the Artin-Rees property (is strict). Then $\widehat{f}\colon \widehat{M} \to \widehat{N}$ has the Artin-Rees property (is also strict).

Proof. Let $\xi' \in \widehat{f}(\widehat{M}) \cap F_n\widehat{N}$. Then there exists $\xi \in \widehat{M}$ such that $\widehat{f}(\xi) = \xi'$. Next, since $\varphi_M(M)$ is everywhere dense in $\widehat{M}$, there exists $e \in M$ such that $\xi - \varphi_M(e) \in F_n\widehat{M}$. It follows that $\widehat{f}(\varphi_M(e)) \in F_n\widehat{N}$. But $\widehat{f}(\varphi_M(e)) = \varphi_N(f(e))$ and, by Theorem 3.4.2., $\varphi_N^{-1}(F_n\widehat{N}) = F_nN$. Thus $f(e) \in F_nN$ and hence $f(e)$ is in $f(M) \cap F_nN \subseteq f(F_{n+c}M)$ for some $c \in \mathbb{Z}$ because f has the Artin-Rees property. We now see that there is an element $x \in F_{n+c}M$ such that $f(x) = f(e)$ and therefore $\widehat{f}(\varphi_M(x)) = \widehat{f}(\varphi_M(e))$. Thus $\varphi_M(x) - \varphi_M(e)$ belongs to $\operatorname{Ker}\widehat{f}$ and so $\xi = (\xi - \varphi_M(e)) + \varphi_M(x) + (\varphi_M(e) - \varphi_M(x))$ is in $F_{n+c}\widehat{M} + \operatorname{Ker}\widehat{f}$. Finally, $\xi' = \widehat{f}(\xi) \in \widehat{f}(F_{n+c}\widehat{M} + \operatorname{Ker}\widehat{f}) = \widehat{f}(F_{n+c}\widehat{M}$. Hence the corollary has been proved.   □.

Theorem 3.4.5. shows that each continuous homomorphism $M \to N$ gives rise to a well defined continuous homomorphism between a given completion of M and a given completion of N. From here on the latter mapping will automatically be designated by putting a^over the symbol for the original mapping.

7. Theorem. Let M , N and K be filtered R-modules with filtrations FM, FN and FK respectively, and let $\widehat{M}$, $\widehat{N}$ and $\widehat{K}$ be their respectively given completions. Suppose that $f\colon M \to N$ and $g\colon N \to K$ are continuous R-homomorphisms. Then $g \circ f\colon M \to K$ is a continuous R-homomorphism and $\widehat{g \circ f} = \widehat{g} \circ \widehat{f}$.

Proof. This is obvious.   □

8. Corollary. Let M and N be filtered R-modules with filtrations FM and FN respectively. If $f\colon M \to N$ is a continuous isomorphism of filtered R-modules, i.e., f has an inverse continuous isomorphism $g\colon N \to M$, then $\widehat{f}\colon \widehat{M} \to \widehat{N}$ is a continuous isomorphism and its inverse is $\widehat{g}$.

Now we are ready to mention the following

9. Theorem. Let M be a filtered R-module with filtration FM. Suppose that $\widehat{M}$ with filtration $F\widehat{M}$ and $\widehat{M}'$ with filtration $F\widehat{M}'^{\cdot}$ are two completions of M. Then there exists one and only one filtered morphism $\omega\colon \widehat{M} \to \widehat{M'}$ such that the diagram

$$
\begin{array}{ccc}
 & M & \\
\varphi_M \swarrow & & \searrow \varphi'_M \\
\widehat{M} & \underset{\omega}{\longrightarrow} & \widehat{M'}
\end{array}
$$

is commutative, where φ_M and φ'_M are the canonical homomorphisms associated with the given completions. The mapping ω has the further properties that it is an isomorphism of the R-module $\widehat{M}$ onto the R-module $\widehat{M'}$ and $\omega(F_n\widehat{M}) = F_n\widehat{M'}$ for all $n \in \mathbb{Z}$.

Let M and N be filtered R-modules with their respective filtrations FM and FN. Let $f\colon M \to N$ be an R-morphism and $r \in R$. In general the mapping $\varphi\colon M \to N$ defined by $\varphi(e) = rf(e)$ is not a homomorphism because we are not assuming that R is commutative. However, in the special case where r belongs to the centre of R, φ will be R-linear. This observation prepared us for:

10. Theorem. Let M and N be in R-filt with filtrations FM and FN respectively, and let $\widehat{M}$ and $\widehat{N}$ be their respectively given completions. Furthermore, let f, f_1 and f_2 be continuous homomorphisms of M into N and let γ belong to the centre of R and $\gamma \in F_0 R$. Then $f_1 + f_2$, $f_1 - f_2$ and γf are continuous homomorphisms of M into N and $\widehat{f_1 + f_2} = \hat{f}_1 + \hat{f}_2$, $\widehat{f_1 - f_2} = \hat{f}_1 - \hat{f}_2$ and $\widehat{\gamma f} = \gamma \hat{f}$.

Proof. Note that the diagram

$$
\begin{array}{ccc}
M & \xrightarrow{f_1 + f_2} & N \\
\varphi_M \downarrow & & \downarrow \varphi_N \\
\widehat{M} & \xrightarrow{\hat{f}_1 + \hat{f}_2} & \widehat{N}
\end{array}
$$

is commutative (where φ_M and φ_N are the canonical mappings), it follows from Theorem 3.4.5. that $\widehat{f_1 + f_2} = \hat{f}_1 + \hat{f}_2$. The other two equalities may be established in a similar way. $\square$

11. Corollary. Suppose that $f\colon M \to N$ is a null homomorphism. Then so is $\hat{f}\colon \widehat{M} \to \widehat{N}$.

Suppose now that M, N, K are filtered R-modules with filtrations FM, FN and FK respectively, and that f, g are filtered morphisms

$$ M \xrightarrow{f} N \xrightarrow{g} K $$

this gives rise to the filtered morphisms

$$ \widehat{M} \xrightarrow{\hat{f}} \widehat{N} \xrightarrow{\hat{g}} \widehat{K} $$

(we are assuming that each of the given modules possesses a completion).

Next, for each $n \in \mathbb{Z}$, we have a commutative diagram

$$(*) \qquad \begin{array}{ccccc} M/F_nM & \xrightarrow{f_n} & N/F_nN & \xrightarrow{g_n} & K/F_nK \\ {\scriptstyle\cong}\downarrow & & {\scriptstyle\cong}\downarrow & & {\scriptstyle\cong}\downarrow \\ \widehat{M}/F_n\widehat{M} & \xrightarrow[\hat{f}_n]{} & \widehat{N}/F_n\widehat{N} & \xrightarrow[\hat{g}_n]{} & \widehat{K}/F_n\widehat{K} \end{array}$$

here the mappings in the upper row are induced by f and g whereas those in the lower row arise similarly from $\hat{f}$ and $\hat{g}$; The vertical mappings, on the other hand, come from the canonical mappings of M, N and K into their respective completions. Hence, by Corollary 3.4.4., all the vertical mappings are isomorphisms.

12. Lemma. Let the situation be as described above and suppose, in addition, that f has the Artin-Rees property and

$$M/F_nM \longrightarrow N/F_nN \longrightarrow K/F_nK$$

is exact for every $n \in \mathbb{Z}$. Then the sequence

$$\widehat{M} \xrightarrow{\hat{f}} \widehat{N} \xrightarrow{\hat{g}} \widehat{K}$$

is also exact.

Proof. The properties of the foregoing diagram $(*)$ show that

$$\widehat{M}/F_n\widehat{M} \longrightarrow \widehat{N}/F_n\widehat{N} \longrightarrow \widehat{K}/F_n\widehat{K}$$

is exact for every $n \in \mathbb{Z}$. Also, by Corollary 3.4.6., $\hat{f}$ has the Artin-Rees property. This means that for the purposes of the proof we may assume that M, N and K are themselves complete and by deducing that

$$M \xrightarrow{f} N \xrightarrow{g} K$$

is exact, establish the lemma.

Let $x \in M$. Since the result of composing $M/F_nM \xrightarrow{f_n} N/F_nN$ and $N/F_nN \xrightarrow{g_n} K/F_nK$ is a null map, it follows that $gf(x) \in F_nK$ for all n. But $\cap_{n \in \mathbb{Z}} F_nK = 0$ because K is complete. Hence $gf(x) = 0$ and therefore gf is a null homomorphism.

Next, let $\xi \in \text{Ker}g$. Then $\xi + F_nN \subseteq \text{Ker}g_n = \text{Im}f_n$ for all $n \in \mathbb{Z}$ by assumption. Since f has the Artin-Rees property, there is an integer $c \geq 0$ such that $F_nN \cap \text{Im}f \subseteq f(F_{n+c}M)$ for all $n \in \mathbb{Z}$. Take any $\eta_0 \in M$ such that $f(\eta_0) - \xi \in F_{-c}N$. Since $\xi + F_{-c-1}N \subseteq \text{Ker}g_{-c-1} = \text{Im}f_{-c-1}$, there is another $\eta^* \in M$ such that $f(\eta^*) - \xi \in F_{-c-1}N$. Hence

$f(\eta^* - \eta_0) \in F_{-c}N \cap \operatorname{Im} f \subseteq f(F_0 M)$. Let $f(\eta^* - \eta_0) = f(x_0)$ with $x_0 \in F_0 M$. If we put $\eta_1 = \eta_0 + x_0$, then $\eta_1 - \eta_0 \in F_0 M$. Since $f(\eta_1) = f(\eta_0) + f(x_0) = f(\eta_0) + f(\eta^*) - f(\eta_0) = f(\eta^*)$ we also have $f(\eta_1) - \xi \in F_{-c-1}N$. By an inductive argument we may construct a sequence $\eta_0, \eta_1, \eta_2, \cdots$ of elements of M such that

(a). $f(\eta_n) - \xi \in F_{-c-n}N$,

(b). $\eta_{n+1} - \eta_n \in F_{-n}M$.

Hence by Proposition 3.3.3. and above (b). it follows that $\lim \eta_n = \eta$ (say) exists. Finally from above (a). and the fact that f is continuous we obtain $\xi = \lim f(\eta_n) = f(\lim \eta_n) = f(\eta)$ as required. □

13. Theorem. Let M, N and K be in R-filt with filtrations FM, FN and FK respectively, and let $\widehat{M}$, $\widehat{N}$ and $\widehat{K}$ be the given completions of M, N and K respectively. Furthermore, let

$$(*) \qquad\qquad M \xrightarrow{\ f\ } N \xrightarrow{\ g\ } K$$

be an exact sequence in which f and g are strict filtered morphisms. Then for each $n \in \mathbb{Z}$, the induced sequence

$$(**) \qquad\qquad M/F_n M \longrightarrow N/F_n N \longrightarrow K/F_n K$$

is exact. In addition, $\widehat{f}$ and $\widehat{g}$ are strict filtered morphisms and the sequence

$$\widehat{M} \xrightarrow{\ \widehat{f}\ } \widehat{N} \xrightarrow{\ \widehat{g}\ } \widehat{K}$$

is also exact.

Proof. It is only necessary to show that the sequence $(**)$ is an exact sequence for then the theorem will follow by virtue of Corollary 3.4.6. and Lemma 3.4.12.. But the exactness of $(**)$ follows immediately from the exactness of $(*)$ and the strictness of g. □

14. Proposition. Let M be in R-filt with two filtrations FM and $F'M$. Suppose that FM and $F'M$ are topologically equivalent, then the identity map $i: M \to M$ induces the isomorphism $\widehat{i}: \widehat{M} \to \widehat{M}'$, where $\widehat{M}$ resp. $\widehat{M}'$ is the completion of M with respect to FM resp. to $F'M$ (at the moment we are assuming that the completion of M exists).

Proof. This is a direct result of Corollary 3.4.8.. □

15. Proposition. Let $f: M \to N$ be a filtered morphism in R-filt, where M and N have their rspective filtrations FM and FN. Suppose that f has the Artin-Rees Property, then the sequence of filtered morphisms

$$0 \longrightarrow \widehat{\operatorname{Ker} f} \longrightarrow \widehat{M} \xrightarrow{\ \widehat{f}\ } \widehat{N} \longrightarrow \widehat{N/f(M)} \longrightarrow 0$$

is exact, where $\mathrm{Ker}f$ has the filtration induced by FM and $N/f(M)$ has the quotient filtration given by FN, and moreover, the mapping $\widehat{\mathrm{Ker}f} \to \widehat{M}$ is inducedby the inclusion mapping $\mathrm{Ker}f \hookrightarrow M$ and the mapping $\widehat{N} \to N/\widehat{f(M)}$ is induced by the natural mapping $N \to N/f(M)$.

Proof. Since the exact sequence

$$0 \longrightarrow \mathrm{Ker}f \longrightarrow M \overset{f}{\longrightarrow} N \longrightarrow N/f(M) \longrightarrow 0$$

may be decomposed into the following exact sequence

$$
\begin{array}{ccccccccc}
 & & & & & 0 & & & \\
 & & & & & \downarrow & & & \\
0 & \to & \mathrm{Ker}f & \to & M & \to & M/\mathrm{Ker}f & \overset{f^*}{\to} & f(M) & \to & 0 \\
 & & & & & & & & i\downarrow & & \\
 & & & & 0 & \to & f(M) & \to & N & \to & N/f(M) & \to & 0 \\
 & & & & & & & & \downarrow & & \\
 & & & & & & & & 0 & & \\
\end{array}
$$

where f^* is the canonical isomorphism induced by f, and i is the identity map. If $f(M)$ is endowed with the filtrations $\{f(F_nM),\ n \in \mathbb{Z}\}$ and $\{f(M) \cap F_nN,\ n \in \mathbb{Z}\}$, then since f has the Artin-Rees property, f^* is a strict filtered morphism and i is a continuous isomorphism. Now the proof can be completed by using Theorem 3.4.7., Theorem 3.4.13. and Proposition 3.4.14.. $\square$

From the above proposition we immediately arrive at the following

16. Theorem. Consider in R-filt the exact sequence

$$M \overset{f}{\longrightarrow} N \overset{g}{\longrightarrow} K$$

where M, N and K have their respective filtrations FM, FN and FK, if f and g have the Artin-Rees property then the sequence

$$\widehat{M} \overset{\widehat{f}}{\longrightarrow} \widehat{N} \overset{\widehat{g}}{\longrightarrow} \widehat{K}$$

is exact.

Remark. From the argumentation of the foregoing results we have seen that the Artin-Rees property on f may be replaced by the condition: the filtrations $\{f(F_nM),\ n \in \mathbb{Z}\}$ and $\{f(M) \cap F_nN,\ n \in \mathbb{Z}\}$ on $f(M)$ are topologically equivalent. Similarly, this can be done for g.

3.5. The existence of completions

In the last section we established a number of properties of the completion associated with a filtered module, but we still have to show that a filtered module always possesses a completion.

Let R be a filtered ring with filtration $FR = \{F_nR,\ n \in \mathbb{Z}\}$ and $M \in R$-filt with filtration $FM = \{F_nM,\ n \in \mathbb{Z}\}$. Consider the projective system

$$\{M/F_pM,\ \psi_q^p : M/F_pM \to M/F_qM,\ p \leq q\}$$

we have the (canonical) inverse limit, denoted $\widehat{M} = \varprojlim M/F_pM$, which, of course, is an additive group. Put

$$\pi_p : \qquad \widehat{M} \to M/F_pM,\ \text{the canonical morphisms,}$$

$$u_p : \qquad M \to M/F_pM,\ \text{the natural morphisms,}$$

$$\varphi_M : \qquad M \to \widehat{M},\ \text{the canonical morphism with } \varphi_M(x) = (u_p(x))_{p \in \mathbb{Z}}.$$

Then one easily sees that

(a). for any $\xi \in \widehat{M}$, $\xi = (u_p(x_p))_{p \in \mathbb{Z}}$, $x_p \in M$ such that $x_p - x_{p+1} \in F_{p+1}M$;

(b). $\pi_p \circ \varphi_M = u_p$, $p \in \mathbb{Z}$;

(c). π_p is surjective, $p \in \mathbb{Z}$;

(d). $\operatorname{Ker}\pi_p = \{\xi = (u_k(x_k))_{k \in \mathbb{Z}},\ x_k \in F_pM\} = \varprojlim_{r \leq p} F_pM/F_rM \subseteq \operatorname{Ker}\pi_{p+1}$;

(e). Put $F_n\widehat{M} = \operatorname{Ker}\pi_n$, $n \in \mathbb{Z}$. Let $\xi = (u_p(x_p))_{p \in \mathbb{Z}}$ be an element of $\widehat{M}$. If $p_0 \in \mathbb{Z}$, for the element x_{p_0} there exists a $k \in \mathbb{Z}$ such that $x_{p_0} \in F_kM$ (since we always assume that FM is exhaustive). If $k \leq p_0$ then $x_{p_0} \in F_{p_0}M$ and consequently $u_{p_0}(x_{p_0}) = 0$, hence $\pi_{p_0}(\xi) = 0$ and thus $\xi \in F_{p_0}\widehat{M}$. If $p_0 \leq k$, we get $x_k \in F_kM$ and consequently $\xi \in F_k\widehat{M}$. Thus $\cup_{n \in \mathbb{Z}} F_n\widehat{M} = \widehat{M}$ and hence $\{F_n\widehat{M},\ n \in \mathbb{Z}\}$ forms an (exhaustive) filtration on $\widehat{M}$.

(f). Let $\xi = (u_p(x_p))_{p \in \mathbb{Z}}$ be an element of $\widehat{M}$, $r \in R$. Define

$$r\xi = \begin{cases} (u_{p+k}(rx_p))_{p \in \mathbb{Z}}, & \text{if } r \in F_kR - F_{k-1}R; \\ 0, & \text{if } r \in \cap_{n \in \mathbb{Z}} F_nR, \end{cases}$$

then $\widehat{M}$ becomes a filtered R-module and φ_M becomes an R-linear morphism.

1. Proposition. With notations as above, $\widehat{M}$ has the following properties:

(a). $\operatorname{Ker}\varphi_M = \cap_{n \in \mathbb{Z}} F_nM$;

(b). φ_M is a strict filtered morphism;

(c). $\varphi_M(M)$ is everywhere dense in $\widehat{M}$;

(d). $\widehat{M}$ is complete (in the sense of Definition 3.3.2.) with respect to its filtration topology. Hence $(\widehat{M}, \varphi_M)$ is a completion of M in the sense of Definition 3.4.1..

Proof. (a). Obvious.

(b). It is easy to see that φ_M is a filtered morphism. Let $\xi \in \varphi_M(M) \cap F_n\widehat{M}$, then by definition $\xi = \varphi_M(x) = (u_p(x_p))_{p \in \mathbf{Z}} = (u_k(x_k))_{k \in \mathbf{Z}}$, $x \in M$, $x_k \in F_nM$ for all $k \in \mathbb{Z}$. Hence $x \in F_nM$, i.e., $\xi \in \varphi_M(F_nM)$. This shows that φ_M is strict.

(c). Let $\xi = (u_p(x_p))_{p \in \mathbf{Z}}$ be an element of $\widehat{M}$. Let us denote $\xi_i = \varphi_M(x_{-i})$ for all $i \geq 1$. As $\pi_{-i}(\xi - \xi_i) = 0$ one deduces $\xi - \xi_i \in F_{-i}\widehat{M}$. Therefore $\xi = \lim_{i \to \infty} \xi_i$, and consequently $\varphi_M(M)$ is dense in $\widehat{M}$.

(d). Obviously $\cap_{n \in \mathbf{Z}} F_n\widehat{M} = 0$. Let $(\xi_i)_{i \geq 1}$ be a Cauchy sequence of $\widehat{M}$. This sequence is equivalent to a sequence $(\eta_i)_{i \geq 1}$ with the following property: $\eta_{i+1} - \eta_i \in F_{-i}\widehat{M}$ for any $i \geq 0$. Indeed, there exists a natural number N_i such that $\xi_r - \xi_s \in F_{-i}\widehat{M}$ for $r, s \geq N_i$. We get the ascending sequence $N_1 < N_2 < \cdots$. Let $\eta_i = \xi_{N_i}$, then the sequence $(\eta_i)_{i \geq 1}$ satisfies the required condition. Therefore in order to show that $\widehat{M}$ is complete it will be sufficient to establish that all sequences $(\xi_i)_{i \geq 1}$, with the property $\xi_{i+1} - \xi_i \in F_{-i}\widehat{M}$, converge. Let $\xi_i = (u_p(x_p^i))_{p \in \mathbf{Z}}$, $(i \geq 1)$. So $x_{p+1}^i - x_p^i \in F_{p+1}M$, $p \in \mathbb{Z}$, $i \geq 1$. As $\xi_{i+1} - \xi_i \in F_{-i}\widehat{M}$, then $x_{-i}^{i+1} - x_{-i}^i \in F_{-i}M$ for all $i \geq 1$. As $x_{p+1}^{i+1} - x_p^{i+1} \in F_{p+1}M$, then for $p = -i - 1$ we get $x_{-i}^{i+1} - x_{-(i+1)}^{i+1} \in F_{-i}M$ and so $x_{-i}^i - x_{-(i+1)}^{i+1} \in F_{-i}M$. If we now denote $\xi = (y_i)_{i \in \mathbf{Z}}$ where $y_{-i} = u_{-i}(x_{-i}^i)$ for $i \geq 1$ and $y_i = u_j(x_{-i}^1)$ for $j \geq 0$, it follows that $\xi \in \widehat{M}$. On the other hand, $\xi - \xi_i \in F_{-i}\widehat{M}$ for $i \geq 1$, therefore $\xi = \lim_{i \to \infty} \xi_i$. Consequently $\widehat{M}$ is complete. $\square$

2. Remark. There is a more down-to-earth description of $\widehat{M}$ (we leave the details to the interested reader). Assume $\cap_{n \in \mathbf{Z}} F_nM = 0$. If $x \in M$, and $x \neq 0$, there exists k with $x \in F_kM - F_{k-1}M$; define $\|x\| = 2^k$; If $x = 0$, define $\|x\| = 0$. It is easy to see that $\|x - y\|$ is a metric on M. It turns out that $\widehat{M}$ is the completion of M in the sense of metric spaces.

In order to make $\widehat{R}$ into a filtered ring and $\widehat{M}$ a filtered $\widehat{R}$-module, we first recall the R-module structure of $\widehat{M}$: for any $r \in R$ and $\xi \in \widehat{M}$ with $\xi = (u_p(x_p))_{p \in \mathbf{Z}}$, $r\xi = (u_{p+k}(rx_p))_{p \in \mathbf{Z}}$ if $r \in F_kR - F_{k-1}R$, and $r\xi = 0$ if $r \in \cap_{n \in \mathbf{Z}} F_nR$.

Let $\xi = (u_p(x_p))_{p \in \mathbf{Z}} \in \widehat{R}$ and $\eta = (u_p(y_p))_{p \in \mathbf{Z}} \in \widehat{M}$. Then $\xi = \lim \varphi_R(x_i)$ for some $x_i \in R$, and for any $p \geq 0$, there exists an integer $N(p) > 0$ such that $\xi - \varphi_R(x_s) \in F_{-p}\widehat{R}$ whenever $s \geq N(p)$. If s, $t \geq N(p)$, then

$$\varphi_R(x_s - x_t) = \varphi_R(x_s) - \varphi_R(x_t) \in F_{-p}\widehat{R} \cap \varphi_R(R) = \varphi_R(F_{-p}R)$$

It follows that $x_s - x_t - y \in \mathrm{Ker}\varphi_R = \cap_{n \in \mathbb{Z}} F_n R$, where $y \in F_{-p}R$, and hence $x_s - x_t \in F_{-p}R$. If $\eta \in F_d\widehat{M}$ then for $p \geq 0$ such that $p + d \geq 0$, there is an integer $N(p + d) > 0$ such that whenever s, $t \geq N(p + d)$, $x_s - x_t \in F_{-p-d}R$ and consequently $x_s\eta - x_t\eta = (x_s - x_t)\eta \in F_{-p-d}RF_d\widehat{M} \subseteq F_{-p}\widehat{M}$. This shows that $(x_i\eta)_{i \geq 1}$ is a Cauchy sequence in $\widehat{M}$. Also if $(x_i')_{i \geq 1}$ is another sequence of elements of R for which $\xi = \lim\varphi_R(x_i')$, then one can easily check that $(x_i\eta - x_i'\eta)_{i \geq 1}$ converges to zero. It is therefore in order to put

(∗)
$$\xi\eta = \lim(x_i\eta).$$

From this definition we immediately obtain

(a). $\varphi_R(1)\eta = \eta$ for all $\eta \in \widehat{M}$, and
(b). $\varphi_R(x)\varphi_M(y) = \varphi_M(xy)$ for all $x \in R$, $y \in M$.
Moreover, if $\alpha \in F_d\widehat{R}$, $\beta \in F_m\widehat{R}$ and $\eta \in F_n\widehat{M}$, then we claim that

$$\alpha(\beta\eta) = (\alpha\beta)\eta.$$

Indeed, choose sequences (x_i) and (y_i) of elements of R, so that $\alpha = \lim\varphi_R(x_i)$ and $\beta = \lim\varphi_R(y_i)$. Then since $x_i \in F_d R$ and $\beta\eta = \lim(y_i\eta)$, for $p \geq 0$ such that $p + d \geq 0$, there is $N(p + d) > 0$ such that $\beta\eta - y_s\eta \in F_{-p-d}\widehat{M}$ whenever $s \geq N(p + d)$. Accordingly $x_s(\beta\eta) - (x_s y_s)\eta \in F_{-p}\widehat{M}$. By definition $\alpha(\beta\eta) = \lim(x_i(\beta\eta))$. Also $\varphi_R(x_i y_i) = \varphi_R(x_i)\varphi_R(y_i)$ tends to $\alpha\beta$. Hence $(x_i y_i)\eta \to (\alpha\beta)\eta$. It follows that $\alpha(\beta\eta) - (\alpha\beta)\eta \in F_{-p}\widehat{M}$ for all $p \geq 0$. This shows that $\alpha(\beta\eta) = (\alpha\beta)\eta$. For $\alpha \in \widehat{R}$, β, $\eta \in \widehat{M}$, the verification of the formulas $\alpha(\beta + \eta) = \alpha\beta + \alpha\eta$ and $(\beta + \eta)\alpha = \beta\alpha + \eta\alpha$ are also quite straightforward and will be left to the reader.
Therefore, we have just proved the following

3. Proposition. With notations as above, we have
(1). the filtered R-module $\widehat{R}$ is a filtered ring and the canonical map φ_R is a filtered ring homomorphism;
(2). for any $M \in R$-filt with filtration FM, the filtered R-module $\widehat{M}$ is also a filtered $\widehat{R}$-module.

4. Remark. (1). If $F_0 R = R$, then the subgroups $F_n R$, $n \in \mathbb{Z}$ are two-sided ideals of R. Hence the multiplication on $\widehat{R}$ may be simplified as follows: if ξ, $\eta \in \widehat{R}$ with $\xi = (u_p(x_p))_{p \in \mathbb{Z}}$

and $\eta = (u_p(y_p))_{p \in \mathbb{Z}}$ then $\xi\eta = (u_p(x_p y_p))_{p \in \mathbb{Z}}$. In this case the subgroups $F_n \hat{R}$, $n \in \mathbb{Z}$, are two-sided ideals of $\hat{R}$. Moreover, if $F_0 M = M$, then the module structure of $\hat{M}$ is given by the following definition: if $\alpha = (u_p(y_p))_{p \in \mathbb{Z}}$ is an element of $\hat{M}$, then $\xi\alpha = (u_p(x_p y_p))_{p \in \mathbb{Z}}$.
(2). From the discussion of section 3.4., in particular from the proof of Theorem 3.4.5., we see that the application $M \to \hat{M}$ defines a functor $\widehat{(-)}$: R-filt $\to \hat{R}$-filt.

Let us finish this section with two useful lemmas.

5. Lemma. Let R be a filtered ring with filtration FR.
(1). Suppose $F_0 R = R$. Denote by Ω the subset of R consisting of all elements x such that x^n converges to zero as n tends to infinity. Then Ω is closed with respect to the filtration topology of R.
(2). If R is complete then $F_{-1} R \subseteq J(F_0 R)$, where $J(F_0 R)$ denotes the Jacobson radical of the subring $F_0 R$.

Proof. (1). Let y be an element in the closure $\overline{\Omega}$ of Ω. Then for any $k \in \mathbb{Z}$, $y + F_k R$ meets Ω and therefore $y - x \in F_k R$ for some $x \in \Omega$. But $F_k R$ is now a two-sided ideal of R. Consequently $y^n - x^n \in F_k R$ for all $n \geq 1$. However $x^n \in F_k R$ for all large n and therefore $y^n \in F_k R$ when n is large. Thus $y^n \to 0$, that is to say $y \in \Omega$ as desired.
(2). Let $x \in F_{-1} R$, then $x^n \to 0$ and therefore, by Proposition 3.3.4., the series $1 + x + x^2 + x^3 + \cdots$ converges. Let y be its sum. Then since

$$(1 + x + x^2 + x^3 + \cdots + x^n)(1 - x) = 1 - x^{n+1}$$

for all $n \geq 0$, it follows that $y(1 - x) = 1$. Since $F_{-1} R$ is a two-sided ideal of $F_0 R$ we have $F_{-1} R \subseteq J(F_0 R)$. □

The next lemma is a classical result concerning topological rings. We refer the reader to [Sj] or [NVO1] for a complete proof of it.

6. Lemma. Let R be a complete topological ring and consider a fundamental set of neighbourhoods of zero, consisting of additive subgroups. If φ: $R \to S$ is a ring homomorphism such that every $x \in \mathrm{Ker}\varphi$ is topologically nilpotent (i.e., the sequence $(x^n)_{n \geq 1}$ is a Cauchy sequence converging to zero), then every idempotent element of $\mathrm{Im}\varphi$ may be lifted to R.

§4. Filtrations and Associated Gradations

4.1. Preliminaries on gradations

For a general theory of graded rings we refer to [NVO1]. Here we only recall some basic notions and facts on graded rings.

Let R be an associative ring with identity 1. R is said to be a **graded ring of type** G (or a G-graded ring), where G is any group, if $R = \oplus_{\sigma \in G} R_\sigma$ where the R_σ, $\sigma \in G$, are additive subgroups of R satisfying $R_\sigma R_\tau \subseteq R_{\sigma\tau}$ for all σ, $\tau \in G$. If $R_\sigma R_\tau = R_{\sigma\tau}$ for all $\sigma, \tau \in G$, then R is called a **strongly graded ring of type** G (or a strongly G-graded ring).

From the definition it follows that $1 \in R_e$, and R_e is a subring of R where e is the neutral element of G.

Let R be a graded ring of type G. An R-module M is said to be a **graded** R-module if $M = \oplus_{\sigma \in G} M_\sigma$ for additive subgroups M_σ, $\sigma \in G$, of M such that $R_\sigma M_\tau \subseteq M_{\sigma\tau}$ for all σ, $\tau \in G$.

The elements of $h(R) = \cup_{\sigma \in G} R_\sigma$ resp. $h(M) = \cup_{\sigma \in G} M_\sigma$ are called **homogeneous elements** of R resp. of M. If $m \neq 0$ and $m \in M_\sigma$, $\sigma \in G$, then m is said to be a **homogeneous element of degree** σ, written $\deg m = \sigma$. Any nonzero $m \in M$ may be written in a unique way as a finite sum $m = m_{\sigma_1} + \cdots + m_{\sigma_n}$ where $m_{\sigma_j} \neq 0$, $j = 1, \cdots, n$ and $m_{\sigma_j} \in M_{\sigma_j}$. The elements m_{σ_j} are called the **homogeneous components** of m.

Let R be a graded ring of type G, where G is an **ordered group**. A graded R-module M is said to be **left limited (resp. right limited)** if there is a $\sigma_0 \in G$ such that $M_\sigma = 0$ for all $\sigma < \sigma_0$ (resp. for all $\sigma > \sigma_0$). If $M_\sigma = 0$ for each $\sigma < e$ then M is said to be **positively graded** and if $M_\sigma = 0$ for all $\sigma > e$ then M is **negatively graded**. It is clear that a strongly graded ring of type G cannot be left (or right) limited (note that finite groups and groups containing torsion elements cannot be ordered).

It is obvious how to define graded right modules and graded bimodules.

Let R be a graded ring of type G and M a graded R-module. An R-submodule N of M is said to be a **graded submodule** if $N = \oplus_{\sigma \in G}(M_\sigma \cap N)$; or equivalently, if for every $x \in N$ the homogeneous components of x are again in N. If N is a graded R-submodule of M then $M/N = \oplus_{\sigma \in G}(M_\sigma + N)/N$ is again a G-graded R-module.

Let R and S be two G-graded rings. A ring homomorphism $g: R \to S$ is said to be a **graded ring homomorphism of degree** τ if $g(R_\sigma) \subseteq S_{\sigma\tau}$ for all $\sigma \in G$.

Let R be a G-graded ring and consider graded R-modules M and N. An R-linear mapping $f: M \to N$ is said to be a **graded morphism of dgree** τ if $f(M_\sigma) \subseteq M_{\sigma\tau}$ for all $\sigma \in G$. Morphisms of degree τ form an additive subgroup of $\mathrm{Hom}_R(M, N)$ which we will denote by $\mathrm{HOM}_R(M, N)_\tau$, and $\mathrm{HOM}_R(M, N) = \oplus_{\tau \in G} \mathrm{HOM}_R(M, N)_\tau$ is a graded abelian group of type G.

The category R-gr consists of **graded (left)** R-modules and the graded morphisms

of degree e. We will write $\mathrm{Hom}_{R\text{-gr}}(M, N) = \mathrm{HOM}_R(M, N)_e$. This category is an abelian category.

Forgetting the graded structure defines a **forgetful functor** U: R-gr $\to$ R-mod. We will sometimes write $U(M) = \underline{M}$ but on most occasions we omit this distinction and just write M hoping that it will be clear from the context whether the graded or the ungraded module is being considered.

For each $\tau \in G$, one may define the **shift functor** $T(\tau)$: R-gr $\to$ R-gr, associating to $M \in R$-gr the graded R-module obtained by defining on $\underline{M}$ a new gradation given by $(T(\tau)M)_\sigma = M_{\tau\sigma}$. We will write $M(\tau)$ for $T(\tau)M$.

The family $\{R(\tau),\ \tau \in G\}$ is a family of generators for R-gr and consequently the graded R-module $\oplus_{\tau \in G} R(\tau)$ is a generator of R-gr. It follows that R-gr **has enough injective objects**.

A graded R-module F is said to be **gr-free** if it has a basis consisting of homogeneous elements, or equivalently if there is a family $\{\tau_i \in G,\ i \in I\}$ such that $F \cong \oplus_{i \in I} R(\tau_i)$ where the isomorphism is in R-gr. Any graded R-module is ismorphic to a quotient of a gr-free R-module, hence it follows that R-gr **has enough projective objects**. Note that a graded R-module L which is a free R-module need not be gr-free!

Let R be a G-graded ring. The following two lemmas from [NVO1] are fundamental in the study of graded rings.

1. Lemma. Let M and N be graded R-modules and suppose that M is finitely generated, then $\mathrm{HOM}_R(M, N) = \mathrm{Hom}_R(\underline{M}, \underline{N})$.

2. Lemma. Let M, N and P be graded R-modules and suppose that we are given a commutative diagram of R-linear maps:

$$\begin{array}{ccc} M & \xrightarrow{h} & N \\ & \searrow{\scriptstyle f}\swarrow{\scriptstyle g} & \\ & P & \end{array}$$

where f is a graded morphism of degree e. If g (h) has degree e then there exists a graded morphism h' (g') of degree e too, such that $f = g \circ h'$ $(f = g' \circ h)$.

3. Corollary. (a). A graded R-module P is a projective object in R-gr if and only if $\underline{P}$ is a projective R-module.

(b). Let M be a graded R-module. A graded R-submodule N of M is a direct summand in R-gr if and only if $\underline{N}$ is a direct summand of $\underline{M}$ in R-mod.

(c). The graded projective dimension of a graded R-module M, denoted $\mathrm{gr.p.dim}_R M$, is equal to $\mathrm{p.dim}_R M$, where the latter one is the usual projective dimension of M in R-mod.

Unfortunately, an injective object of R-gr need not be injective as an ungraded module (unlike what happens for projective objects! (cf. Remark 3.3.11. of [NVO1]). However we still have the following

4. Lemma. If $E \in R$-gr is such that $\underline{E}$ is injective then E is injective in R-gr.

Let R be a G-graded ring. Consider $\mathbb{Z}$ as a G-graded ring with the trivial gradation, i.e., $\mathbb{Z}_e = \mathbb{Z}$ and $\mathbb{Z}_\sigma = 0$ for all $\sigma \neq e$. For $M \in$ gr-R (the category of graded right R-modules), $N \in R$-gr we may consider the abelian group $\underline{M} \otimes_R \underline{N}$ and define a $\mathbb{Z}$-gradation on it by putting:

$$(\underline{M} \otimes_R \underline{N})_\tau = \left\{ \sum_{\alpha,\beta} x^{(\alpha)} \otimes y^{(\beta)}, \ x^{(\alpha)} \in M_\alpha, \ y^{(\beta)} \in N_\beta; \ \alpha\beta = \tau \right\}$$

The object obtained in $\mathbb{Z}$-gr is denoted by $M \otimes_R N$ and we call it the **tensor-product** of the graded modules M and N.

Let R° be the opposite ring of R, i.e., the abelian group of R with multiplication reversed. Then we also have the functor $\otimes_R$: R°-gr $\times$ R-gr $\to$ $\mathbb{Z}$-gr. The functor $- \otimes_R M$: R°-gr $\to$ $\mathbb{Z}$-gr is right exact. We say that M is **gr-flat** if the latter functor is also left exact.

5. Proposition. (a). For any α, $\beta \in G$, $T(\alpha\beta)(M \otimes_R N) = T(\alpha)(M) \otimes_R T(\beta)(N)$. When R and S are G-graded rings whereas $M \in$ gr-R, $N \in R$-gr-S, then $M \otimes_R N$ is a graded right S-module.

(b). If $M \in$ gr-R, $P \in$ gr-S, then there is a natural isomorphism: $\mathrm{HOM}_S(M \otimes_R N, P) \cong \mathrm{HOM}_R(M, \mathrm{HOM}_S(N, P))$.

(c). If $P \in$ gr-R, $N \in S$-gr, $M \in S$-gr-R then there exists a canonical morphism φ: $P \otimes \mathrm{HOM}_S(M, N) \to \mathrm{HOM}_S(\mathrm{HOM}_R(P, M), N)$, defined by $\varphi(p \otimes f)(g) = (f \circ g)(p)$ for $p \in P$, $f \in \mathrm{HOM}_S(M, N)$, $g \in \mathrm{HOM}_R(P, M)$.

6. Lemma. A graded R-module M is gr-flat if and only if $\underline{M}$ is flat as an R-module. If we denote the gr-flat dimension of M in R-gr by gr.w.dim$_R M$, then gr.w.dim$_R M = $ w.dim$_R \underline{M}$, where the latter one is the usual flat dimension of M in R-mod.

The following proposition concerns a strongly G-graded ring R.

7. Proposition. Let $R = \oplus_{\sigma \in G} R_\sigma$ be a strongly G-graded ring.
(1). Every graded R-module $M = \oplus_{\sigma \in G} M_\sigma$ is strongly graded in the sense that $R_\sigma M_\tau = M_{\sigma\tau}$ for all σ, $\tau \in G$.
(2). The functors $R \dot{\otimes}_{R_e} -$: R_e-mod $\to$ R-gr with $M \mapsto R \otimes_{R_e} M$, $(R \otimes_{R_e} M)_\sigma = R_\sigma \otimes_{R_e} M$, $\sigma \in G$, and $(-)_e$: R-gr $\to$ R_e-mod with $M \mapsto M_e$, define an equivalence between R-gr and R_e-mod.

(3). If M is a graded R-module and N is a graded R-submodule of M, then $N = RN_e$, where N_e is the part of degree e of N.

Finally, Let us recall that for a G-graded ring R, the **graded Jacobson radical** of R, denoted $J^g(R)$, is the largest proper graded ideal of R such that for all $a \in J^g(R) \cap R_e$ it follows that $1 + ar$, $r \in R_e$, is a unit in R_e. In other words, $J^g(R)$ is the largest proper graded ideal of R such that its intersection with R_e is contained in the Jacobson radical of R_e, where e is the neutral element of G.

8. Lemma. (Graded version of Nakayama's lemma) If $M \in R$-gr is finitely generated and I is any graded ideal of R contained in $J^g(R)$, then $IM = M$ if and only if $M = 0$.

Caution. In general, the graded Jacobson radical $J^g(R)$ of a graded ring R does not coincide with the Jacobson radical $J(R)$ of R. For example, let R be the commutative polynomial ring $k[t]$ over a field in variable t, if we consider the natural gradation on R: $R_n = kt^n$, $n \geq 0$, then $(t) = J^g(R) \neq J(R) = 0$.

4.2. Associated graded rings and modules

Let R be a filtered ring with filtration FR and let M be a filtered R-module with filtration FM.
Define the following abelian groups:

$$G(R) = \oplus_{n \in \mathbb{Z}} F_n R / F_{n-1} R,$$
$$G(M) = \oplus_{n \in \mathbb{Z}} F_n M / F_{n-1} M.$$

To an $x \in F_n M$, $n \in \mathbb{Z}$, we may associate its image $x_{(n)} = x + F_{n-1}M$ in $G(M)_n = F_n M / F_{n-1} M$. If $r \in F_i R$, $m \in F_j M$ then we define $r_{(i)} m_{(j)} = (rm)_{(i+j)}$ and extend this to a $\mathbb{Z}$-bilinear mapping $\mu_M: G(R) \times G(M) \to G(M)$. In case $M = R$, μ_R makes $G(R)$ into a $\mathbb{Z}$-graded ring; in general $G(M)$ is made into a $\mathbb{Z}$-graded $G(R)$-module by μ_M.

1. Definition. $G(R)$ resp. $G(M)$, as defined above with gradation: $G(R)_n = F_n R / F_{n-1} R$, $n \in \mathbb{Z}$, resp. $G(M)_n = F_n M / F_{n-1} M$, $n \in \mathbb{Z}$, is called the **associated graded ring** of R with respect to the given filtration FR resp. the **associated graded module** of M with respect to the given filtration FM.

If $m \in F_n M - F_{n-1} M$ for some $n \in \mathbb{Z}$ then we write $\sigma(m) = m_{(n)}$ and we say that $\sigma(m)$ is the **principal part** of m; We refer to σ as the **principal symbol** (map). Observe that σ is neithermultiplicative nor additive! Nevertheless if $a \in R$, $m \in M$ are such that $\sigma(a)\sigma(m) \neq 0$, then $\sigma(am) = \sigma(a)\sigma(m)$ does hold.

If we take an arbitrary filtered morphism $f \in \mathrm{Hom}_{FR}(M, N)$ for some filtered R-module M and N, then f induces canonical mappings $f_n \colon F_n M / F_{n-1} M \to F_n N / F_{n-1} N$ such that $x_{(n)} \mapsto f(x)_{(n)}$. It is clear that $G(f) = \oplus_{n \in \mathbb{Z}} f_n$ defines a graded morphism of degree zero from $G(M)$ to $G(N)$, i.e., $G(f) \in \mathrm{Hom}_{G(R)\text{-gr}}(G(M), G(N))$, and if $g \in \mathrm{Hom}_{FR}(N, K)$ is another filtered morphism then $G(g \circ f) = G(g) \circ G(f)$. Moreover if 1_M is the identity morphism in $\mathrm{Hom}_{FR}(M, M)$ then $G(1_M) = 1_{G(M)}$. Hence G defines a functor: R-filt $\to$ $G(R)$-gr.

One of the main tools in the study of filtered rings and modules is the "lifting" of information from the associated graded objects and morphisms. First we state some general properties of the functor G.

2. Proposition. (1). If $M \in R$-filt with separated filtration FM, then $M = 0$ if and only if $G(M) = 0$.

(2). If $M \in R$-filt with filtration FM, then for every $n \in \mathbb{Z}$, $G(T(n)M) = T(n)G(M)$, here we use the same $T(n)$ to represent the n-th shift functor on both R-filt and $G(R)$-gr.

(3). If $M \in R$-filt with discrete filtration FM, then $G(M)$ is left limited in the sense that $G(M)_i = 0$ for all $i < p$ for some $p \in \mathbb{Z}$.

(4). The functor G commutes with (filtered-)direct sums, (filtered-)products and (filtered-)inductive limits.

(5). If $M \in R$-filt with filtration FM, then the canonical (filtered) morphism $\varphi_M \colon M \to \widehat{M}$ induces a graded isomorphism $G(\varphi_M) \colon G(M) \to G(\widehat{M})$, where $\widehat{M}$ is the completion of M with respect to FM and $\widehat{M}$ has the canonical filtration as defined in §3..

Proof. The last assertion follows from Corollary 3.4.4.. The others may easily be verified, we leave this as an exercise for the reader. □

3. Definition. A sequence

$$L \xrightarrow{f} M \xrightarrow{g} N$$

in R-filt is called **strict exact** if it is an exact sequence of filtered R-modules such that both f and g are strict filtered morphisms (recall: all filtrations are exhaustive!).

4. Theorem. Consider the following sequence such that $g \circ f = 0$ in R-filt:

$$(*) \qquad L \xrightarrow{f} M \xrightarrow{g} N$$

where L, M and N have filtrations FL, FM and FN respectively, and consider the associated sequence in $G(R)$-gr:

$$G(*) \qquad G(L) \xrightarrow{G(f)} G(M) \xrightarrow{G(g)} G(N)$$

(1). If $(*)$ is strict exact then $G(*)$ is exact.

(2). If $G(*)$ is exact then g is strict.

(3). If $G(*)$ is exact, L is complete with rspect to its filtration topology and FM is separated then f is strict.

(4). If $G(*)$ is exact and FM is discrete then f is strict.

(5). If L is complete with respect to its filtration topology and FM is separated, or if FM is discrete then $(*)$ is strict exact if and only if $G(*)$ is exact.

Proof. (1). That $G(g) \circ G(f) = 0$ is obvious. If $x \in F_n M$ is such that $G(g)(x_{(n)}) = 0$ then $(g(x))_{(n)} = 0$ or $g(x) \in F_{n-1}N$. Since g is strict there is an $x' \in F_{n-1}M$ such that $g(x) = g(x')$, i.e., $x - x' = f(y)$ for some $y \in F_n L$. This entails: $G(f)(y_{(n)}) = x_{(n)}$ and therefore $\mathrm{Im}\,G(f) = \mathrm{Ker}\,G(g)$.

(2). Take $y \in (F_n N \cap \mathrm{Im}\,g) - F_{n-1}N$. There is an $x \in M$ such that $g(x) = y$, say $x \in F_{n+s}M$ for some $s \geq 0$. If $s = 0$ then there is nothing to prove. In case $s > 0$, $G(g)(x_{(n+s)}) = 0$ and the exactness of $G(*)$ implies that $x_{(n+s)} = G(f)(z_{(n+s)})$ for some $z \in F_{n+s}L$. Then $x - f(z) \in F_{n+s-1}M$ and $y = g(x) = g(x - f(z)) = g(x')$ with $x' \in F_{n+s-1}M$. repetition of this procedure leads to an $m \in F_n M$ such that $y = g(m)$.

(3). Look at $y \in F_n M \cap \mathrm{Im}\,f$. Exactness of $G(*)$ yields: $G(g)(y_{(n)}) = g(y)_{(n)} = 0$ and thus $y_{(n)} = G(f)(x^n_{(n)})$ for some $x^n \in F_n L$. Hence, $y - f(x^n) \in \mathrm{Im}\,f \cap F_{n-1}M$. By induction we obtain a sequence x^n, x^{n-1}, $\cdots$, x^{n-s} with $x^{n-s} \in F_{n-s}L$ such that $y - f(x^n) - \cdots - f(x^{n-s}) \in \mathrm{Im}\,f \cap F_{n-s-1}M$. Since L is complete we may define an element $x = \sum_{s=0}^{\infty} x^{n-s}$ in $F_n L$. Now we arrive at: $y - f(x) = y - \lim_{s \to \infty} f(x^n + x^{n-1} + \cdots + x^{n-s}) = 0$, the latter follows from the separatedness of FM. Therefore we obtain $y \in f(F_n L)$ and $f(F_n L) \subseteq F_n M \cap \mathrm{Im}\,f$ is obvious.

(4). Along the lines of (3).

(5). Strict exactness of $(*)$ implies exactness of $G(*)$ because of (1). On the other hand, if $G(*)$ is exact and $y \in M$, $y \neq 0$, is such that $g(y) = 0$ then $y \in F_n M - F_{n-1}M$ for some $n \in \mathbb{Z}$. Thus we obtain $G(g)(y_{(n)}) = 0$ and $y_{(n)} = G(f)(x^n_{(n)}) = f(x^n)_{(n)}$ for some $x^n \in F_n L$. Consequently $y - f(x^n) \in F_{n-1}M$ and by induction we arrive at $x^{n-s} \in F_{n-s}L$ such that $y - f(x^n + x^{n-1} + \cdots + x^{n-s}) \in F_{n-s-1}M$. If FM is discrete then $F_{n-s-1}M = 0$ for some s, accordingly we have $y = f(x^n + x^{n-1} + \cdots + x^{n-s})$ and $(*)$ will be exact. If L is complete and FM is separated then we can define $x = \sum_{s=0}^{\infty} x^{n-s}$ and get $y = f(x)$. The fact that both f and g are strict filtered morphisms now follows readily from (2) and (3). $\square$

Later we will establish a generalization of some of the results in the theorem, considerably enlarging the utility of the result, in particular for Zariskian filtrations.

5. Corollary. Let $f: M \to N$ be a morphism in R-filt. Suppose that FM and FN are separated.

(1). $G(f)$ is injective if and only if f is injective and strict.

(2). If M is complete with respect to its filtration topology, then $G(f)$ is an isomorphism in $G(R)$-gr if and only if f is an isomorphism in R-filt.

(3). If either M is complete with respect to its filtration topology or FM is discrete, then $G(f)$ is surjective if and only if f is surjective and f is strict.

6. Remark. Let us dwell somewhat longer on some properties of the functor G, pointing out some problems where caution is necessary.

(1). Given a filtered morphism $f: M \to N$ in R-filt with associated graded morphism $G(f): G(M) \to G(N)$ in $G(R)$-gr. We may define two filtrations on $\text{Im} f$ that are related to FN and FM (as we did in the proof of Proposition 3.4.15.), i.e., $F_n(\text{Im} f) = \text{Im} f \cap F_n N$, $F'_n(\text{Im} f) = f(F_n M)$, $n \in \mathbb{Z}$. Both filtrations give rise to associated graded mordules:

$$G_F(\text{Im} f) \cong \oplus_{n \in \mathbb{Z}}(\text{Im} f \cap F_n N + F_{n-1} N)/F_{n-1} N,$$
$$G_{F'}(\text{Im} f) = \oplus_{n \in \mathbb{Z}} f(F_n M)/f(F_{n-1} M).$$

On the other hand, for the graded morphism $G(f)$ we have

$$\text{Im} G(f) = \oplus_{n \in \mathbb{Z}} f(F_n M) + F_{n-1} N/F_{n-1} N$$

and so it is obvious that $\text{Im} G(f) = G_F(\text{Im} f)$ when f is strict, but this need not hold in general.

(2). If we are given a sequence

(*) $$M \xrightarrow{f} N \xrightarrow{g} T$$

in R-filt but it is not necessarily a complex, then the exactness of the associated sequence $G(*)$ does not learn us much about the exactness of $(*)$ and in general we cannot deduce that $(*)$ is a complex (except when $g \circ f$ is strict and FM is separated, but we have seen in section 3.2. that $g \circ f$ need not be strict even if g and f are strict).

4.3. Dehomogenizations of gradings to filtrations Rees rings (modules)

In projective algebraic geometry, homogeneous coordinate rings appear together with a suitable dehomogenization. For example, if $V(I)$ is a projective variety determined by a homogeneous prime ideal I of the polynomial ring $k[x_0, \cdots, x_n]$, where k is an algebraically closed field, and R is the graded coordinate ring $k[x_n, \cdots, x_n]/I$, then $A = R/(1 - \overline{x}_0)R$ is isomorphic to the coordinate ring of the open affine subvariety complementary to the hyperplane "at infinity" (defined by the vanishing of x_0) in $V(I)$. In a similar way every determinantal ring is a dehomogenization of a Schubert cycle (being the graded coordinate ring of a Schubert variety) and this dehomogenization principle is the basis for the study of determinantal rings (cf. [BV] for detail). We now extend this principle to general $\mathbb{Z}$-graded rings such that, on the one hand, every filtered ring is the dehomogenization of its Rees ring with respect to a canonical element; and on the other hand, every suitable dehomogenization of a $\mathbb{Z}$-graded ring is a filtered ring with the original graded ring as its Rees ring.

To begin with, Let $S = \oplus_{n \in \mathbb{Z}} S_n$ be any $\mathbb{Z}$-graded ring and T a **homogeneous element of degree 1** in S. Since for any homogeneous element $s_n \in S_n$ and $j > 0$ we have $s_n = T^j s_n + (1 - T^j) s_n$, it follows that the quotient ring $R = S/\langle 1 - T\rangle$, where $\langle 1 - T\rangle$ is the ideal of S generated by $1 - T$, may be made into a filtered ring by putting:

$$F_n R = S_n + \langle 1 - T\rangle/\langle 1 - T\rangle, \qquad n \in \mathbb{Z}.$$

Similarly, if $\overline{M} = \oplus_{n \in \mathbb{Z}} \overline{M}_n$ is a graded S-module, then the quotient module $M = \overline{M}/ < 1 - T > \overline{M}$ may be made into a filtered R-module with filtration:

$$F_n M = \overline{M}_n + \langle 1 - T\rangle\overline{M}/\langle 1 - T\rangle\overline{M}, \qquad n \in \mathbb{Z}.$$

1. Definition. With notations as above, the filtered ring R is called the **dehomogenization of** S with respect to T and, similarly, the filtered R-module M is called the **dehomogenization of** $\overline{M}$ with respect to T.

Obviously, for any $\overline{M} \in S$-gr, the filtration defined on the dehomogenization M of $\overline{M}$ above is **exhaustive**.

2. Lemma. With notations as above, suppose that T is a **regular homogeneous element of degree 1**, then
(1). $(1 - T)S \cap S_n = 0$, $n \in \mathbb{Z}$, and if $\overline{M}$ is a T-torsionfree graded S-module (i.e., $Tu = 0$ implies $u = 0$ for any element $u \in \overline{M}$) then $(1 - T)\overline{M} \cap \overline{M}_n = 0$ for all $n \in \mathbb{Z}$;
(2). if T is also a normalizing element (i.e., $TS = ST$) then T is in the centre of S if and only if $(1 - T)S$ is an ideal of S.

Proof. (1). This can be checked in a straightforward way. To prove (2)., let us consider the natural morphism $\pi: S \to S/(1 - T)S$. Since S is a graded ring and T is a normalizing element, if we look at the image of $Ts = s'T$, under the morphism π for any homogeneous element s in S, then property (1). yields the required equivalence immediately.   □

From now on, we fix the $\mathbb{Z}$-graded ring $S = \oplus_{n \in \mathbb{Z}} S_n$ and its homogeneous element T of degree 1; Moreover, we always assume that T is a **regular element contained in the**

centre of S. Let the quotient ring $R = S/(1-T)S$ have the filtration FR as described above.

3. Proposition. With notations and conventions as above,

(1). $G(R) \cong S/TS$ as graded rings;

(2). the localization of S at the (central) Ore set $\{1, T, T^2, \cdots\}$ exists, it is denoted by $S_{(T)}$, and the natural homomorphism $\pi\colon S \to R$ factors through $S_{(T)}$ in a canonical way. Accordingly there is a commutative diagram of ring homomorphisms:

$$
\begin{array}{ccc}
S & \longrightarrow & S_{(T)} \\
& \searrow_{\pi} \quad \swarrow_{\psi} & \\
& R &
\end{array}
$$

where $S_{(T)}$ is still a graded ring: $S_{(T)} = \oplus_{n \in \mathbb{Z}}[S_{(T)}]_n$, $[S_{(T)}]_n = \{T^j s, \ j \in \mathbb{Z}, \ s \in S_{n-j}\}$, and $\psi(T^j s) = s + (1-T)$. Note that $S_{(T)}$ is a strongly $\mathbb{Z}$-graded ring and its structure is particularly simple (cf. [NVO] Ch.1.): $S_{(T)} \cong [S_{(T)}]_0[t, t^{-1}]$, where the latter one is the ring of finite Laurent series over $[S_{(T)}]_0$ in the variable t and t corresponds to T;

(3). with notations as in (2), the homomorphism ψ maps $[S_{(T)}]_0$ isomorphically onto R.

Proof. By Lemma 4.3.2. and results of [NVO1] concerning strongly graded ring, (2) and (3) may be verified easily. We only include the proof of (1) here.
By definition

$$
\begin{aligned}
G(R) &= \oplus_{n \in \mathbb{Z}}(S_n + (1-T)S)/(S_{n-1} + (1-T)S), \\
S/TS &= \oplus_{n \in \mathbb{Z}}(S_n + TS)/TS.
\end{aligned}
$$

For each n one may define an isomorphism of additive groups $\varphi_n\colon G(R)_n \to (S_n + TS)/TS$ as follows: $s_n + S_{n-1} + (1-T)S \mapsto s_n + TS$. Indeed, since for every $s_{n-1} \in S_{n-1}$ we have $s_{n-1} = T s_{n-1} + (1-T)s_{n-1}$ it follows that φ_n is well defined. Moreover, if $s_n \in TS$ then $s_n = T s_{n-1}$ for some $s_{n-1} \in S_{n-1}$, hence $s_n = T s_{n-1} = s_{n-1} - (1-T)s_{n-1}$. This shows that φ_n is an isomorphism. Hence we can define the required graded ring isomorphism by using all φ_n in the obvious way. $\quad\square$

In a similar way one may derive the following

4. Proposition. Let $\overline{M}$ be a T-torsionfree graded S-module and $M = \overline{M}/(1-T)\overline{M}$ the dehomogenization of $\overline{M}$ with the filtration as described in the begining. Then

(1). $G(M) \cong \overline{M}/T\overline{M}$ as graded $G(R)$-modules;

(2). the localization of $\overline{M}$ at the (central) Ore set $\{1, T, T^2, \cdots\}$ exists, it is denoted by $\overline{M}_{(T)}$, and the natural homomorphism $\pi\colon \overline{M} \to M$ factors through $\overline{M}_{(T)}$ in a canonical way.

Accordingly there is a commutative diagram of S-module homomorphisms:

$$\begin{array}{ccc} \overline{M} & \longrightarrow & \overline{M}_{(T)} \\ & \searrow_{\pi} \quad \swarrow_{\psi} & \\ & M & \end{array}$$

where $\overline{M}_{(T)}$ is a graded $S_{(T)}$-module: $\overline{M}_{(T)} = \oplus_{n\in\mathbb{Z}}[\overline{M}_{(T)}]_n$, $[\overline{M}_{(T)}]_n = \{T^j\overline{m}, \ j \in \mathbb{Z}, \ \overline{m} \in \overline{M}_{n-j}\}$, and $\psi(T^j\overline{m}) = \overline{m} + (1 - T)\overline{M}$. Since $S_{(T)}$ is a strongly graded ring it follows that $S_{(T)} \otimes_{[S_{(T)}]_0} [\overline{M}_{(T)}]_0 \cong \overline{M}_{(T)}$.

(3). with notations as in (2)., the homomorphism ψ maps $[\overline{M}_{(T)}]_0$ isomorphically onto M.

Now, let us turn to filtered rings. Let R be an arbitrary filtered ring with filtration FR. Consider the abelian group

$$\tilde{R} = \oplus_{n\in\mathbb{Z}} F_n R$$

given by the filtration FR, it becomes a $\mathbb{Z}$-graded ring if we define the gradation and multiplication as follows: for $n \in \mathbb{Z}$, $\tilde{R}_n = F_n R$, $r_n \cdot r_m = (r_n r_m)_{n+m}$ for $r_n \in F_n R$, $r_m \in F_m R$, and $(r_n r_m)_{n+m}$ is the product $r_n r_m$ in R viewed as an element in $\tilde{R}_{n+m}$. If $M \in R$-filt with filtration FM, then one may make the abelian group

$$\tilde{M} = \oplus_{n\in\mathbb{Z}} F_n M$$

into a graded $\tilde{R}$-module in an obvious way.

5. Definition. The graded ring $\tilde{R}$ resp. graded $\tilde{R}$-module $\tilde{M}$ defined above is called the **Rees ring** resp. **Rees module** of R with respect to FR resp. of M with respect to FM.

If M and N are two filtered R-modules with filtrations FM and FN respectively and $f \in \mathrm{Hom}_{FR}(M, N)$, then $f(F_n M) \subseteq F_n N$ for all $n \in \mathbb{Z}$. Consequently there is a graded morphism of degree zero $\tilde{f}: \tilde{M} \to \tilde{N}$ defined in a natural way. Hence one easily sees that we have defined a functor $\widetilde{(-)}: R$-filt $\to \tilde{R}$-gr with $M \mapsto \tilde{M}$.

6. Observations.
(a). Let $S = \oplus_{n\in\mathbb{Z}} S_n$ be a $\mathbb{Z}$-graded ring and $T \in S_1$ a regular central homogeneous element of degree 1. If $R = S/(1 - T)S$ is the dehomogenization of S with the filtration $F_n R = S_n + (1 - T)S/(1 - T)S$, $n \in \mathbb{Z}$, then from Lemma 4.3.2. it is clear that $\tilde{R} \cong S$ as graded rings. Similarly, if $\overline{M} \in S$-gr is a T-torsionfree graded S-module and $M = \overline{M}/(1 - T)\overline{M}$ is the dehomogenization of $\overline{M}$ with the filtration $F_n M = \overline{M}_n + (1 - T)\overline{M}/(1 - T)\overline{M}$, $n \in \mathbb{Z}$, then $\tilde{M} \cong \overline{M}$ as graded $S(\cong \tilde{R})$-modules.

(b). Let R be an arbitrary filtered ring with filtration FR, and let $\tilde{R}$ be the Rees ring of R with respect to FR. As $\tilde{R}_n = F_n R$, for $r \in F_n R$, we write the homogeneous element represented by r in $\tilde{R}_n$ as $(\tilde{r})_n$; Note that $(\tilde{1})_1 \in \tilde{R}_1$ is a regular central element, we denote this **canonical element** by X. Accordingly, we have the homomorphisms of abelian groups: $\varphi_s^n \colon \tilde{R}_n \to \tilde{R}_{n+s}$, $s \geq 1$, with $\varphi_s^n((\tilde{r})_n) = X^s(\tilde{r})_n = (\tilde{r})_{n+s}$. It follows that the map $\vartheta \colon R \to \tilde{R}/(1 - X)\tilde{R}$ defined by $\vartheta(r) = (\tilde{r})_n + (1 - X)\tilde{R}$ for $r \in F_n R$, is a ring isomorphism. This shows that:

Every filtered ring R with filtration FR is the dehomogenization of its Rees ring $\tilde{R}$ with respect to the canonical element X; Similarly, every filtered R-module M with filtration FM is the dehomogenization of its Rees module $\widetilde{M}$ with respect to X.

(c). Again let $S = \oplus_{n \in \mathbb{Z}} S_n$ be a $\mathbb{Z}$-graded ring and $T \in S_1$ a regular central element. If $R = S/(1 - T)S$ is the dehomogenization of S with the filtration $F_n R = S_n + (1 - T)S/(1 - T)S$, $n \in \mathbb{Z}$, then $1 + (1 - T)S = T + (1 - T)S \in F_0 R$. Considering the Rees ring $\tilde{R}$ of R, then by the definition of X in (b). it follows that for every filtered R-module M with filtration FM the Rees module $\widetilde{M}$ of M is a T-torsionfree graded S-module.

Summing up, We have the following

7. Proposition. Let R be filtered ring with filtration FR, and let X be the canonical homogeneous element of degree 1 of $\tilde{R}$ as described in observation (b). above. Let M be a filtered R-module with filtration FM.

(1). $\tilde{R}/X\tilde{R} \cong G(R)$, $\widetilde{M}/X\widetilde{M} \cong G(M)$.

(2). $\tilde{R}/(1 - X)\tilde{R} \cong R$, $\widetilde{M}/(1 - X)\widetilde{M} \cong M$.

(3). The class of X-torsionfree graded $\tilde{R}$-modules is a full subcategory of $\tilde{R}$-gr, denoted $\mathcal{F}_X$, and the functor $\widetilde{(-)} \colon R\text{-filt} \to \tilde{R}\text{-gr}$ given by $M \mapsto \widetilde{M}$ defines an equivalence of categories between R-filt and $\mathcal{F}_X$.

(4). If we write $\tilde{R}_{(X)}$ to be the localization of $\tilde{R}$ at the (central) Ore set $\{1, X, X^2, \cdots\}$ then $\tilde{R}_{(X)} \cong R[t, t^{-1}]$; Similarly let $\widetilde{M}_{(X)}$ be the localization of $\widetilde{M}$ at $\{1, X, X^2, \cdots\}$ then $\widetilde{M}_{(X)} \cong R[t, t^{-1}] \otimes_R M = M[t, t^{-1}]$.

(5). The functor $\tilde{G} = \tilde{R}/X\tilde{R} \otimes_{\tilde{R}} -$ is exact on $\mathcal{F}_X$ and the functor $D = \tilde{R}/(1 - X)\tilde{R} \otimes_{\tilde{R}} -$ $\tilde{R}$-gr $\to R$-filt is exact.

Comparing with the functor G, the functor $\widetilde{(-)}$ acquires some extra properties.

8. Proposition. Let R be a filtered ring with filtration FR.

(1). If $f \in \mathrm{Hom}_{FR}(M, N)$ is a strict filtered morphism for M, $N \in R$-filt, then $\mathrm{Ker}\tilde{f}$ and $\mathrm{Coker}\tilde{f}$ are in $\mathcal{F}_X$.

(2). If

$$(*) \qquad\qquad M \xrightarrow{f} N \xrightarrow{g} K$$

is a sequence in R-filt, where M, N and K have their respective filtrations FM, FN and FK, and

$$\widetilde{(*)} \qquad\qquad \widetilde{M} \xrightarrow{\tilde{f}} \widetilde{N} \xrightarrow{\tilde{g}} \widetilde{K}$$

is the associated sequence in $\tilde{R}$-gr, then $(*)$ is exact and f is strict when $\widetilde{(*)}$ is exact; Conversely, if $(*)$ is exact and f is strict then $\widetilde{(*)}$ is exact.

Proof. Exercises. $\square$

The situation of filtered R-modules in connection with the associated graded structures may be best summed up in the following diagram of functors:

where $\tilde{G}$ and D are as described in Proposition 4.3.7., ι is the inclusion functor and $\#$ is the forgetful functor defined by forgetting the gr-(filt-)structure.

Although the Rees ring (in noncommutative case) has appeared in several papers on filtered rings (this ring has been called different names a.o. the formally graded ring in [Shar] etc. ...), it seems that the category equivalence of R-filt and $\mathcal{F}_X$, the dehomogenization of a $\mathbb{Z}$-graded ring to filtrations have been fully exploited for the first time in [AVVO], [LVO1] and [LVO5].

9. Examples.

(a). Let $A_n(k)$ be the n-th Weyl algebra over a field k of characteristic zero. Consider the Bernstein filtration on $A_n(k)$ as defined in §2. example I., then it is easily seen that the Rees ring of $A_n(k)$ with respect to this filtration is isomorphic to the graded (proper) subring $k[x_1t, \cdots, x_nt, y_1t, \cdots, y_nt, t]$ of the polynomial ring $A_n(k)[t]$ over $A_n(k)$ in t, where $A_n(k)[t]$ has the (natural) gradation defined by the power of t.

Next, consider the Σ-filtration on $A_n(k)$ (see §2. example I.). Since we have the well known ring isomorphism (cf. [M-R] Ch.1.): $A_n(k) \cong U(\mathbf{g})/(1-z)U(\mathbf{g})$, where $U(\mathbf{g})$ is the enveloping algebra of the $2n+1$-dimensional Heisenberg Lie algebra $\mathbf{g}$ over k with basis $\{x_1, \cdots, x_n, y_1, \cdots, y_n, z\}$ and products $[x_i, y_i] = z$ for $i = 1, \cdots, n$, all other products being zero. We claim that $U(\mathbf{g})$ is the Rees algebra of $A_n(k)$ with respect to the Σ-filtration. Indeed, if we put $\deg x_i = 0$, $\deg y_i = 1$ for $i = 1, \cdots, n$ in the tensor algebra $T(\mathbf{g})$ determined by $\mathbf{g}$, then $U(\mathbf{g})$ becomes a graded k-algebra such that z is a regular central homogeneous element of degree 1, and the filtration we defined on the dehomogenization $U(\mathbf{g})/(1-z)U(\mathbf{g})$ of $U(\mathbf{g})$ as before is actually the Σ-filtration. Consequently $\widetilde{A_n(k)} \cong U(\mathbf{g})$ by Observation 4.3.6.(a).. Since the associated graded ring of $A_n(k)$ with respect to both the Bernstein filtration and the Σ-filtration are the same, that is the polynomial ring in $2n$ variables over k, we have $\widetilde{A_n(k)}/X\widetilde{A_n(k)} \cong k[t_1, \cdots, t_{2n}]$ by Proposition 4.3.3..

(b). Let $R = \oplus_{n \in \mathbb{Z}} R_n$ be an arbitrary $\mathbb{Z}$-graded ring. Consider the polynomial ring $R[T]$ over R in commuting variable T and put the **"mixed"** gradation on $R[T]$:

$$R[T]_n = \Big\{ \sum_{i+j=n} r_i T^j, \; r_i \in R_i \Big\}, \quad n \in \mathbb{Z},$$

then we easily see that

(1). $T \in R[T]_1$, R is a graded subring of $R[T]$, and the gradation so defined above will be still positive or left limited if the gradation on R is positive or left limited;

(2). the dehomogenization $A = R[T]/(1-T)R[T]$ of $R[T]$ with respect to T has the filtration FA determined by the "mixed" gradation on $R[T]$. Hence by Proposition 4.3.3. and Observation 4.3.6.(a). we have $G(A) \cong R[T]/TR[T] \cong R$ as graded rings (note here R has the original gradation!), and $\tilde{A} \cong R[T]$ as graded rings; But on the other hand, A is obviously isomorphic to R as ungraded rings.

(c). Let R be a ring and I an ideal of R. Consider an overring S of R such that there is an R-bisubmodule J of S such that $IJ = JI = R$, then I is said to be **invertible** (in S) and one writes $J = I^{-1}$. For such an I, if we consider the usual I-adic filtration on R: $F_nR = I^{-n}$ for $n < 0$, $F_nR = R$ for $n \geq 0$, there is a related graded ring

$$R(I) = \oplus_{n \geq 0} I^n t^n \subseteq R[t]$$

where $R[t]$ is the usual polynomial ring over R in the commuting variable t. $R(I)$ is, as a graded ring, isomorphic to the negative part $\tilde{R}^- = \oplus_{n \leq 0} I^{-n}$ of the Rees ring $\tilde{R}$ of R with respect to the I-adic filtration; if we consider the filtered overring of R defined as

$$S(I) = \cup_{n \in \mathbb{Z}} I^n \subseteq S$$

with $F_n S(I) = I^{-n}$, $n \in \mathbb{Z}$, there is another related graded ring

$$\check{R}(I) = \oplus_{n \in \mathbb{Z}} I^n t^n \subseteq S[t, t^{-1}]$$

where $S[t, t^{-1}]$ is the ring of finite Laurent series over S in variable t. Obviously, $\check{R}(I)$ is isomorphic to the Rees ring of $S(I)$. $\check{R}(I)$ is called the **generalized Rees ring** of R with respect to I. All of those rings defined above play important roles in [LV] and [LVV].

Note that $\check{R}(I)$ has gradation: $\check{R}(I)_n = I^{-n} t^{-n}$, $n \in \mathbb{Z}$, in particular $\check{R}(I)_0 = R$, and $R \subseteq I^{-1}$. Moreover, one sees that $t^{-1} \in \check{R}(I)_1$ is a central homogeneous element of degree 1 in $\check{R}(I)$. Put $y = t^{-1}$, then in view of Proposition 4.3.3. and Observation 4.3.6.(a). we have the following

Observations. Consider the dehomogenization $A = \check{R}(I)/(1 - y)\check{R}(I)$ of $\check{R}(I)$ with respect to y, where A has the filtration defined by the gradation of $\check{R}(I)$ as before, and let $\check{R}(I)_{(y)}$ be the localization of $\check{R}(I)$ at the (central) Ore set $\{1, y, y^2, \cdots\}$, then

(1). $A \cong S(I) \cong (\check{R}(I)_{(y)})_0$ where $(\check{R}(I)_{(y)})_0$ is the part of degree zero of the graded ring $\check{R}(I)_{(y)}$;

(2). $G(A) \cong \check{R}(I)/y\check{R}(I)$, in particular $G(A)_0 \cong R/I$;

(3). $G(A)^- = \oplus_{n \leq 0} G(A)_n \cong \oplus_{n \geq 0} I^n/I^{n+1} = G_I(R)$, where $G_I(R)$ denotes the associated graded ring of R with respect to the I-adic filtration on R.

(d). Rees rings of grading filtrations.

Let $R = \oplus_{n \in \mathbb{Z}} R_n$ be an arbitrary $\mathbb{Z}$-graded ring and FR be one of the grading filtrations (as defined in §2. example L.) on R. Then it is easy to see that that $G(R) \cong R$ as graded rings and FR is separated.

In order to describe the Rees rings for grading filtrations on R we first introduce two special gradations on the polynomial ring extension $R[t]$ of R.

Indeed, we may start with a G-graded ring, where G is any group. Let $Z(G)$ denote the centre of G. Take any submonoid H of $Z(G)$ containing the neutral element e of G and consider the semigroup ring $R[H]$ over R. For each $g \in G$ if we put

$$R[H]_g = \oplus_{h \in H} R_{gh} \cdot h$$

then a direct verification yields the following

10. Lemma. (i). $R[H] = \oplus_{g \in G} R[H]_g$ is a G-graded ring and $R[H]_e \cong R^{(H)} = \oplus_{h \in H} R_h$.

(ii). When R is strongly G-graded then $R[H]$ is strongly G-graded too, so in particular there is an equivalence of categories (see subsection 4.1.): $R^{(H)}$-mod $\leftrightarrow R[H]$-gr, where the latter one is the category of graded left $R[H]$-modules with respect to the gradation defined above.

Now let R be a $\mathbb{Z}$-graded ring. Consider the polynomial ring $R[t]$ over R in commuting variable t. For each $k \in \mathbb{Z}$, put

$$R[t]_k^+ = \oplus_{n \geq 0} R_{k+n} t^n$$
$$R[t]_k^- = \oplus_{n \geq 0} R_{k-n} t^n$$

then from the above lemma we may derive the following

Corollary. (i). $R[t] = \oplus_{k \in \mathbb{Z}} R[t]_k^+ = \oplus_{k \in \mathbb{Z}} R[t]_k^-$.
(ii). $\{R[t]_k^+\}_{k \in \mathbb{Z}}$ resp. $\{R[t]_k^-\}_{k \in \mathbb{Z}}$ is a $\mathbb{Z}$-gradation on $R[t]$; $R[t]_0^+ = \oplus_{n \geq 0} R_n t^n \cong R^+ = \oplus_{n \geq 0} R_n$; $R[t]_0^- = \oplus_{n \geq 0} R_{-n} t^n \cong R^- = \oplus_{n \geq 0} R_{-n}$.
(iii). if R is strongly $\mathbb{Z}$-graded the $R[t]$ is strongly $\mathbb{Z}$-graded with respect to the $+$-gradation resp. to the $-$-gradation defined above.

For a $\mathbb{Z}$-graded ring R the $\pm$-gradations on $R[t]$ are called **sign gradations**.

Concerning with the Rees ring $\tilde{R}$ of R, where R is a $\mathbb{Z}$-graded ring, we have the following

11. Theorem. With notations as defined above, let $\tilde{R}^{(1)}$ resp. $\tilde{R}^{(2)}$ be the Rees ring of R with respect to the grading filtration $F^{(1)}R$ resp. to the grading filtration $F^{(2)}R$, then we have the isomorphism of graded rings: $\tilde{R}^{(1)} \cong R[t]$, where $R[t]$ has the $-$-gradation; $\tilde{R}^{(2)} \cong R[t]$, where $R[t]$ has the $+$-gradation.

Proof. By definition we have for each $k \in \mathbb{Z}$ the isomorphism of additive groups:

$$\oplus_{n \geq 0} R_{k-n} t^n = R[t]_k^- \xrightarrow{\cong} \tilde{R}_k^{(1)} = F_k^{(1)} R = \oplus_{i \leq k} R_i.$$

Hence a graded ring isomorphism $R[t] \xrightarrow{\cong} \tilde{R}^{(1)}$ may be defined in an obvious way. Similarly, we also obtain the second graded ring isomorphism. $\qquad \square$

As a typical example let us look at the ring of differential polynomials $R = A[X][Y, D]$, where $A[X]$ is the polynomial ring over a domain A of characteristic zero in commuting variable X, $D = \partial/\partial X$ on $A[X]$, then $YX - XY = 1$ in R (i.e., R is the usual skew polynomial ring $A[X][Y, \sigma, D]$ with $\sigma = 1_{A[X]}$). R has a natural $\mathbb{Z}$-gradation defined by giving X-degree -1 and Y-degree $+1$; in other words, for $n \in \mathbb{Z}$ we have $R_n = \sum_{j-i=n} AX^i Y^j$ where $i \geq 0$,

$j \geq 0$. Now taking the grading filtration $F^{(1)}R$ on R, it follows that $\tilde{R}^{(1)} \cong A[X][Y, D][t]$. In particular, with respect to the Fuchsian filtration (§2. Example 2.3.L.) on the $n + 1$-st Weyl algebra $A_{n+1}(k)$ over a field k of characteristic zero, we have $\widetilde{A_{n+1}(k)} \cong A_{n+1}(k)[t]$.

4.4. Finite intersection property and faithful filtrations

First of all, let us recall the following

1. Definition. Let R be a ring and I an ideal of R.
(1). I is said to have the **finite intersection property** if for any finitely generated left R-module M it follows that

$$\bigcap_{n \geq 1} I^n M = \{m \in M, \ (1 - a)m = 0 \text{ for some } a \in I\}.$$

(2). I is said to have the **Artin-Rees property** if for any finitely generated left R-module M, any submodule N of M and any natural number n, there exists an integer $h(n) \geq 0$ such that $I^{h(n)}M \cap N \subseteq I^n N$. In other words, I has the Artin-Rees property if the I-adic topology of N coincides with the topology induced on N by the I-adic topology of M.

It is easy to see that if an ideal I of a given ring R has the Artin-Rees property then I has the finite intersection property (indeed, if $x \in \cap_{n \geq 1} I^n M$ then it is sufficient to consider the submodule Rx of M).

2. Definition. Let R be a ring and $a \in R$.
(1). We say that a is a **normalizing element** of R if $aR = Ra$. The set of normalizing elements of R is denoted by $N(R)$.
(2). Let I be an ideal of R. The subset $\{a_1, \cdots, a_n\}$ is a **normalizing (centralizing) set of generators** of I if

    a). the ideal generated by $\{a_1, \cdots, a_n\}$, denoted $(a_1, \cdots, a_n)$, is equal to I,

    b). $a_1 \in N(R)$ $\left(a_1 \in Z(R), \text{ the centre of } R\right)$,

    c). $a_i + (a_1, \cdots, a_{i-1}) \in N(R/(a_1, \cdots, a_{i-1}))$ $\left(a_i + (a_1, \cdots, a_{i-1}) \in N(R/(a_1, \cdots, a_{i-1}))\right)$
       for $i = 2, \cdots, n$.

3. Lemma. Let R be a left Noetherian ring and X a normalizing element of R. If M is a finitely generated X-torsionfree R-module, then

$$\bigcap_{n \geq 1} X^n M = \{m \in M, \ (1 - a)m = 0 \text{ for some } a \in XR\}.$$

Proof. Put $S = \{m \in M, (1-a)m = 0$ for some $a \in XR\}$, then obviously $S \subseteq \cap_{n \geq 1} X^n M$. Conversely, let $y \in \cap_{n \geq 1} X^n M$. Then $y = Xm_1 = X^2 m_2 = \cdots$, where $X^k m_k \in X^k M$. Since M is X-torsionfree we have $m_k = Xm_{k+1}$, $k = 1, 2, \cdots$, and consequently $Rm_1 \subseteq Rm_2 \subseteq Rm_3 \subseteq \cdots$. Since R is Noetherian and M is finitely generated there is a positive integer n_0 such that $Rm_n = Rm_{n+1}$ for all $n \geq n_0$. It follows that $m_{n+1} = rm_n = rXm_{n+1}$ and hence $X^{n+1} m_{n+1} = X^{n+1} r X m_{n+1} = X r' X^{n+1} m_{n+1}$ for some r, $r' \in R$ since X is a normalizing element by assumption. Therefore $0 = (1 - Xr')X^{n+1} m_{n+1} = (1 - Xr')y$, i.e., $y \in S$. This shows that $\cap_{n \geq 1} X^n M \subseteq S$ and hence the equality follows. □

4. Lemma. Let R and X be as in Lemma 4.4.3.. Then for any finitely generated left R-module M, either $\cap_{n \geq 1} X^n M = 0$ or $\cap_{n \geq 1} X^n M$ is an X-torsionfree R-submodule of M.

Proof. Obviously $\cap_{n \geq 1} X^n M$ is an R-submodule of M. Suppose $\cap_{n \geq 1} X^n M \neq 0$. By the assumption on R, X and M, the X-torsion part of M, denoted $t(M) = \{m \in M, X^w m = 0$ for some integer $w > 0\}$, is a finitely generated R-submodule of M such that $X^k t(M) = 0$ for large k. It follows that $X^k M$ has no X-torsion element. Hence $\cap_{n \geq 1} X^n M$ is X-torsionfree. □

5. Proposition. Let R and X be as in Lemma 4.4.3.. Then the ideal XR of R has the finite intersection property.

Proof. Let M be any finitely generated left R-module. Then there is an exact sequence of R-modules:
$$0 \longrightarrow t(M) \longrightarrow M \longrightarrow M/t(M) \longrightarrow 0$$
where $t(M)$ is the X-torsion part of M. Since $M/t(M)$ is X-torsionfree it follows from Lemma 4.4.3. and Lemma 4.4.4. that

$$\bigcap_{n \geq 1} X^n M \cong \left(\cap_{n \geq 1} X^n M + t(M)\right)/t(M)$$
$$\subseteq \cap_{n \geq 1} \left(X^n M + t(M)\right)/t(M)$$
$$= \cap_{n \geq 1} X^n \left(M/t(M)\right)$$
$$= \{\overline{m} \in M/t(M), (1-a)\overline{m} = 0 \text{ for some } a \in XR\}$$

This shows that if $y \in \cap_{n \geq 1} X^n M$ then there is an $a \in XR$ such that $(1-a)y = 0$ as desired. □

6. Proposition. (see [PS] Corollary 2.8. or [NVO1] Ch.D. Proposition V.1.) Let R be a left Noetherian ring, I an ideal generated by a centralizing system in the sense of Definition 4.4.2.. Then I has the Artin-Rees property.

7. Proposition. (see [NVO2] Lemma 6.5.2.) Let R be a left Noetherian ring and let P be an invertible ideal of R. Then P has the Artin-Rees property.

8. Corollary. Let R be a left Noetherian ring and X a normalizing element of R. If X is also regular in R then XR has the Artin-Rees property.

Proof. This follows from Proposition 4.4.7. since by the assumption XR is an invertible ideal of R.   □

9. Proposiiton. (1). Let R be a left Noetherian ring and I an ideal of R. If I is contained in the Jacobson radical $J(R)$ of R and R/I is Artinian, then the following statements are equivalent:
(a). I has the Artin-Rees property;
(b). $\cap_{n\geq 1} I^n M = 0$ for every finitely generated left R-module M, i.e., the I-adic filtration on every finitely generated R-module M is separated.

(2). Let R be a ring and I an ideal of R generated by a centralizing system. If R/I is left Noetherian then the following statements are equivalent:
(a). I has the finite intersection property;
(b). I has the Artin-Rees property.

10. Remark. The first part of Proposition 4.4.9. is ([PS] Theorem 2.5.); The second part of Proposition 4.4.9. is ([PS] Theorem 2.9.) but note that the condition "R/I is Noetherian" has to be added to the statement of the original theorem in order to make the proof correct.

Now let $S = \oplus_{n\in \mathbb{Z}} S_n$ be a $\mathbb{Z}$-graded ring and T a regular central homogeneous element of degree 1 in S.

11. Lemma. With notations as above, consider the dehomogenization $R = S/(1-T)S$ of S with respect to T, where R has the filtration FR determined by the gradation of S as before, then
(1). $\cap_{n\geq 1} T^n \overline{M} = 0$ if and only if $\cap_{n\in \mathbb{Z}} F_n M = 0$, where $\overline{M}$ is any T-torsionfree graded S-module and M is the dehomogenization of $\overline{M}$ with filtration FM on M determined by the gradation of $\overline{M}$ as before, in other words, the TS-adic filtration on $\overline{M}$ is separated if and only if the filtration FM on M is separated;
(2). $T \in J^g(S)$ if and only if $F_{-1}R \subseteq J(F_0R)$, where $J^g(S)$ denotes the graded Jacobson radical of S and $J(F_0R)$ denotes the Jacobson radical of F_0R.

Proof. (1). this is easily verified.
(2). Suppose $T \in J^g(S)$. It follows from the definition (given in subsection 4.1) that

$TS_{-1} \subseteq J^g(S) \cap S_0$, i.e., $1 - Ts_{-1}$ is invertible in S_0 for every $s_{-1} \in S_{-1}$. But then it follows from $s_{-1} = Ts_{-1} + (1 - T)s_{-1}$, $s_{-1} \in S_{-1}$, that

$$S_{-1} + (1 - T)S/(1 - T)S = F_{-1}R \subseteq J(F_0R).$$

The inverse implication follows from Lemma 4.3.2.. □

12. Corollary. Let R be a filtered ring with filtration FR and let M be a filtered R-module with filtration FM. If X is the canonical element of $\tilde{R}$ defined in subsection 4.3., then

(1). $\cap_{n \geq 1} X^n \tilde{M} = 0$ if and only if $\cap_{n \in \mathbf{Z}} F_nM = 0$;

(2). $X \in J^g(\tilde{R})$ if and only if $F_{-1}R \subseteq J(F_0R)$.

13. Definition. Let R be a filtered ring with filtration FR. If $F_{-1}R \subseteq J(F_0R)$ then FR is said to be **faithful**.

14. Theorem. Let R be a filtered ring with filtration FR. Suppose that FR is faithful and $\tilde{R}$ is left Noetherian, then for any $M \in R$-filt with filtration FM such that $\tilde{M}$ is finitely generated in $\tilde{R}$-gr we have $\cap_{n \in \mathbf{Z}} F_nM = 0$, i.e., FM is separated.

Proof. Considering the canonical element X in $\tilde{R}$, the theorem follows from Corollary 3.4.8. and Corollary 4.4.12. because $X \in J^g(\tilde{R})$, and $\cap_{n \geq 1} X^n \tilde{M}$ is now a graded submodule of $\tilde{M}$. □

§5. Good Filtrations

Let R be a filtered ring with filtration FR.

5.1. Definition. Let M be in R-filt with filtration FM. If there exist $m_1, \cdots, m_s \in M$, $k_1, \cdots, k_s \in \mathbb{Z}$, such that for all $n \in \mathbb{Z}$

$$F_n M = \sum_{i=1}^{s} F_{n-k_i} R m_i,$$

then FM is called a **good filtration** on M.

5.2. Remark. From the definition it is clear that if M has a good filtration FM then it is a finitely generated R-module (note that we always assume that all filtrations are exhaustive). On the other hand, if M is a finitely generated R-module and $\{m_1, \cdots, m_s\}$ is a generating system of M, then one can always define a good filtration FM on M as follows: take any $k_1, \cdots, k_s \in \mathbb{Z}$, and put $F_n M = \sum_{i=1}^{s} F_{n-k_i} R m_i$, $n \in \mathbb{Z}$, then it is obvious that $\cup_{n \in \mathbb{Z}} F_n M = M$, $F_n M \subseteq F_{n+1} M$ and $F_m R F_n M \subseteq F_{m+n} M$, $m, n \in \mathbb{Z}$.
However, one should also be careful with the following two points:
(1). If M is a finitely generated R-module with filtration FM, FM is not necessarily good. For example, consider the trivial filtration FR on R and let M be a non-Artinian R-module that is however a finitely generated R-module (of course this exists if R is not Artinian); define a filtration FM by taking an infinite descending chain of submodules of M and check that this filtration fails to be good.
(2). If $M \in R$-filt has a good filtration FM and N is any submodule of M, then the quotient filtration on M/N defined by $F_n(M/N) = F_n M + N/N$, $n \in \mathbb{Z}$, is again a good filtration. But the filtration induced on N by FM need not be good at all; in fact if R is not Noetherian and N is a left ideal that is not finitely generated then there is no chance at all that $FR \cap N$ is a good filtration on N.

From the remark given above it is clear that a finitely generated R-module may have many different good filtrations. But fortunately we have the following:

5.3. Lemma. Any two good filtrations on an R-module M are equivalent in the sense of subsection 3.2..

Proof. Since there exist good filtrations on M it must be a finitely generated R-module and we may write, for all $n \in \mathbb{Z}$,

$$\begin{aligned}
F_n M &= F_{n-d_1} R m_1 + \cdots + F_{n-d_r} R m_r, \\
F'_n M &= F_{n-e_1} R m'_1 + \cdots + F_{n-e_s} R m'_s.
\end{aligned}$$

Choose N to be an integer such that all $m_i \in F_N' M$ and N' such that all $m_j' \in F_{N'} M$, and put $|N''| = \max\{|e_j|, |d_i|, j, i\}$, $w = |N| + |N'| + |N''|$, then one checks that $F_{n-w} M \subseteq F_n' M \subseteq F_{n+w} M$, $n \in \mathbb{Z}$. □

Now let us focus to the associated gradations of a good filtration.

5.4. Lemma. Let M be in R-filt with filtration FM.

(1). FM is good if and only if $\widetilde{M}$ is finitely generated in $\widetilde{R}$-gr.

(2). If FM is good then $G(M)$ is a finitely generated $G(R)$-module.

Proof. (1). Suppose that FM is good, say $F_n M = \sum_{i=1}^{s} F_{n-k_i} R m_i$, $n \in \mathbb{Z}$, for some $k_i \in \mathbb{Z}$, $m_i \in M$. By the definition of $\widetilde{M}$ it is clear that $\widetilde{M}_n = \sum_{i=1}^{s} \widetilde{R}_{n-k_i} \cdot (m_i)_{k_i}$, $n \in \mathbb{Z}$, where $(m_i)_{k_i}$ are the homogeneous elements represented by m_i in $\widetilde{M}_{k_i}$, hence $\widetilde{M} = \sum_{i=1}^{s} \widetilde{R} \cdot (m_i)_{k_i}$. Conversely, let $\widetilde{M} = \sum_{i=1}^{s} \widetilde{R} \cdot (m_i)_{k_i}$, then $\widetilde{M}_n = \sum_{i=1}^{s} \widetilde{R}_{n-k_i} \cdot (m_i)_{k_i}$, $n \in \mathbb{Z}$. it follows from Proposition 4.3.7. that $F_n M = \sum_{i=1}^{s} F_{n-k_i} R m_i$, $n \in \mathbb{Z}$, hence FM is good.

(2). This follows from (1) and Proposition 4.3.7.. □

5.5. Corollary. Suppose that the Rees ring $\widetilde{R}$ of R with respect to FR is left Noetherian.

(1). Good filtrations in R-filt induce good filtrations on R-submodules.

(2). If FM is a good filtration on $M \in R$-filt and $F'M$ is another filtration on M which is equivalent to FM, then $F'M$ is also a good filtration on M.

(3). If FR is also faithful, i.e., $F_{-1} R \subseteq J(F_0 R)$, then for any $M \in R$-filt with good filtration FM and any submodule $N \subseteq M$ we have

$$N = \cap_{n \in \mathbb{Z}} (N + F_n M),$$

or in other words, N is closed with respect to the filtration topology of M and any good filtration in R-filt is separated. Consequently, if $G(M)$ may be generated by n homogeneous generators then M may be generated by n generators. Moreover, if $N \subseteq P \subseteq M$ are submodules of M with the filtrations induced by FM, then $G(N) = G(P)$ implies $N = P$.

Proof. Since $\widetilde{R}$ is Noetherian, (1) and (2) are easily obtained by Lemma 5.4.. It remains to prove (3). By the assumptions on R it follows from Theorem 4.4.14. that good filtrations are separated. If we consider the quotient filtration on M/N given by FM then it is still good and hence separated. Accordingly N is closed. Now let $G(M) = \sum_{i=1}^{s} G(R)\sigma(u_i)$, where $u_i \in F_{k_i} M - F_{k_i - 1} M$, $\sigma(u_i) = u_i + F_{k_i - 1} M$ is the principal part of u_i (see subsection 4.2.).

Then for each $n \in \mathbb{Z}$,

$$G(M)_n = \sum_{i=1}^{s} G(R)_{n-k_i} \sigma(u_i),$$

$$F_n M = \sum_{i=1}^{s} F_{n-k_i} R u_i + F_{n-1} M.$$

It follows that $M \subseteq \cap_{k \in \mathbb{Z}} (\sum_{i=1}^{s} R u_i + F_k M) = \sum_{i=1}^{s} R u_i \subseteq M$, hence $M = \sum_{i=1}^{s} R u_i$. The last assertion may be verified in a similar way and we leave it to the reader (or see Ch.II. §3.).   □

5.6. Proposition. Assume that $G(R)$ is left Noetherian. Let $F'M$, FM be equivalent filtrations on the R-module M. If $G_F(M)$, the associated graded module of M with respect to FM, is finitely generated in $G(R)$-gr, then $G_{F'}(M)$ is also finitely generated in $G(R)$-gr, where $G_{F'}(M)$ is the associated graded module of M with respect to $F'M$.

Proof. (The proof given here is due to T.A. Springer, see [VE2]) Since FM and $F'M$ are euqivalent filtrations there exists an integer $w > 0$ such that $F_{n-w} M \subseteq F'_n M \subseteq F_{n+w} M$ for all $n \in \mathbb{Z}$. Put $\Lambda_n = F_{n-w} M$, $\Gamma_n = F'_n M$, $n \in \mathbb{Z}$, and $c = 2w$. Then $\Lambda_n \subseteq \Gamma_n \subseteq \Lambda_{n+c}$ for all $n \in \mathbb{Z}$. Observe that $G_\Lambda(M)$ is finitely generated we want to prove that $G_\Gamma(M)$ is finitely generated.

Define $T_i = \oplus_{n \in \mathbb{Z}} (\Gamma_n \cap \Lambda_{n+i} / \Gamma_{n-1} \cap \Lambda_{n+i})$ for all $0 \leq i \leq c$, then $T_c = \oplus_{n \in \mathbb{Z}} \Gamma_n / \Gamma_{n-1} = G_\Gamma(M)$. we use induction on c to prove that $G_\Gamma(M)$ is finitely generated. First consider T_0. Note that $T_0 = \oplus_{n \in \mathbb{Z}} (\Lambda_n / \Gamma_{n-1} \cap \Lambda_n)$ and $\Lambda_{n-1} \subseteq \Gamma_{n-1} \cap \Lambda_n$ we obtain an exact sequence $G_\Lambda(M) \to T_0 \to 0$. Hence T_0 is finitely generated. Now consider the canonical map

$$T_i = \bigoplus_{n \in \mathbb{Z}} \frac{\Gamma_n \cap \Lambda_{n+i}}{\Gamma_{n-1} \cap \Lambda_{n+i}} \xrightarrow{\varphi_i} \bigoplus_{n \in \mathbb{Z}} \frac{\Gamma_n \cap \Lambda_{n+i+1}}{\Gamma_{n-1} \cap \Lambda_{n+i+1}} = T_{i+1}$$

then

$$\mathrm{Ker}\varphi_i = \bigoplus_{n \in \mathbb{Z}} \frac{\Gamma_n \cap \Lambda_{n+i} \cap \Gamma_{n-1} \cap \Lambda_{n+i+1}}{\Gamma_{n-1} \cap \Lambda_{n+i}} = \bigoplus_{n \in \mathbb{Z}} \frac{\Gamma_{n-1} \cap \Lambda_{n+i}}{\Gamma_{n-1} \cap \Lambda_{n+i}} = 0,$$

and

$$\mathrm{Coker}\varphi_i \cong \bigoplus_{n \in \mathbb{Z}} \frac{\Gamma_n \cap \Lambda_{n+i+1}}{\Gamma_n \cap \Lambda_{n+i} + \Gamma_{n-1} \cap \Lambda_{n+i+1}}.$$

Using the exact sequence

$$U_i = \bigoplus_{n \in \mathbb{Z}} \frac{\Gamma_n \cap \Lambda_{n+i+1}}{\Gamma_n \cap \Lambda_{n+i}} \longrightarrow \bigoplus_{n \in \mathbb{Z}} \frac{\Gamma_n \cap \Lambda_{n+i+1}}{\Gamma_n \cap \Lambda_{n+i} + \Gamma_{n-1} \cap \Lambda_{n+i+1}} \longrightarrow 0$$

we find that Cokerφ_i will be finitely generated if U_i is finitely generated. However, $U_i \hookrightarrow V_i = \oplus_{n \in \mathbb{Z}} \Lambda_{n+i+1} / \Lambda_{n+i}$ and V_i is a $G(R)$-module isomorphic to $G_\Lambda(M)$. Consequently V_i is finitely generated and hence U_i and Cokerφ_i are finitely generated. Finally, using the exact sequence

$$0 \longrightarrow T_i \xrightarrow{\varphi_i} T_{i+1} \longrightarrow \text{Coker}\varphi_i \longrightarrow 0$$

and induction it follows that T_c is finitely genrated as desired. $\qquad\square$

5.7. Theorem. Suppose that R is a complete filtered ring with respect to FR and $M \in R$-filt with separated filtration FM, then FM is good if and only if $G(M)$ is a finitely generated $G(R)$-module. Furthermore if $G(M)$ is generated by s homogeneous elements as a $G(R)$-module then M can be generated by s (or less) elements as an R-module.

Proof. Suppose that FM is good then $G(M)$ is finitely generated by Lemma 5.4.. onversely, let $G(M)$ be finitely generated as a graded $G(R)$-module, say $G(M) = \sum_{i=1}^s G(R)\sigma(u_i)$, where $u_i \in F_{k_i}M - F_{k_i-1}M$ for some $k_i \in \mathbb{Z}$ and $\sigma(u_i) = u_i + F_{k_i-1}M$ is the principal part of u_i in the sense of subsection 4.2., then for every $n \in \mathbb{Z}$,

$$G(M)_n = \sum_{i=1}^s G(R)_{n-k_i} \sigma(u_i),$$

$$\bar{F}_n M = \sum_{i=1}^s F_{n-k_i} R u_i + F_{n-1}M.$$

If $u \in F_n M$, then $u = \sum_{i=1}^s a_{n-k_i} u_i + u_{n-1}$ with $a_{n-k_i} \in F_{n-k_i}R$, $u_{n-1} \in F_{n-1}M$, and $u - \sum_{i=1}^s a_{n-k_i} u_i = u_{n-1} \in F_{n-1}M$. Similarly we have $u_{n-1} = \sum_{i=1}^s a_{n-k_i-1} u_i + u_{n-2}$ with $a_{n-k_i-1} \in F_{n-k_i-1}R$, $u_{n-2} \in F_{n-2}M$ and $u_{n-1} - \sum_{i=1}^s a_{n-k_i-1} u_i = u_{n-2} \in F_{n-2}M$. Inductively for each $q \geq 0$ we obtain

$$u - \sum_{i=1}^s \left(\sum_{t=1}^q a_{n-k_i-t} \right) u_i = u_{n-q-1} \in F_{n-q-1}M$$

with $a_{n-k_i-t} \in F_{n-k_i-t}R$. Since R is complete we may define $a_i = \sum_{t=1}^\infty a_{n-k_i-t}$ for $i = 1, \cdots, s$, then obviously $a_i \in F_{n-k_i}R$ and

$$u - \sum_{i=1}^s a_i u_i = u - \sum_{i=1}^s \left(\sum_{t=1}^\infty a_{n-k_i-t} \right) u_i$$

$$= u - \sum_{i=1}^s \left(\sum_{t=1}^q a_{n-k_i-t} \right) u_i - \sum_{i=1}^s \left(\sum_{t=q+1}^\infty a_{n-k_i-t} \right) u_i$$

$$= u_{n-q-1} - \sum_{i=1}^s \left(\sum_{t=q+1}^\infty a_{n-k_i-t} \right) u_i$$

$$\in F_{n-q-1}M.$$

But FM is separated by assumption, it follows that $u = \sum_{i=1}^{s} a_i u_i \in \sum_{i=1}^{s} F_{n-k_i} R u_i$. Hence $F_n M = \sum_{i=1}^{s} F_{n-k_i} R u_i$ and $M = \sum_{i=1}^{s} R u_i$. This completes the proof.   □

§6. Projective and Injective Objects in R-filt

Let R be a filtered ring with filtration FR.

6.1. Definition. An object $L \in R$-filt is said to be **filt-free** if it is free as an R-module and has a basis $(e_i)_{i \in J}$ consisting of elements with the property that there exists a family $(k_i)_{i \in J}$ of integers such that:

(1). $F_n L = \sum_{i \in J} F_{n-k_i} Re_i = \oplus_{i \in J} F_{n-k_i} Re_i, n \in \mathbb{Z}$,

(2). $e_i \notin F_{k_i-1} L, i \in J$.

An object $P \in R$-filt is said to be **filt-projective (or projective in R-filt)** if it is a direct summand of a filt-free R-module in R-filt, in other words, there exists L_0, L in R-filt with filtrations FL_0, FL respectively such that L is filt-free, $L \cong P \oplus L_0$ and moreover $F_n L \cong F_n P \oplus F_n L_0$ for all $n \in \mathbb{Z}$.

6.2. Lemma. Let L be a filtered R-module with filtration FL.

(1). L is filt-free with filt-basis $(e_i)_{i \in J}$ and associated family $(k_i)_{i \in J}$ if and only if $L \cong \oplus_{i \in J} T(-k_i) R$ in R-filt, where $T(-k_i)$ is the $-k_i$-th shift functor (see subsection 2.7.).

(2). If L is filt-free with filt-basis $(e_i)_{i \in J}$ and associated family $(k_i)_{i \in J}$, then $G(L)$ is a gr-free$G(R)$-module with homogeneous basis $\{(e_i)_{(k_i)} = \sigma(e_i), i \in J\}$.

(3). If $G(L)$ is gr-free in $G(R)$-gr with homogeneous basis $\{(e_i)_{(k_i)} = \sigma(e_i), i \in J\}$ where $e_i \in F_{k_i} L - F_{k_i-1} L$, then if FL is discrete it follows that L is filt-free with filt-basis $(e_i)_{i \in J}$ and associated family $(k_i)_{i \in J}$.

(4). If $M \in G(R)$-gr is gr-free then there exists a filt-free $L \in R$-filt with filtration FL such that $G(L) \cong M$ as graded modules.

(5). if L is filt-free with filt-basis $(e_i)_{i \in J}$ and associated family $(k_i)_{i \in J}$ and if $f: (e_i)_{i \in J} \to M$ is a function into an $M \in R$-filt such that $f(e_i) \in F_{s+k_i} M$ for some $s \in \mathbb{Z}$ and all $i \in J$, then there is a unique filtered morphism of degree s, say $g: L \to M$, extending the function f to L.

(6). If L is filt-free and $M \in R$-filt are such that we have a graded morphism of degree s, say $g: G(L) \to G(M)$, then there is a filtered morphism of degree s, say $f: L \to M$, such that $G(f) = g$.

(7). Let $L \in R$-filt be filt-free with filt-basis $(e_i)_{i \in J}$ and associated family $(k_i)_{i \in J}$, then FL is separated if and only if FR is separated; If FR is discrete and $(k_i)_{i \in J}$ is bounded below then FL is discrete; If J is finite and R is complete with respect to FR then FL is complete.

Proof. (1), (2), (3), (4), (5) can easily be checked and we leave it to the reader.

(6). If $L = \oplus_{i \in J} Re_i$ and $G(L) = \oplus_{i \in J} G(R)\sigma(e_i)$ and if $g(\sigma(e_i)) = \overline{\xi}_i \in G(M)$, then define f by putting $f(e_i) = \xi_i$.

(7). The first two assertions may be checked directly. Now suppose $L = \oplus_{i=1}^{s} Re_i$, then on the one hand, as a direct sum in R-filt, L has the filtration $F'_n L = \oplus_{i=1}^{s} F_n Re_i$, $n \in \mathbb{Z}$, and on the other hand L has the filtration $F_n L = \oplus_{i=1}^{s} F_{n-k_i} Re_i$, $n \in \mathbb{Z}$. But these filtrations on L are obviously equivalent, consequently L is complete with respect to FL because, as is easily seen, it is complete with respect to $F'L$.   □

As a consequence of Lemma 6.2.(4). it is clear that for every $M \in R$-filt there is a filt-free R-module L and a **strict exact sequence**

$$L \longrightarrow M \longrightarrow 0.$$

In particular we have the following

6.3. Corollary. (1). Let M be in R-filt with filtration FM. Then FM is a good filtration if and only if there exists a filt-free $L \in R$-filt of finite rank and a strict filtered epimorphism $f: L \to M$.
(2). Let $M \in R$-filt have good filtration FM. Suppose that R is complete with respect to FR and FM is separated, then M is complete with respect to FM.

Proof. (1). This is obvious.
(2). Since FM is a good filtration it follows from (1) that there is a filt-free R-module L of finite rank and a strict exact sequence in R-filt

$$0 \longrightarrow K \longrightarrow L \xrightarrow{\pi} M \longrightarrow 0$$

where $K = \text{Ker}\pi$ has the filtration FK induced by FL. By the assumption FM is separated, it follows that K is closed with respect to the filtration topology of L and consequently K is complete with respect to FK since L is complete by Lemma 6.2.(7).. Now take the corresponding completions of M, L and K and consider the commutative diagram of exact sequences in R-filt:

$$
\begin{array}{ccccccccc}
 & & 0 & & 0 & & & & \\
 & & \downarrow & & \downarrow & & & & \\
0 & \to & K & \longrightarrow & L & \xrightarrow{\pi} & M & \to & 0 \\
 & & \downarrow & & \downarrow & & \downarrow & & \\
0 & \to & \widehat{K} & \longrightarrow & \widehat{L} & \xrightarrow{\hat{\pi}} & \widehat{M} & \to & 0 \\
 & & \downarrow & & \downarrow & & & & \\
 & & 0 & & 0 & & & & \\
\end{array}
$$

where the vertical mappings are canonical, we obtain $M \cong \widehat{M}$, hence M is complete.   □

6.4. Lemma. (1). A filtered R-module M with filtration FM is filt-free if and only if $\widetilde{M}$ is gr-free in $\widetilde{R}$-gr.

(2). A filtered R-module P with filtration FP is filt-projective if and only if $\widetilde{P}$ is a projective $\widetilde{R}$-module.

(3). Suppose that R is complete with respect to FR. Then a filtered R-module L with separated filtration FL is filt-free of finite rank if and only if $G(L)$ is a gr-free $G(R)$-module of finite rank.

Proof. (1) and (2) may be checked by using Proposition 4.3.7., Proposition 4.3.8. and Corollary 4.1.3..

(3). If L is filt-free of finite rank then it follows from Lemma 6.2.(1). that $G(L)$ is gr-free of finite rank in $G(R)$-gr. Conversely, let $G(L)$ be gr-free of finite rank, say $G(L) = \oplus_{i=1}^{s} G(R)\sigma(e_i)$, where $e_i \in F_{k_i}L - F_{k_i-1}L$ and $\sigma(e_i) = e_i + F_{k_i-1}L$ is the principal part of e_i for some $k_i \in \mathbb{Z}$. Then a similar argumentation as in the proof of Theorem 5.7. gives rise to $F_n L = \sum_{i=1}^{s} F_{n-k_i} Re_i$, $n \in \mathbb{Z}$, and hence $L = \sum_{i=1}^{s} Re_i$. Moreover, since FR is separated one easily checks that $e_1, \cdots, e_s$ are R-linear independent. Hence L is filt-free in R-filt. $\square$

If P is a filt-projective object in R-filt and $f: M \to M'$ is a **strict epimorphism** of filtered R-modules, where M and M' have their respective filtrations FM and FM', then for a given filtered morphism $g: P \to M'$ we may find a filtered morphism $h: P \to M$ such that $f \circ h = g$. Indeed, the strictness of f yields that $\widetilde{f}$ is surjective and the projectivity of $\widetilde{P}$ in $\widetilde{R}$-gr leads to a graded morphism $\widetilde{h}: \widetilde{P} \to TM$ such that $\widetilde{f} \circ \widetilde{h} = \widetilde{g}$, but then we obtain h by applying the functor D (see subsection 4.3.). This leads to:

6.5. Proposition. For $P \in R$-filt we have the following properties:

(1). if P is filt-projective then $G(P)$ is projective in $G(R)$-gr;

(2). P is filt-projective if and only if for a given strict exact sequence in R-filt $M \xrightarrow{f} M' \to 0$ and filtered morphism $g: P \to M'$ there exists a filtered morphism $h: P \to M$ such that the following diagram is commutative in R-filt:

$$
\begin{array}{ccc}
 & & P \\
 & {}^{h}\swarrow & \downarrow {}^{g} \\
M & \xrightarrow{f} & M' \to 0
\end{array}
$$

Proof. (1). This follows from $G(P) \cong \widetilde{P}/X\widetilde{P}$ and Lemma 6.4..

(2). This follows from the remark preceding the proposition. $\square$

Of course, the theory of projective modules is particularly useful in connection with the homological dimension(s) of rings. As a prerequisite of such applications we have to clarify the structure of the HOM-functor and the tensor product of filtered modules, and to compare the filtered theory with the associated graded thory.

First note that our conventions of considering exhaustive filtrations only will not cause any problems when we study $\text{HOM}_R(M, N) \in \mathbb{Z}$-filt (see §2.) because the latter will be again exhaustively filtered.

6.6. Proposition. (Comparing with Lemma 4.1.1.) Let M, $N \in R$-filt have filtrations FM and FN respectively. Suppose that FM is good then $\text{HOM}_R(M, N) = \text{Hom}_R(M, N)$.

Proof. Let $m_1, \cdots, m_s$ be R-generators for M such that for each $n \in \mathbb{Z}$, $F_n M = \sum_{i=1}^s F_{n-k_i} R m_i$, where $k_i \in \mathbb{Z}$. Take an $f \in \text{Hom}_R(M, N)$. Select an $t \in \mathbb{Z}$ such that $f(m_i) \in F_{k_i+t}N$ for $i = 1, \cdots, s$, then $f \in F_t\text{HOM}_R(M, N)$ and hence certainly $f \in \text{HOM}_R(M, N)$. Therefore $\text{HOM}_R(M, N) = \text{Hom}_R(M, N)$.   □

6.7. Proposition. Let M, N be in R-filt with filtrations FM and FN respectively.
(1). If FN is separated then $F\text{HOM}_R(M, N)$ is separated.
(2). If M is finitely generated and FN is discrete then $F\text{HOM}_R(M, N)$ is discrete.
(3). If FN is complete with respect to its filtration topology then $\text{HOM}_R(M, N)$ is complete with respect to its filtration topology.

Proof. (1). If $f \in \cap_{n \in \mathbb{Z}} F_n\text{HOM}_R(M, N)$, then $f(F_n M) \subseteq \cap_{s \in \mathbb{Z}} F_{n+s}N = 0$, so for all $n \in \mathbb{Z}$, $f(F_n M) = 0$ and thus $f = 0$.
(2). Let $m_1, \cdots, m_s$ be R-generators for M and let $s_0 \in \mathbb{Z}$ be such that $\{m_1, \cdots, m_s\} \subseteq F_{s_0} M$. Since $F_n N = 0$ for all $n < n_0$ for some $n_0 \in \mathbb{Z}$ we obtain for any $f \in F_j\text{HOM}_R(M, N)$, where $j < n_0 - s_0$, that $f(m_i) = 0$. hence for all $j < n_0 - s_0$, $F_j\text{HOM}_R(M, N) = 0$.
(3). Let $p < q \in \mathbb{Z}$ and consider the projective system

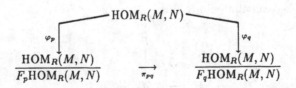

then we claim that

$$\text{HOM}_R(M, N) = \varprojlim_p \text{HOM}_R(M, N)/F_p\text{HOM}_R(M, N) = \widehat{\text{HOM}_R}(M, N).$$

Indeed, let $(f^{(p)})_{p \in \mathbb{Z}}$ be an element of $\widehat{\mathrm{HOM}_R}(M, N)$ such that $\pi_{pq}(f^{(p)}) = f^{(q)}$ for $p < q$. Let $f^{(p)} = \varphi_p(g^{(p)})$ with $g^{(p)} \in \mathrm{HOM}_R(M, N)$. If $p < q$ then $g^{(p)} - g^{(q)} \in F_q\mathrm{HOM}_R(M, N)$ and therefore $(g^{(p)} - g^{(q)})(F_s M) \subseteq F_{s+q}N$ for $s \in \mathbb{Z}$. Since FM is exhaustive, for any $m \in M$ there exists some $t \in \mathbb{Z}$ such that $m \in F_t M$ and thus $g^{(p)}(m) - g^{(q)}(m) \in F_{t+q}N$. Consequently the sequence $(g^{(p)}(m))_{p \in \mathbb{Z}}$ is a Cauchy sequence in N. Completeness of N entails that the mapping $f: M \to N$ defined by $f(m) = \lim_{p \to \infty} g^{(p)}(m)$, is well defined. If $m \in F_t M$, then the fact that $g^{(p)}(m) \in g^{(q)}(m) + F_{t+q}N$ for $p \leq q$ implies that $f(m) \in g^{(q)}(m) + F_{t+q}N$. Therefore $f - g^{(q)} \in F_q\mathrm{HOM}_R(M, N)$. Hence $f \in \mathrm{HOM}_R(M, N)$ and $\varphi_q(f) = f^{(q)}$ for selected q. This shows that the canonical mapping of $\mathrm{HOM}_R(M, N)$ to $\widehat{\mathrm{HOM}_R}(M, N)$ is an isomorphism. $\qquad \square$

6.8. Corollary. Let $M \in R$-filt be such that M is complete with respect to its filtration FM, then
(1). the ring $\mathrm{HOM}_R(M, M)$ is complete with respect to its filtration topology;
(2). the ring $\mathrm{Hom}_{FR}(M, M)$ is complete with respect to its filtration topology;
(3). suppose that $f \in \mathrm{Hom}_{FR}(M, M)$ is such that $G(f)^2 = G(f)$, then there exists a strict filtered morphism $g: M \to M$ in R-filt satisfying: $g^2 = g$ and $G(g) = G(f)$.

Proof. (1). Directly follows from Proposition 6.7..
(2). Since $\mathrm{Hom}_{FR}(M, M) = F_0\mathrm{HOM}_R(M, M)$ it is closed by property 3.1.(a).. Hence $\mathrm{Hom}_{FR}(M, M)$ is complete.
(3). Consider the morphism

$$\varphi: \quad \mathrm{Hom}_{FR}(M, M) \longrightarrow \mathrm{Hom}_{G(R)\text{-gr}}(G(M), G(M))$$

given by $\varphi(f) = G(f)$, then φ is a ring homomorphism. We see that $\mathrm{Ker}\varphi = F_{-1}\mathrm{HOM}_R(M, M)$. Hence if $f \in \mathrm{Ker}\varphi$ then $f^n \in F_{-n}\mathrm{HOM}_R(M, M)$ and we find that $\lim_{n \to \infty} f^n = 0$. Now the assertion follows from Lemma 3.5.6. and Corollary 3.2.3.. $\qquad \square$

For any M, $N \in R$-filt there is a natural map:

$$\psi = \psi(M, N): \quad G(\mathrm{HOM}_R(M, N)) \longrightarrow \mathrm{HOM}_{G(R)}(G(M), G(N))$$

given by associating to $f \in F_p\mathrm{HOM}_R(M, N)$ the graded $G(R)$-morphism $\psi(f_{(p)})$ such that for $x \in F_q M$, $\psi(f_{(p)})(x_{(q)}) = f(x)_{(p+q)}$.

6.9. Lemma. For every M, $N \in R$-filt, $\psi(M, N)$ is a graded monomorphism. If M is filt-projective then $\psi = \psi(M, N)$ is an isomorphism.

Proof. Clearly ψ is a graded $G(R)$-morphism. If $\psi(f_{(p)}) = 0$ for some $f \in F_p\mathrm{HOM}_R(M, N)$ then the definition of ψ implies that $f(x)_{(p+q)} = 0$ for every $x \in F_qM$. It follows that $f(F_qM) \subseteq F_{p+q-1}N$ for all $q \in \mathbb{Z}$ and hence $f \in F_{p-1}\mathrm{HOM}_R(M, N)$. Consequently $f_{(p)} = 0$. If M is filt-projective then there is a $Q \in R$-filt such that $L = M \oplus Q$ in R-filt is a filt-free R-module. Now since HOM_R commutes with finite direct sums it will suffice to establish the second statement in case M is filt-free. So assume that $(m_i)_{i \in I}$ is a filt-basis for M with associated family $(d_i)_{i \in I}$. In this case $(m_{i(d_i)})_{i \in I}$ is a homogeneous $G(R)$-basis for $G(M)$. Giving $g \in \mathrm{HOM}_{G(R)}(G(M), G(N))_p$ then $g(m_{i(d_i)}) = x^i_{p+d_i}$, $i \in I$. Define $f \colon M \to N$ by putting $f(m_i) = x^i$, one checks that $f \in F_p\mathrm{HOM}_R(M, N)$ and that $\psi(f_{(p)}) = g$. $\square$

6.10. Definition. A filtered R-module M with filtration FM is said to be **filt-injective**, or **injective in R-filt**, if for every left ideal I of R equiped with the filtration FI induced by FR, any filtered morphism $f \colon I \to M$ extends to a filtered morphism $f' \colon R \to M$, or equivalently, the canonical morphism

$$\mathrm{HOM}_R(i, 1_M) \colon \quad \mathrm{HOM}_R(R, M) \to \mathrm{HOM}_R(I, M)$$

is an epimorphism.

It is clear from this definition that a filtered R-module Q such that Q is injective in R-mod is also filt-injective.

6.11. Lemma. If R is complete with respect to FR and $G(R)$ is left Noetherian then a filtered R-module Q that is filt-injective is also injective in R-mod.

Proof. Apply Theorem 5.7. to the left ideals of R (note that the induced filtrations are separated), so it follows that R is left Noetherian. Then one can use Proposition 6.6. and derive from the surjectivity of $\mathrm{HOM}_R(i, 1_M)$ as above, the surjectivity of $\mathrm{Hom}_R(i, 1_M)$ and this allows to deduce from the filt-injectivity of Q the injectivity of Q in R-mod. $\square$

6.12. Theorem. Assume that $Q \in R$-filt is complete with respect to its filtration FQ. If $G(Q)$ is injective in $G(R)$-gr then Q is filt-injective.

Proof. Let I be a left ideal of R with the filtration FI induced by FR. Since $i \colon I \hookrightarrow R$ is a strict filtered morphism we obtain a commutative diagram of graded $G(R)$-morphisms:

$$
\begin{array}{ccc}
G(\mathrm{HOM}_R(R, Q)) & \xrightarrow{\Delta} & G(\mathrm{HOM}_R(I, Q)) \\
\psi(R,Q) \downarrow & & \downarrow \psi(I,Q) \\
\mathrm{HOM}_{G(R)}(G(R), G(Q)) & \xrightarrow[\theta]{} & \mathrm{HOM}_{G(R)}(G(I), G(Q))
\end{array}
$$

where $\Delta = G(\text{HOM}_R(i, 1_Q))$ and $\Theta = \text{HOM}_{G(R)}(G(i), 1_{G(Q)})$. Since $G(i)$ is a monomorphism, the injectivity of $G(Q)$ entails that $\text{HOM}_{G(R)}(G(i), 1_{G(Q)})$ is an epimorphism. Hence $\psi(I, Q)$ is epimorphic and therefore an isomorphism. Finally, surjectivity of $G(\text{HOM}_R(i, 1_Q))$ and Proposition 6.7. and Theorem 4.2.4. lead to the surjectivity of $\text{HOM}_R(i, 1_Q)$. □

Let us finish this section by introducing the tensor product of filtered R-modules.

If R is a filtered ring with filtration FR and $M \in R$-filt, $N \in$ filt-R (the category of filtered right R-modules) with filtrations FM and FN respectively, then the abelian group $\underline{N} \otimes_R \underline{M}$, where $\underline{N}$ and $\underline{M}$ denote the abelian groups N and M without filtrations respectively, has a filtration $F(\underline{N} \otimes_R \underline{M})$ that may be obtained by putting $F_n(\underline{N} \otimes_R \underline{M})$ equal to the subgroup of $\underline{N} \otimes_R \underline{M}$ generated by all elements of type $x \otimes y$, $x \in F_t N$, $y \in F_s M$, such that $s + t \leq n$. The filtered $\mathbb{Z}$-module thus defined is called the **tensor product** of N and M in R-filt, we denote it by $N \otimes_R M$ and call $F(N \otimes_R M)$ the **tensor filtration**. The tensor product defines a functor $\otimes_R$: filt-$R \times R$-filt $\rightarrow \mathbb{Z}$-filt.

6.13. Lemma. With conventions and notations as above, we have

(1). $F(N \otimes_R M)$ is exhaustive;

(2). if FN and FM are discrete then so is $F(N \otimes_R M)$;

(3). if L is a filt-free R-module with filt-basis $(e_i)_{i \in J}$ and associated family $(k_i)_{i \in J}$, then $\hat{R} \otimes_R L$ is, with respect to the filtration $F(\hat{R} \otimes_R L)$, a filt-free $\hat{R}$-module with filt-basis $(1 \otimes e_i)_{i \in J}$ and associated family $(k_i)_{i \in J}$;

(4). for all $m, n \in \mathbb{Z}$, $T(n)N \otimes_R T(m)M = T(m+n)(N \otimes_R M)$, where $T(n)N$ resp. $T(m)M$ is the n-shift of N resp. the m-shift of M;

(5). in R-filt and filt-R, $R \otimes_R M \cong M$, $N \otimes_R R \cong N$;

(6). the functor $\otimes_R$ commutes with direct sums and inductive limits;

(7). for $M \in$ filt-R, $N \in R$-filt-S, $P \in$ filt-S, where S is another filtered ring with filtration FS, there is a filtered isomorphism of filtered abelian groups: $\text{HOM}_S(M \otimes_R N, P) \cong \text{HOM}_R(M, \text{HOM}_S(N, P))$.

Proof. The statements (1), (2), (3), (4), (5) are easy enough.

(6). Let $N = \oplus_{\alpha \in I} N_\alpha$, $M = \oplus_{\beta \in J} M_\beta$, then we claim $\oplus_{(\alpha, \beta) \in I \times J}(N_\alpha \otimes_R M_\beta) \cong N \otimes_R M$. Indeed, let φ: $N \otimes_R M \rightarrow \oplus_{(\alpha, \beta) \in I \times J}(N_\alpha \otimes_R M_\beta)$ be the R-module homomorphism which is given by $\varphi((\sum_{\alpha \in I} x_\alpha) \otimes (\sum_{\beta \in J} y_\beta)) = \sum_{(\alpha, \beta) \in I \times J} x_\alpha \otimes y_\beta$, then it is well known that φ is an isomorphism of abelian groups. Let $x \otimes y \in F_n(N \otimes_R M)$ where $x \in F_i N$, $y \in F_j M$ and $i + j = n$. If $x = \sum_{\alpha \in I} x_\alpha$, $y = \sum_{\beta \in J} y_\beta$ then $x_\alpha \in F_i N_\alpha$ and $y_\beta \in F_j M_\beta$. It follows immediately that $x_\alpha \otimes y_\beta \in F_n(N_\alpha \otimes_R M_\beta)$, hence $\varphi(x \otimes y) \in F_n(\oplus_{(\alpha, \beta) \in I \times J}(N_\alpha \otimes_R M_\beta))$. In a similar way one establishes that φ^{-1} is actually a morphism in $\mathbb{Z}$-filt. The assertion for inductive limits may be checked in a similar way.

(7). Define τ: $\mathrm{HOM}_S(M \otimes_R N, P) \to \mathrm{HOM}_R(M, \mathrm{HOM}_S(N, P))$ by $\tau(f)(x)(y) = f(x \otimes y)$, then one may define a map ψ: $\mathrm{HOM}_R(M, \mathrm{HOM}_S(N, P)) \to \mathrm{HOM}_S(M \otimes_R N, P)$ by putting $\psi(g)(x \otimes y) = g(x)(y)$. If $f \in F_n\mathrm{HOM}_S(M \otimes_R N, P)$ whereas $x \in F_tM$, $y \in F_sN$ then from $x \otimes y \in F_{t+s}(M \otimes_R N)$ it follows that $f(x \otimes y) \in F_{n+t+s}P$. Hence $\tau(f)(x) \in F_{n+t}\mathrm{HOM}_S(N, P)$ and $\tau(f) \in F_n\mathrm{HOM}_R(M, \mathrm{HOM}_S(N, P))$. So we have established that τ is a filtered morphism in $\mathbb{Z}$-filt. In a similar way one may verify that ψ is a filtered morphism and that it is the inverse of τ. $\square$

If $N \in R$-filt, $M \in$ filt-R with filtrations FN and FM respectively, then we may define a graded morphism

$$\varphi(M, N) : \quad G(M) \otimes_{G(R)} G(N) \longrightarrow G(M \otimes_R N)$$

by putting $x_{(s)} \otimes y_{(t)} \mapsto (x \otimes y)_{s+t}$, where $x_{(s)} = \sigma(x)$, $y_{(t)} = \sigma(y)$ for $x \in F_sM - F_{s-1}M$, $y \in F_tN - F_{t-1}N$.

It is obvious that $\varphi(M, N)$ is a well defined graded morphism and moreover $\varphi(M, N)$ **is an epimorphism**.

6.14. Lemma. With notations as above, if either M is filt-free as a right R-module or N is filt-free as a left R-module then $\varphi(M, N)$ is an isomorphism.

Proof. The functors $\otimes_R$ and G commute with direct sums so we may reduce the proof to the case where $M = T(n)R \in$ filt-R, and in view of Lemma 6.13.(4). we may reduce further to the case where $M = R$ in filt-R so that we arrive at a commutative diagram of graded $G(R)$-morphisms

$$
\begin{array}{ccc}
G(R_R) \otimes_{G(R)} G(N) & \xrightarrow{\varphi(R,N)} & G(R \otimes_R N) \\
\cong \downarrow & & \downarrow \cong \\
G(N) & \xrightarrow[G(1_N)]{} & G(N)
\end{array}
$$

It follows that $\varphi(R, N)$ is also an isomorphism. $\square$

It is equally straightforward to prove the following

6.15. Lemma. Let A and B be filtered rings with filtrations FA and FB respectively. Suppose that f: $A \to B$ is a filtered ring homomorphism of degree zero (note that for a ring homomorphism f we always assume $f(1_A) = 1_B$). If $N \in A$-filt has a good filtration FN then the tensor product filtration on the B-module $B \otimes_A N$ is a good filtration (where the A-module structure on B is defined by using the ring homomorphism f in a classical way). In particular, if N is filt-free of finite rank with filt-basis $\{e_1, \cdots, e_s\}$ and associated family $\{k_1, \cdots, k_s\}$, then $B \otimes_A N$ is a filt-free B-module with filt-basis $\{1 \otimes e_1, \cdots, 1 \otimes e_s\}$ and associated family $\{k_1, \cdots, k_s\}$.

§7. Krull Dimension and Global Dimension of Filtered Rings

7.1. Krull dimension of filtered rings

Let us first recall the Krull dimension of modules as it has been introduced by Gabriel and Rentschler in [G-R] and extended by [Krau] for not necessarily finite ordinal numbers.

Let $(E, \leq)$ be an ordered set. For a, $b \in E$ we put $[a, b] = \{x \in E, a \leq x \leq b\}$, $\Gamma(E) = \{(a, b), a \leq b\}$. By transfinite recurrence we define on $\Gamma(E)$ the following filtration: $\Gamma_{-1}(E) = \{(a, b) \in \Gamma(E), a = b\}$, $\Gamma_0(E) = \{(a, b) \in \Gamma(E), [a, b]$ is Artinian $\}$, if $\Gamma_\beta(E)$ has been defined for all $\beta < \alpha$, then put $\Gamma_\alpha(E) = \{(a, b) \in \Gamma(E)$, for all $b \geq b_1 \geq \cdots \geq b_n \geq \cdots \geq a$, there is a positive integer n such that $[b_{i+1}, b_i] \in \Gamma_\beta(E)$ for all $i \geq n\}$. We have the ascending chain: $\Gamma_{-1}(E) \subseteq \Gamma_0(E) \subseteq \cdots \subseteq \Gamma_\alpha(E) \subseteq \cdots$ and there exists an ordinal ξ such that $\Gamma_\xi(E) = \Gamma_{\xi+1}(E) = \cdots$. When $\Gamma(E) = \Gamma_\alpha(E)$ for some ordinal α then we say that E has Krull dimension. The smallest ordinal α such that $\Gamma(E) = \Gamma_\alpha(E)$ is called the Krull dimension of E and we write $\mathrm{K.dim}\, E = \alpha$.

1. Lemma. (cf. [G-R]) Let E and F be ordered sets.

(1). If $f\colon E \to F$ is a strictly increasing mapping and F has Krull dimension then E has Krull dimension and $\mathrm{K.dim}\, E \leq \mathrm{K.dim}\, F$.

(2). If E and F have Krull dimension then $E \times F$ has Krull dimension and $\mathrm{K.dim}(E \times F) = \mathrm{Sup}\{\, \mathrm{K.dim}\, E, \mathrm{K.dim}\, F\}$ where $E \times F$ has the product ordering.

If $\mathcal{C}$ is any abelian category and $M \in \mathcal{C}$ then we let $E(M)$ be the set of all subobjects of M ordered by inclusion. If $E(M)$ has Krull dimension then M is said to **have Krull dimension** and we denote it by $\mathrm{K.dim}_\mathcal{C} M$ or simply $\mathrm{K.dim}\, M$.

If $M \in \mathcal{C}$ has $\mathrm{K.dim}\, M = \alpha$ then M is said to be α-**critical** if $\mathrm{K.dim}(M/M') < \alpha$ for every nonzero subobject M' of M. In particular, M is 0-critical if and only if M is a simple object in the category $\mathcal{C}$. Clearly if M is α-critical then any nonzero subobject of M is α-critical, too.

If $\mathcal{C} = R$-mod for some ring R and $M \in R$-mod has Krull dimension then we denote it by $\mathrm{K.dim}_R M$; in particular, $\mathrm{K.dim}_R R$ is called the (left) Krull dimension of R and we denote it by $\mathrm{K.dim}\, R$.

2. Proposition. Let R be a complete filtered ring with filtration FR, and let $M \in R$-filt with separated filtration FM. If $G(M)$ is left Noetherian then M is left Noetherian too and $\mathrm{K.dim}_R M \leq \mathrm{K.dim}_{G(R)} G(M)$. Moreover if $\mathrm{K.dim}_R M = \mathrm{K.dim}_{G(R)} G(M) = \alpha$ and $G(M)$ is α-critical then M is an α-critical R-module.

Proof. Let N be any submodule of M with the filtration FN induced by FM, then

$G(N) \hookrightarrow G(M)$ by Theorem 4.2.4.. Accordingly $G(N)$ is a finitely generated $G(R)$-module. In view of Theorem 5.7. it follows that N is a finitely generated R-submodule and therefore M is left Noetherian. If $N \subseteq P \subseteq M$ are R-submodules filtered by the induced filtrations and such that $G(N) = G(P)$ then $N = P$ follows again from Theorem 5.7. (in fact it follows from the proof of this result). Now the foregoing Lemma yields K.dim$_R M \leq$ K.dim$_{G(R)} G(M)$ because the map $N \mapsto G(N)$ determines a strictly increasing map on the lattice of subobjects of M to the lattice of subobjects of $G(M)$.

Finally, assume that K.dim$_R M =$ K.dim$_{G(R)} G(M) = \alpha$ and that $G(M)$ is α-critical as a $G(R)$-module. If N is a nonzero R-submodule of M then we may filter M/N by the quotient filtration of FM and by Theorem 4.2.4. we obtain an exact sequence in $G(R)$-gr:

$$0 \longrightarrow G(N) \longrightarrow G(M) \longrightarrow G(M/N) \longrightarrow 0$$

where N has the filtration FN induced by FM. Since $G(N) \neq 0$ (note that FM is separated) the Krull dimension of $G(M/N)$ is less than α, hence we obtain K.dim$_R(M/N) \leq$ K.dim$_{G(R)} G(M/N) < \alpha$, or M is α-critical. □

3. Theorem. Let R be a filtered ring with filtration FR. Suppose that $\tilde{R}$ is left Noetherian and FR is faithful, i.e., $F_{-1} R \subseteq J(F_0 R)$.

(1). Let M be in R-filt with good filtration FM. If the Krull dimension of $G(M)$ is well defined then the Krull dimension of M is defined and K.dim$_R M \leq$ K.dim$_{G(R)} G(M)$.

(2). Let M be as in (1). If K.dim$_R M =$ K.dim$_{G(R)} G(M) = \alpha$ and if $G(M)$ is α-critical then M is an α-critical R-module.

Proof. (1). This follows from Corollary 5.5. and Lemma 7.1.1..

(2). Let N be a nonzero R-submodule of M. Considering the filtration FN on N induced by FM and the quotient filtration on M/N given by $F_n M + N/N$, $n \in \mathbb{Z}$, we have the exact sequence in $G(R)$-gr:

$$0 \longrightarrow G(N) \longrightarrow G(M) \longrightarrow G(M/N) \longrightarrow 0$$

by Theorem 4.2.4.. Since FM is separated by Theorem 4.4.14. we have $G(N) \neq 0$, hence the Krull dimension of $G(M/N)$ is less than α. Consequently by (1) we obtain K.dim$_R(M/N) \leq$ K.dim$_{G(R)} G(M/N) < \alpha$. This shows that M is α-critical. □

7.2. Global dimension of filtered rings

We refer the reader to [Rot] for the fundamental theory concerning the (homological) global dimension of rings.

First recall that a ring R is called **left regular** if every finitely generated left R-module has finite projective dimension, or equivalently, that every cyclic R-module has finite projective dimension. Similarly, for any (group)G-graded ring R, one may define the **left gr-regularity** in terms of objects in R-gr; we also write gr.gl.dim$R$ for the (left) graded global dimension of R in the sense of [NVO1], and here it suffices to know that the definition is the obvious one expressed intrinsically in terms of objects of R-gr.

Now let R be a filtered ring with filtration FR. We write gl.dim$R$ for the **left global dimension** of R (it is well known that the left global dimension and right global dimension of R are equal when R is a left and right Noetherian ring). If $M \in R$-filt, we denote by p.dim$_R M$ the **projective dimension** of M.

1. Proposition. Let P be in R-filt with separated filtration FP. Suppose that R is complete with respect to its filtration topology and $G(P)$ is finitely generated. If $G(P)$ is projective in $G(R)$-gr then P is filt-projective in R-filt.

Proof. Since $G(P)$ is finitely generated it follows from Theorem 5.7. that FP is good. Accordingly there is a strict exact sequence in R-filt:

$$0 \longrightarrow K \longrightarrow L \longrightarrow P \longrightarrow 0$$

such that L is filt-free of finite rank and K has the filtration FK induced by FL. By Lemma 6.4. $\widetilde{L}$ is a graded free $\widetilde{R}$-module of finite rank and hence $\widetilde{L}/X\widetilde{L}$ is a graded free $\widetilde{R}/X\widetilde{R}$-module of finite rank, where X is the canonical homogeneous element of degree 1 in $\widetilde{R}$. Note that $\widetilde{P}/X\widetilde{P} \cong G(P)$ and $\widetilde{P}$ is X-torsionfree it follows from the assumption that the sequence

$$0 \longrightarrow \widetilde{K}/X\widetilde{K} \longrightarrow \widetilde{L}/X\widetilde{L} \longrightarrow \widetilde{P}/X\widetilde{P} \longrightarrow 0$$

is exact and spliting, i.e., $\widetilde{L}/X\widetilde{L} \cong \widetilde{P}/X\widetilde{P} \oplus \widetilde{K}/X\widetilde{K}$. Put $T = P \oplus K$ in R-filt by equiping T with the filtration $F_n T = F_n P \oplus F_n K$, $n \in \mathbb{Z}$. Then $\widetilde{T} \cong \widetilde{P} \oplus \widetilde{K}$ by Proposition 4.3.8.. Moreover we have

$$\widetilde{T}/X\widetilde{T} \cong (\widetilde{P} \oplus \widetilde{K})/X(\widetilde{P} \oplus \widetilde{K}) \cong \widetilde{P}/X\widetilde{P} \oplus \widetilde{K}/X\widetilde{K} \cong \widetilde{L}/X\widetilde{L}$$

Hence $G(T)$ is gr-free of finite rank in $G(R)$-gr since $G(L) \cong \widetilde{L}/X\widetilde{L}$ is gr-free of finite rank in $G(R)$-gr. From the construction of T we see that FT is separated, it follows from Lemma 6.4.(3). that T is filt-free of finite rank in R-filt. Therefore P is filt-projective in R-filt. $\square$

2. Corollary. Suppose that R is complete with respect to its filtration topology and that $G(R)$ is left Noetherian. Then R is left regular in case $G(R)$ is left gr-regular; moreover we always have gl.dim$R \leq$ gr.gl.dim$G(R)$.

Proof. If $\text{gr.gl.dim}\,G(R) = \infty$ then there is nothing to prove. Suppose that $\text{gr.gl.dim}\,G(R) = n$ is finite. Let M be an arbitrary finitely generated R-module. Take a good filtraion, say FM, on M. Then by Corollary 6.3. there is a strict exact sequence in R-filt:

$$0 \longrightarrow K \longrightarrow L \longrightarrow M \longrightarrow 0$$

where L is filt-free of finite rank and K has the filtration FK induced by FL. Since $G(R)$ is left Noetherian $G(K)$ is finitely generated by Lemma 5.4.. Note that FR is separated it follows that FL is separated and hence FK is separated, too. Then by Theorem 5.7. FK is good. By a repetition of this procedure for K and so on we get a strict exact sequence in R-filt:

$$0 \longrightarrow K_n \longrightarrow L_{n-1} \longrightarrow \cdots \longrightarrow L_0 \longrightarrow M \longrightarrow 0$$

where all L_i are filt-free of finite rank and K_n has the separated filtration FK_n induced by FL_{n-1} which is also a good filtration. Now Theorem 4.2.4. yields the following exact sequence in $G(R)$-gr:

$$0 \longrightarrow G(K_n) \longrightarrow G(L_{n-1}) \longrightarrow \cdots \longrightarrow G(L_0) \longrightarrow G(M) \longrightarrow 0$$

such that all $G(L_i)$ are gr-free of finite rank (Lemma 6.2.) and $G(K_n)$ is a finitely generated projective $G(R)$-module. It follows from Proposition 7.2.1. that K_n is filt-projective in R-filt, hence $\text{p.dim}_R M \leq \text{p.dim}_{G(R)} G(M) \leq n$. Accordingly we have proved $\text{gl.dim}\,R \leq n$. From the proof it is clear that if $G(R)$ is left gr-regular then R is left regular. $\qquad\square$

Before considering the filtered rings without the completeness assumption, we first prove a lemma which has independent interest in general ring theory.

3. Lemma. Let A be a left and right Noetherian ring and T a central regular element. Let $A_{(T)}$ be the localization of A at the (central) Ore set $\{1,\ T,\ T^2,\ \cdots\}$. If M is a finitely generated T-torsionfree A-module such that M/TM is a projective A/TA-module and $M_{(T)} = A_{(T)} \otimes_A M$ is a projective $A_{(T)}$-module, then M is a projective A-module (the converse holds obviously).

Proof. Let

$$\cdots \longrightarrow L_1 \longrightarrow L_0 \longrightarrow M \longrightarrow 0$$

be a free resolution of M, where each L_j is a free A-module of finite rank. Let $Z[T]$ denote the subring of A generated by 1 and T, then $Z[T]$ is contained in $Z(A)$, the centre of A. From ([Bass] Proposition 4.5., or a direct verification) it follows that for any A-module N there is an isomorphism of complexes of $Z[T]_{(T)}$-modules:

$$0 \to Z[T]_{(T)} \otimes \mathrm{Hom}_A(M,N) \;\to\; Z[T]_{(T)} \otimes \mathrm{Hom}_A(L_0,N) \;\to\; Z[T]_{(T)} \otimes \mathrm{Hom}_A(L_1,N)$$

$$\Big\downarrow \cong \qquad\qquad\qquad\qquad \Big\downarrow \cong \qquad\qquad\qquad\qquad \Big\downarrow \cong$$

$$0 \to \mathrm{Hom}_{A_{(T)}}(M_{(T)},N_{(T)}) \;\to\; \mathrm{Hom}_{A_{(T)}}(L_{0(T)},N_{(T)}) \;\to\; \mathrm{Hom}_{A_{(T)}}(L_{1(T)},N_{(T)})$$

where $Z[T]_{(T)}$ is the localization of $Z[T]$ at the Ore set $\{1,\, T,\, T^2,\, \cdots\}$. Hence there is an isomorphism of $Z[T]_{(T)}$-modules:

$$Z[T]_{(T)} \otimes \mathrm{Ext}_A^i(M,N) \cong \mathrm{Ext}_{A_{(T)}}^i(M_{(T)},N_{(T)})$$

for each $i \geq 0$. Now $M_{(T)}$ is projective by the assumption so we obtain for $i \geq 1$: $\mathrm{Ext}_{A_{(T)}}^i(M_{(T)},N_{(T)}) = 0$ and thus $\mathrm{Ext}_A^i(M,N)$ is T-torsion for all $i \geq 1$. On the other hand, the projective dimension of M/TM as a left A-module equals one by the well known first "change of rings" theorem (see also later Ch.II. Theorem 5.1.5.). Since M is T-torsionfree we have an exact sequence in A-mod:

$$0 \longrightarrow M \xrightarrow{\mu_T} M \longrightarrow M/TM \longrightarrow 0$$

where μ_T is multiplication by T and we have a long exact sequence

$$\to \mathrm{Ext}_A^2(M/TM,N) \longrightarrow \mathrm{Ext}_A^2(M,N) \xrightarrow{\mu_T^*} \mathrm{Ext}_A^2(M,N) \longrightarrow \mathrm{Ext}_A^3(M/TM,N) \to$$

Since $\mathrm{Ext}_A^i(M,N)$ is independent of the choice of projective resolution for M it follows (cf. [Rot] Theorem 7.16.) that μ_T^* is again multiplication by T and we arrive at an isomorphism $\mu_T^*\colon \mathrm{Ext}_A^2(M,N) \to \mathrm{Ext}_A^2(M,N)$. Combined with the fact that $\mathrm{Ext}_A^2(M,N)$ is T-torsion, the latter states that $\mathrm{Ext}_A^2(M,N) = 0$ for any left A-module N. Hence p.dim$_A M \leq 1$. Now if we have an exact sequence of left A-modules:

$$0 \longrightarrow K \longrightarrow L \longrightarrow N \longrightarrow 0$$

where L is free and N is finitely generated, then $\mathrm{Ext}_A^1(M,A) = 0$ entails that $\mathrm{Ext}_A^1(M,N) = 0$ and therefore M is projective. That $\mathrm{Ext}_A^1(M,A) = 0$ follows from the exact sequence of right A-modules:

$$\mathrm{Ext}_A^1(M,A) \xrightarrow{\mu_T^*} \mathrm{Ext}_A^1(M,A) \longrightarrow \mathrm{Ext}_A^2(M/TM,A) = 0$$

The surjectivity of μ_T combined with the fact that $\mathrm{Ext}_A^1(M,A)$ is finitely generated as a right A-module but also a T-torsion module, yields that $\mathrm{Ext}_A^1(M,A) = 0$. $\square$

4. Theorem. Let A be a left and right Noetherian ring and T a central regular element of A.

(1). If A/TA and $A_{(T)}$ are left regular then A is left regular, where $A_{(T)}$ is defined as in Lemma 3..

(2). gl.dim$A \leq$ max$\{1+$ gl.dim(A/TA), gl.dim$A_{(T)}\}$; In case gl.dim(A/TA) is finite then we always have an equality gl.dim$A =$ max$\{1+$ gl.dim(A/TA), gl.dim$A_{(T)}\}$.

Proof. (1) will follow from the proof of (2) below.

(2). If max$\{1+$ gl.dim(A/TA), gl.dim$A_{(T)}\} = \infty$ then there is nothing to prove. So we may assume max$\{1+$ gl.dim(A/TA), gl.dim$A_{(T)}\} = n$.

For a finitely generated left A-module M we let $t(M)$ stand for the T-torsion submodule of M and we have an exact sequence of A-modules:

$$(*) \qquad\qquad 0 \longrightarrow t(M) \longrightarrow M \longrightarrow M/t(M) \longrightarrow 0$$

Since A is Noetherian there is an integer $w > 0$ such that $T^w t(M) = 0$. In view of the well known first "change of rings" theorem (or see also later Ch.II. Theorem 5.1.5.) we may apply induction on w and arrive at

$$\text{p.dim}_A t(M) \leq 1 + \text{gl.dim}(A/TA) \leq n.$$

Now applying the foregoing lemma to the finitely generated T-torsionfree A-module $M/t(M)$ we can conclude that p.dim$(M/t(M)) \leq n$. This establishes the first claim because p.dim$_A M \leq n$ follows from the foregoing and $(*)$. In order to obtain the equality it suffices to note that we also have gl.dim$A \geq 1+$ gl.dim(A/TA) when gl.dim(A/TA) is finite (again see latter Ch.II. Theorem 5.1.5.) since we always have gl.dim$A \geq$ gl.dim$A_{(T)}$. This completes the proof.   □

5. Corollary. Let R be a filtered ring with filtration FR and P a finitely generated R-module that is a projective object in R-mod. Suppose that P is also a filtered R-module with filtration FP and $\tilde{R}$ is left and right Noetherian , then P is filt-projective in R-filt if and only if $G(P)$ is a projective left $G(R)$-module.

Proof. From Lemma 6.4. we know that P is filt-projective if and only if $\tilde{P}$ is (gr-)projective as a left $\tilde{R}$-module. First assume that $G(P)$ is a projective left $G(R)$-module. Putting $T = X$, the canonical homogeneous element of degree 1 in $\tilde{R}$, and $A = \tilde{R}$ we may apply Lemma 7.2.3. to the X-torsionfree $\tilde{R}$-module $\tilde{P}$. So the projectivity of $\tilde{P}_{(X)} \cong P[t, t^{-1}]$ as a left $\tilde{R}_{(X)} \cong R[t, t^{-1}]$-module together with the assumed projectivity of $\tilde{P}/X\tilde{P} \cong G(P)$ as a left $\tilde{R}/X\tilde{R} \cong G(R)$-module lead to the projectivity of $\tilde{P}$ as a left $\tilde{R}$-module hence to the filt-projectivity of P.

Conversely, if $\tilde{P}$ is projective as a left $\tilde{R}$-module then $\tilde{P}/X\tilde{P} \cong G(P)$ is projective as a left $G(R)$-module (an obvious property of the functor $\tilde{R}/X\tilde{R} \otimes_{\tilde{R}} -$) so the claim follows.   □

6. Corollary. Let the situation be as in the foregoing corollary.

(1). If R and $G(R)$ are left regular then $\tilde{R}$ is left regular.

(2). $\mathrm{gl.dim}\tilde{R} \leq \max\{1+ \mathrm{gl.dim}G(R), 1+ \mathrm{gl.dim}R\}$ and the equality holds if $\mathrm{gl.dim}G(R)$ is finite.

Proof. Since $\tilde{R}_{(X)} \cong R[t,t^{-1}]$ we have $\mathrm{gl.dim}\tilde{R}_{(X)} = 1+ \mathrm{gl.dim}R$ and the statement now follows by translating Theorem 7.2.4. according to the dictionary $T = X$, $A = \tilde{R}$, $A/TA = G(R)$.   □

7. Corollary. If R is a filtered ring with filtration FR such that $\tilde{R}$ is left and right Noetherian, then $\mathrm{gl.dim}R < \infty$ plus $\mathrm{gl.dim}G(R) < \infty$ imply that $\mathrm{gl.dim}\tilde{R} < \infty$.

Our next goal is to investigate the case where the Rees ring $\tilde{R}$ of the given filtered ring R is not necessarily two-sided Noetherian but the filtration FR on R is **faithful**. At this extreme we still get decent behaviour of projectivity and certain dimensions.

8. Lemma. Let R be a filtered ring with faithful filtration FR, i.e., $F_{-1}R \subseteq J(F_0R)$, and $M \in R$-filt with good filtration FM. If

$$F \xrightarrow{\ \overline{f}\ } G(M) \longrightarrow 0$$

is an exact sequence in $G(R)$-gr such that F is gr-free of finite rank in $G(R)$-gr, then there is a filt-free R-module L of finite rank such that $G(L) = F$ and there is a strict exact sequence

$$L \xrightarrow{\ f\ } M \longrightarrow 0$$

in R-filt such that $G(f) = \overline{f}$ and $G(\mathrm{Ker}f) = \mathrm{Ker}\overline{f}$.

Proof. From Lemma 6.2. the existence of L and $L \xrightarrow{f} M \to 0$ such that $G(L) = F$ and $G(f) = \overline{f}$ is guaranteed. Since FM is good $\tilde{M}$ is finitely generated in $\tilde{R}$-gr and since $X \in J^g(\tilde{R})$ (Corollary 4.4.12.), where X is the canonical homogeneous element of degree 1 in $\tilde{R}$, we can use the graded version of Nakayama's lemma to derive that every homogeneous generating system of $\tilde{M}/X\tilde{M} = G(M)$ can be lifted to a homogeneous generating system of $\tilde{M}$. This allows to conclude that $\tilde{f}: \tilde{L} \to \tilde{M}$ is a surjective graded morphism and consequently the sequence

$$L \xrightarrow{\ f\ } M \longrightarrow 0$$

is strict exact by Proposition 4.3.8..

Next, if $K = \mathrm{Ker}f$ is equiped with the filtration FK induced by FL (note that L is filt-free) then we have an exact sequence of strict filtered morphisms in R-filt:

$$0 \longrightarrow K \longrightarrow L \xrightarrow{f} M \longrightarrow 0$$

and therefore the associated sequence

$$0 \longrightarrow G(K) \longrightarrow G(L) \xrightarrow{G(f)} G(M) \longrightarrow 0$$

is exact in $G(R)$-gr. From $G(f) = \overline{f}$ it then follows that $G(\mathrm{Ker}f) = G(K) = \mathrm{Ker}G(f) = \mathrm{Ker}\overline{f}$. Hence the proof is completed. □

9. Corollary. Let R be a filtered ring with filtration FR satisfying:

a). good filtrations in R-filt induce good filtrations on R-submodules,

b). FR is faithful.

Consider $M \in R$-filt with good filtration FM, then a finitely generated graded free resolution of $G(M)$ in $G(R)$-gr can be "lifted" to a finitely generated filt-free resolution of M in R-filt such that the lifted sequence is strict exact.

Proof. Any given graded free resolution

$$\cdots \longrightarrow F_1 \xrightarrow{\overline{f}_1} F_0 \xrightarrow{\overline{f}_0} G(M) \longrightarrow 0$$

of the prescribed type may be lifted step by step by considering each short exact sequence

$$0 \longrightarrow K_{i+1} \longrightarrow F_{i+1} \xrightarrow{\overline{f}_{i+1}} K_i \longrightarrow 0$$

By the assumption a). and the foregoing lemma we obtain $K_i = G(N_i)$ where $N_i \in R$-filt has good filtration FN_i. So repeated application of the statement of Lemma 7.2.8. finally completes the proof. □

10. Proposition. Let R be as in Corollary 7.2.9. and let $M \in R$-filt have a good filtration FM. If $G(M)$ is projective in $G(R)$-gr then M is filt-projective in R-filt.

Proof. Since FM is good there is a strict exact sequence in R-filt:

$$(*) \qquad 0 \longrightarrow K \longrightarrow L \longrightarrow M \longrightarrow 0$$

where L is filt-free of finite rank and K has the good filtraiton FK induced by FL (using assumption a).). Then we also have exact sequences of graded morphisms:

$$G(*) \qquad 0 \longrightarrow G(K) \longrightarrow G(L) \longrightarrow G(M) \longrightarrow 0$$

and

$$(\widetilde{*}) \qquad\qquad 0 \longrightarrow \widetilde{K} \longrightarrow \widetilde{L} \longrightarrow \widetilde{M} \longrightarrow 0$$

where $G(L)$ resp. $\widetilde{L}$ is gr-free of finite rank in $G(R)$-gr resp. in $\widetilde{R}$-gr. Put $T = \widetilde{M} \oplus \widetilde{K}$ in $\widetilde{R}$-gr, then T is a finitely generated and X-torsionfree $\widetilde{R}$-module. Since $G(*)$ is split we obtain $\widetilde{L}/X\widetilde{L} = T/XT$ and the graded version of Nakayama's lemma (using assumption b). and Corollary 4.4.12.) allows to lift a homogeneous $G(R)$-basis of $G(L)$ to a homogeneous $\widetilde{R}$-basis of T and therefore T will be gr-free. Since $\widetilde{M}$ is a graded direct summand of T it then follows that $\widetilde{M}$ is gr-projective in $\widetilde{R}$-gr. Hence by Lemma 6.4. M is a filt-projective R-module.   □

Now we are able to prove the theorem containing the best available information concerning the global dimension on this particular situation.

11. Theorem. Let R be satisfying a). and b). as in Corollary 7.2.9., then
(1). for any $M \in R$-filt with good filtration FM we have $\mathrm{p.dim}_R M \leq \mathrm{p.dim}_{G(R)} G(M)$; it follows that if $G(R)$ is left gr-regular then R is left regular;
(2). $\mathrm{gl.dim} R \leq \mathrm{gr.gl.dim} G(R)$.

Proof. Since every finitely generated R-module may be equiped with a good filtration the statement (2) will follow from (1).
To prove (1)., we can assume $\mathrm{p.dim}_{G(R)} G(M) = n < \infty$. Since FM is good we may choose a filt-free resolution in R-filt:

$$(*) \qquad\qquad 0 \longrightarrow K_n \longrightarrow L_{n-1} \xrightarrow{f_{n-1}} \cdots \longrightarrow L_0 \xrightarrow{f_0} M \longrightarrow 0$$

where all L_i are filt-free of finite rank, all f_i are strict filtered morphisms and $K_n = \mathrm{Ker} f_{n-1}$ has the good filtration FK_n induced by FL_{n-1}. Thus $\widetilde{K}_n$ is finitely generated in $\widetilde{R}$-gr (Lemma 5.4.) and by Corollary 4.4.12. we have $G(K_n) \cong \widetilde{K}_n/X\widetilde{K}_n \neq 0$. Accordingly the associated sequence

$$G(*) \qquad 0 \longrightarrow G(K_n) \longrightarrow G(L_{n-1}) \xrightarrow{G(f_{n-1})} \cdots \longrightarrow G(L_0) \xrightarrow{G(f_0)} G(M) \longrightarrow 0$$

is exact and all $G(L_i)$ are gr-free of finite rank in $G(R)$-gr by Theorem 4.2.4. and Lemma 6.2.. Since $\mathrm{p.dim}_{G(R)} G(M) = n$, $G(K_n)$ must be graded projective. But then Proposition 7.2.10. entails that K_n is filt-projective and hence $\mathrm{p.dim}_R M \leq \mathrm{p.dim}_{G(R)} G(M)$.   □

12. Theorem. Let R be a filtered ring with filtration FR. Suppose that $\widetilde{R}$ is left Noetherian and FR is faithful.

(1). If $G(R)$ is left gr-regular then R is left regular.

(2). $\mathrm{gl.dim}R \le \mathrm{gr.gl.dim}G(R) \le \mathrm{gl.dim}G(R)$.

(3). If $\tilde{R}$ is also right Noetherian then $\mathrm{gl.dim}\tilde{R} \le 1+ \mathrm{gl.dim}G(R)$; if $\mathrm{gl.dim}G(R) < \infty$ then we have $\mathrm{gl.dim}\tilde{R} = 1+ \mathrm{gl.dim}G(R)$.

Proof. Use Corollary 5.5., Theorem 7.2.11. and Corollary 7.2.6.. □

CHAPTER II
Zariskian Filtrations

§1. Flatness of Completion

1.1. Artin-Rees property on filtrations

Let R be an arbitrary filtered ring with filtration FR.

Definition. Let M be in R-filt with filtration FM. If for any finitely generated R-submodule $N = \sum_{i=1}^{s} Ru_i$ say, there exists an integer $c \in \mathbb{Z}$ such that for all $n \in \mathbb{Z}$

$$F_n M \cap N \subseteq \sum_{i=1}^{s} F_{n+c} Ru_i$$

then FM is said to have the (left) **Artin-Rees property** (comparing with Definition 3.2.1. of Ch.I.). Similarly, one may define the right Artin-Rees property on filtrations for right R-modules.

2. Remark. (1). From the definition it is clear that if FM is a good filtration and FM induces good filtrations on submodules then FM has the Artin-Rees property.
(2). If FM has the Artin-Rees property and N is any finitely generated submodule of M, then the induced filtration by FM on N is (algebraically) equivalent to any good filtration on N.
(3). In some references the Artin-Rees property on filtrations defined above is called the **comparison condition**, e.g., [Bj2].

The next proposition provides a large class of filtrations having the Artin-Rees property.

Let $A = \oplus_{n \in \mathbb{Z}} A_n$ be a $\mathbb{Z}$-graded ring and $M = \oplus_{n \in \mathbb{Z}} M_n$ an arbitrary graded A-module. With respect to the gradation of M one can define an order function ω on M as follows: for any $x \in M$, let $x = \sum_{i=1}^{s} x_{p_i}$ be the (unique) decomposition of x with $x_{p_i} \in M_{p_i}$ and put

$$\omega(x) = \begin{cases} -\infty, & \text{if } x = 0; \\ p_s, & \text{if } x \neq 0 \text{ and } p_1 < p_2 < \cdots < p_s. \end{cases}$$

Then it is obvious that $\omega(ax) \leq \omega(a) + \omega(x)$ for any $a \in A$, $x \in M$. Moreover, one can define the grading filtration FM on M by putting

$$F_n M = \sum_{k \leq n} M_k, \qquad n \in \mathbb{Z}$$

such that M becomes a filtered A-module, where A has the grading filtration $F^{(1)}R$ as defined in Ch.I. §2..

3. Proposition. With notation as above, then any graded A-module M with the grading filtration FM has the (left) Artin-Rees property if M satisfies one of the following conditions:
(1). M is a left Noetherian A-module;
(2). M is a torsionfree A-module, i.e., for any $0 \neq a \in A$ and $x \in M$, $ax = 0$ implies $x = 0$.

Proof. If M is left Noetherian then argumentation as given in Ch.I. Example 4.3.9.(d). yields $\widetilde{M} \cong M[t] = A[t] \otimes_A M$, where $A[t]$ has the $--$gradation, hence $\widetilde{M}$ is left Noetherian. It follows from Ch.I. Corollary 5.5. (replace $\widetilde{R}$ by $\widetilde{M}$ in the proof given there) and the foregoing remark that M has the left Artin-Rees property.
Now suppose that M satisfies condition (2), then by the definition of ω we have $\omega(ax) = \omega(a) + \omega(x)$ for any $a \in A$ and $x \in M$. Let N be an arbitrary finitely generated A-submodule of M, say $N = \sum_{i=1}^{t} Au_i$, $u_i \in N$. If $x \in F_n M \cap N$ then $x = \sum_{i=1}^{t} a_i u_i$ for some $a_i \in A$. Decompose each a_i resp. each u_i as

$$a_i = \sum_{k=0}^{k_i} a_{i,p_k}, \quad a_{i,p_k} \in A_{p_k}$$

$$u_i = \sum_{r=0}^{r_i} u_{i,q_r}, \quad u_{i,q_r} \in M_{q_r}$$

and we may assume that only nonzero components have been written. Let $p_{s(i)}$ be the largest number in $\{p_0, p_1, \cdots, p_{k_i}\}$ such that $\omega(a_{i,p_{s(i)}} u_{i,q_{r(i)}}) = p_{s(i)} + q_{r(i)} \leq n$ for some $u_{i,q_{r(i)}} \in \{u_{i,q_r}; 1 \leq r \leq r_i\}$. If we assume $p_0 < p_1 < \cdots < p_{k_i}$ (this does not cause any loss of generality!) then it follows from $x \in F_n M$ that

$$x = \sum_{i=1}^{t} a_i u_i = \sum_{i=1}^{t} a_i \sum_{r=0}^{r_i} u_{i,q_r}$$

$$= \sum_{i=1}^{t} \sum_{r=0}^{r_i} \left(\sum_{k=0}^{k_i} a_{i,p_k} u_{i,q_r} \right)$$

$$= \sum_{i=1}^{t} \sum_{r=0}^{r_i} \left(\sum_{k=0}^{p_{s(i)}} a_{i,p_k} u_{i,q_r} \right) = \sum_{i=1}^{t} h_i u_i$$

where $h_i = \sum_{k=0}^{p_s(i)} a_{i,p_k}$. Note that $\omega(h_i) = p_{s(i)} \leq n - q_{r(i)}$ it follows that $h_i \in F_{n-q_{r(i)}}A$ and consequently $x = \sum_{i=1}^{t} h_i u_i \in \sum_{i=1}^{t} F_{n+c} A u_i$, where $c = \text{Sup}\{-q_r; \ 1 \leq r \leq r_i, \ 1 \leq i \leq t\}$. This completes the proof. $\square$

To illustrate the relation between the Noetherian property and the Artin-Rees property we need the following

4. Lemma. Let $M \in R$-filt with filtration FM and N a filtered R-submodule of M in the sense of Ch.I. section 2.5., i.e., N has the given filtration FN such that $F_n N \subseteq F_n M$ for all $n \in \mathbb{Z}$. Writing $F'N$ to be the the filtration on N induced by FM, there are two Rees modules of N associated with FN and $F'N$, denoted $\widetilde{N}^F$, $\widetilde{N}^{F'}$, respectively. With these notations we have $\widetilde{N}^F \subseteq \widetilde{N}^{F'}$ and $\widetilde{N}^{F'}/\widetilde{N}^F$ is an X-torsion $\tilde{R}$-module, where X is the canonical homogeneous element of degree 1 in $\tilde{R}$. Moreover, $F_n N = F'_n N = F_n M \cap N$ holds for all $n \in \mathbb{Z}$ if and only if $\widetilde{M}/\widetilde{N}^F$ is an X-torsionfree $\tilde{R}$-module.

Proof. Since N is a filtered R-submodule of M with filtration FN we have for each $n \in \mathbb{Z}$ the inclusion homomorphism of abelian groups $\vartheta_n\colon F_n N \hookrightarrow F_n M \cap N = F'_n N$. Accordingly, an inclusion morphism $\vartheta\colon \widetilde{N}^F \hookrightarrow \widetilde{N}^{F'}$ of graded $\tilde{R}$-modules is obtained by combining all the ϑ_n, $n \in \mathbb{Z}$. Consider the exact sequence in $\tilde{R}$-gr:

$$0 \longrightarrow \widetilde{N}^F \xrightarrow{\vartheta} \widetilde{N}^{F'} \longrightarrow \widetilde{N}^{F'}/\widetilde{N}^F \longrightarrow 0$$

we have the corresponding exact sequence of localized $\tilde{R}_{(X)}$-modules in $\tilde{R}_{(X)}$-gr (see Ch.I. Proposition 4.3.7. for the notations):

$$0 \longrightarrow (\widetilde{N}^F)_{(X)} \xrightarrow{\vartheta_{(X)}} (\widetilde{N}^{F'})_{(X)} \longrightarrow (\widetilde{N}^{F'}/\widetilde{N}^F)_{(X)} \longrightarrow 0$$

But then it follows from the definition of ϑ and the fact that all filtrations are exhaustive that $\vartheta_{(X)}$ is surjective, hence $(\widetilde{N}^{F'}/\widetilde{N}^F)_{(X)} = 0$. This shows that $\widetilde{N}^{F'}/\widetilde{N}^F$ is an X-torsionfree $\tilde{R}$-module.
If now $F_n N = F'_n N = F_n M \cap N$, $n \in \mathbb{Z}$, then the filtration FN on N is just the filtration induced by FM. It follows that we have a strict exact sequence in R-filt:

$$0 \longrightarrow N \longrightarrow M \longrightarrow M/N \longrightarrow 0$$

where M/N has the quotient filtration. By Ch.I. Proposition 4.3.8. the corresponding sequence

$$0 \longrightarrow \widetilde{N}^F \longrightarrow \widetilde{M} \longrightarrow \widetilde{M/N} \longrightarrow 0$$

is exact in $\tilde{R}$-gr. Hence $\widetilde{M}/\widetilde{N}^F$ is an X-torsionfree $\tilde{R}$-module since $\widetilde{M/N}$ is X-torsionfree by Ch.I. Proposition 4.3.7.. Conversely, if $\widetilde{M}/\widetilde{N}^F$ is X-torsionfree then FN must coincide with $F'N$, because otherwise $\widetilde{N}^{F'}/\widetilde{N}^F$ will be a nonzero X-torsion submodule of $\widetilde{M}/\widetilde{N}^F$ by the previous assertion. This completes the proof. □

5. Theorem. Suppose that the filtration FM on M is separated. Then $\widetilde{M}$ is left Noetherian if and only if M and $G(M)$ are left Noetherian and FM has the left Artin-Rees property.

Proof. If $\widetilde{M}$ is left Noetehrian then it follows from $M \cong \widetilde{M}/(1-X)\widetilde{M}$ and $G(M) \cong \widetilde{M}/X\widetilde{M}$ (Ch.I. Proposition 4.3.7.) that M and $G(M)$ are left Noetherian. Moreover, replacing $\widetilde{R}$ in Ch.I. Corollary 5.5. by $\widetilde{M}$ we claim (from Ch.I. Lemma 5.4.) that FM induces good filtrations on submodules of M. Hence the Artin-Rees property follows from Remark 1.1.2. To establish the converse we only need to show that $\widetilde{M}$ is graded Noetherian because, then it is Noetherian by ([NVO1] Theorem II.3.5., p.88). Let $\widetilde{N}$ be any graded submodule of $\widetilde{M}$. Put $N = \widetilde{N} + (1-X)\widetilde{M}/(1-X)\widetilde{M}$, then this is a submodule of $\widetilde{M}/(1-X)\widetilde{M} \cong M$. We also have $\widetilde{N} \cong \widetilde{N}^F$ if N is endowed with the filtration $F_nN = \widetilde{N}_n + (1-X)\widetilde{M}/(1-X)\widetilde{M}$, $n \in \mathbb{Z}$. Since by assumption FM is separated and M, $G(M)$ are Noetherian, we may choose $x_1, \cdots, x_s \in N$ such that $N = \sum_{i=1}^s Rx_i$ and, with respect to the induced filtration $F'N$ on N, $G(N) = \sum_{i=1}^s G(R)\sigma(x_i)$, where $\sigma(x_i)$ is the principal part of x_i. If $\deg\sigma(x_i) = v_i$, $i = 1, \cdots, s$, then $G(N)_n = \sum_{i=1}^s G(R)_{n-v_i}\sigma(x_i)$, $n \in \mathbb{Z}$, and accordingly

$$(*) \qquad F_n'N \subseteq \bigcap_{k \in \mathbb{Z}} \left(\sum_{i=1}^s F_{n-v_i}Rx_i + F_kM \cap N \right)$$

Note that $N = \sum_{i=1}^s Rx_i$ and the filtration on N defined by $F_n''N = \sum_{i=1}^s F_{n-v_i}Rx_i$, $n \in \mathbb{Z}$, is exhaustive and good, by the Artin-Rees property on FM we have $F_nM \cap N = F_n'N \subseteq \sum_{i=1}^s F_{n+c-v_i}Rx_i$ for some c and all $n \in \mathbb{Z}$ (recall that good filtrations on N are equivalent). Hence $F_{n-c}'N \subseteq \sum_{i=1}^s F_{n-v_i}Rx_i$, $n \in \mathbb{Z}$. But then by $(*)$ we obtain

$$F_n'N \subseteq \sum_{i=1}^s F_{n-v_i}Rx_i + F_{n-c}M \cap N$$

$$= \sum_{i=1}^s F_{n-v_i}Rx_i + F_{n-c}'N$$

$$\subseteq \sum_{i=1}^s F_{n-v_i}Rx_i$$

$$\subseteq F_n'N$$

i.e., $F_n'N = F_n''N$, $n \in \mathbb{Z}$. This shows that the induced filtration $F'N$ on N is good and consequently $\widetilde{N}^{F'} = \sum_{i=1}^s \widetilde{R}\widetilde{x}_i$, where $\widetilde{N}^{F'}$ denotes the Rees module of N with respect to the induced filtration $F'N$. Since $F_n N \subseteq F'_n N$, $n \in \mathbb{Z}$, we have $\widetilde{N}^F \subseteq \widetilde{N}^{F'}$ and $\widetilde{N}^{F'}/\widetilde{N}^F$ is an X-torsion $\widetilde{R}$-module by the foregoing lemma, say $X^w(\widetilde{N}^{F'}/\widetilde{N}^F) = 0$ for some integer $w > 0$. It follows that $X^w \widetilde{N}^{F'} \subseteq \widetilde{N}^F \subseteq \widetilde{N}^{F'}$. Considering the exact sequence of $\widetilde{R}/X\widetilde{R}$-modules

$$0 \longrightarrow X^{w-1}\widetilde{M}/X^w\widetilde{M} \longrightarrow \widetilde{M}/X^w\widetilde{M} \longrightarrow \widetilde{M}/X^{w-1}\widetilde{M} \longrightarrow 0$$

an induction on w shows that $\widetilde{M}/X^w\widetilde{M}$ is Noetherian since $\widetilde{M}/X\widetilde{M} \cong G(M)$ is Noetherian. Since the filtration $F'N$ on N is induced by FM we know from the foregoing lemma that $\widetilde{M}/\widetilde{N}^{F'}$ is X-torsionfree. It follows that the canonical morphism $\widetilde{N}^{F'}/X^w\widetilde{N}^{F'} \to \widetilde{M}/X^w\widetilde{M}$ is injective and hence $\widetilde{N}^{F'}/X^w\widetilde{N}^{F'}$ is Noetherian. Thus $\widetilde{N}^{F'}/X^w\widetilde{N}^{F'}$ is finitely generated. Finally, note that $X^w\widetilde{N}^{F'}$ is finitely generated we see that $\widetilde{N}^F$ is finitely generated and consequently $\widetilde{N}$ is finitely generated because $\widetilde{N} \cong \widetilde{N}^F$ as we have noticed before. This finishes the proof. $\square$

1.2. Flatness of completion

The first two propositions below deal with the flatness of more general filtered ring extensions.

1. Proposition. Let B be a filtered ring with filtration FB. Let A be another filtered ring with filtration FA and $f: A \to B$ a filtered ring homomorphism of degree zero (hence there is the associated graded ring homomorphism $G(f): G(A) \to G(B)$). Assume that the filtered rings B and A satisfy the following conditions:

(a). good filtrations in B-filt are separated;

(b). FA has the left Artin-Rees property;

(c). the right $G(A)$-module $G(B)$ is flat,

then B is a flat right A-module.

Proof. Take a finitely generated left ideal L of A and choose a good filtration $F^{(1)}L$ on L. Since FA has the Artin-Rees property it follows that $F^{(1)}L$ is equivalent to the induced filtration $F^{(2)}L = L \cap FA$ (see Remark 1.1.2.). By Ch.I. Lemma 6.15. we know that $F^{(1)}(B \otimes_A L)$ is a good filtration on $B \otimes_A L$ and also that $F^{(2)}(B \otimes_A L)$ is equivalent to $F^{(1)}(B \otimes_A L)$. Hence by (a), $F^{(1)}(B \otimes_A L)$ is separated, but then the same holds for $F^{(2)}(B \otimes_A L)$. Since $F^{(2)}L$ is induced by FA on L we have a strict exact sequence in A-filt: $0 \to L \xrightarrow{i} A$. Therefore the associated sequence $0 \to G^{F^{(2)}}(L) \xrightarrow{G(i)} G(A)$ is exact in $G(A)$-gr by Ch.I. Theorem 4.2.4.. Consequently, Ch.I. Lemma 6.14. and the flatness of $G(B)$ assumed in (c) entail the existence of an exact commutative diagram of graded $G(A)$-modules:

$$
\begin{array}{ccc}
& & 0 \\
& & \downarrow \\
0 \to G(B) \otimes_{G(A)} G(L) & \longrightarrow & G(B) \otimes_{G(A)} G(A) \\
\varphi(B,L) \downarrow & & \downarrow \varphi(B,A) \\
G^{F^{(2)}}(B \otimes_A L) & \xrightarrow{G(1 \otimes i)} & G(B \otimes_A A) \\
& & \downarrow \\
& & 0
\end{array}
$$

Hence $G(1 \otimes i)$ is injective. Note that $F^{(2)}(B \otimes_A L)$ and $F(B \otimes_A A)$ are separated it follows from Ch.I. Corollary 4.2.5. that $1 \otimes i$ is injective and strict. Of course the injectivity of $1 \otimes i$ amounts to the flatness of B as a right A-module, completing the proof. □

In order to be able to lift the faithful flatness property we have to impose a further condition on FA, in fact the "radical" condition $F_{-1}A \subseteq J(F_0A)$ (or in other words, FA is faithful) will do perfectly.

2. Proposition. Let A and B be as in the foregoing proposition but now assume furthermore that $F_{-1}A \subseteq J(F_0A)$ and $G(B)$ is faithfully flat over $G(A)$, then B is faithfully flat over A.

Proof. Let M be any finitely generated A-module. We may endow M with a good filtration FM say, by giving a strict epimorphism $L \xrightarrow{\varepsilon} M \to 0$ defined on some filt-free A-module L of finite rank (see Ch.I. §6.). Put $\text{Ker}\varepsilon = K$. If K is endowed with the filtration FK induced by FL then we have a strict exact sequence of filtered A-modules:

$$ 0 \longrightarrow K \xrightarrow{i} L \xrightarrow{\varepsilon} M \longrightarrow 0 $$

The strict exactness of this sequence of filtered morphisms entails the exactness of

$$ 0 \longrightarrow G(K) \xrightarrow{G(i)} G(L) \xrightarrow{G(\varepsilon)} G(M) \longrightarrow 0 $$

by Ch.I. Theorem 4.2.4.. Applying Proposition 1.2.1., we may derive the exact sequence of B-modules:

$$ 0 \longrightarrow B \otimes_A K \xrightarrow{1 \otimes i} B \otimes_A L \xrightarrow{1 \otimes \varepsilon} B \otimes_A M \longrightarrow 0 $$

Moreover, since L is filt-free a similar argument as in the proof of Proposition 1.2.1. shows that $1 \otimes i$ is injective and strict. Hence the above sequence is also strict exact. According to Ch.I. Theorem 4.2.4. we arrive at the following exact commutative diagram of $G(B)$-modules:

$$
\begin{array}{ccccccc}
& & 0 & & 0 & & 0 \\
& & \downarrow & & \downarrow & & \downarrow \\
0 \to & G(B) \otimes_{G(A)} G(K) & \longrightarrow & G(B) \otimes_{G(A)} G(L) & \longrightarrow & G(B) \otimes_{G(A)} G(M) & \to 0 \\
& \downarrow & & \downarrow & & \downarrow & \\
0 \to & G(B \otimes_A K) & \longrightarrow & G(B \otimes_A L) & \longrightarrow & G(B \otimes_A M) & \\
& \downarrow & & \downarrow & & \downarrow & \\
& 0 & & 0 & & 0 &
\end{array}
$$

So if $B \otimes_A M = 0$ then $G(B) \otimes_{G(A)} G(M) = 0$ and hence by the faithful flatness of $G(B)$ this leads to $G(M) = 0$. But $G(M) \cong \widetilde{M}/X\widetilde{M}$ and $\widetilde{M}$ is finitely generated in $\widetilde{A}$-gr by Proposition 4.3.7. and Lemma 5.4. of Ch.I., it follows from Ch.I. Corollary 4.4.12. and the graded version of Nakayama's lemma (Ch.I. Lemma 4.1.8.) that $M = 0$. This proves the faithful flatness of B over A. $\square$

Now, let us turn to the flatness of completion.

3. Proposition. Let R be a complete filtered ring with filtration FR. Suppose that $G(R)$ is left Noetherian, then
(1). good filtrations induce good filtrations on R-submodules; in particular, R is Noetherian;
(2). $\widetilde{R}$ is left Noetherian;
(3). good filtrations are separated.

Proof. Since FR is separated and $G(R)$ is Noetherian it follows from Lemma 5.4. and Theorem 5.7. of Ch.I. that FR induces good filtrations on left ideals of R, and hence R is left Noetherian. Furthermore, from the foregoing Remark 1.1.2. we know that FR has the left Artin-Rees property. Thus $\widetilde{R}$ is left Noetherian by Theorem 1.1.5.. Therefore, it follows from Ch.I. Corollary 5.5. that good filtrations in R-filt induce good filtrations on R-submodules and moreover good filtrations are separated. $\square$

4. Theorem. Let R be a filtered ring with filtration FR. Suppose that $G(R)$ is left Noetherian and FR has the left Artin-Rees property. Take the canonical homomorphism of filtered rings $\varphi_R \colon R \to \widehat{R}$, the following statements hold:
(1). $\widehat{R}$ is flat as a right R-module;
(2). If we have in addition that $F_{-1}R \subseteq J(F_0R)$, i.e., FR is faithful, then $\widehat{R}$ is faithfully flat over R.

Proof. (1). The assumptions imply that Proposition 1.2.3. holds for $B = \widehat{R}$ and $A = R$.

Hence it follows from Proposition 1.2.1. that $\widehat{R}$ is a flat right R-module (note that $G(R) \cong G(\widehat{R})!$).

(2). This follows from Proposition 1.2.2. □

In what follows R is a filtered ring with filtration FR.

5. Lemma. Consider an exact sequence in R-filt:

$$L \xrightarrow{f} M \xrightarrow{g} N$$

we suppose that FL, FM and FN are good filtrations, and moreover M and N have the left Artin-Rees property. Then the sequence

$$\widehat{L} \xrightarrow{\widehat{f}} \widehat{M} \xrightarrow{\widehat{g}} \widehat{N}$$

is exact in $\widehat{R}$-filt, where $\widehat{L}$, $\widehat{M}$ and $\widehat{N}$ are the completions corresponding to FL, FM and FN respectively.

Proof. This is a consequence of Ch.I. Theorem3.4.16.. □

Let M be a filtered R-module with filtration FM. Consider the following commutative diagram of R-modules:

$$
(I) \qquad
\begin{array}{ccc}
M & \xrightarrow{\ i\ } & \widehat{R} \otimes_R M \\
& {}_{\varphi_M}\searrow \quad \swarrow{}_{\alpha_M} & \\
& \widehat{M} &
\end{array}
$$

where $\alpha_M(\xi \otimes m) = \xi \varphi_M(m)$ and $\varphi_M = \alpha_M i$ is the canonical morphism, then both i and α_M are filtered morphisms in R-filt and $\widehat{R}$-filt respectively in case $\widehat{R} \otimes M$ is endowed with the tensor product filtration $F(\widehat{R} \otimes M)$ as defined in Ch.I. §6.. Moreover, if $f: M \to N$ is any morphism in R-filt it follows from Theorem 3.4.5. and Remark 3.5.4. of Ch.I. that we have the following commutative diagram in R-filt:

$$
(II) \qquad
\begin{array}{ccc}
M & \xrightarrow{\quad f \quad} & N \\
{}_{\varphi_M}\searrow & & \swarrow{}_{\varphi_N} \\
i\downarrow \quad \widehat{M} & \xrightarrow{\widehat{f}} & \widehat{N} \quad \downarrow i \\
\quad \nearrow{}_{\alpha_M} & & {}_{\alpha_N}\nwarrow \\
\widehat{R} \otimes_R M & \xrightarrow{1 \otimes f} & \widehat{R} \otimes_R N
\end{array}
$$

in particular, the relation $\widehat{f} \circ \alpha_M = \alpha_N \circ (1 \otimes f)$ holds in $\widehat{R}$-filt.

6. Lemma. Let the notations be as given above.

(1). If $M \in R$-filt with filtration FM, then the graded morphism (as defined in Ch.I. §6) $\varphi(\widehat{R}, M)$: $G(\widehat{R}) \otimes_{G(R)} G(M) \to G(\widehat{R} \otimes_R M)$ is an isomorphism, and consequently the associated graded morphism $G(\alpha_M)$: $G(\widehat{R} \otimes_R M) \to G(\widehat{M})$ is an isomorphism.

(2). If $L \in R$-filt is filt-free with filtration FL, then α_L: $\widehat{R} \otimes_R L \to \widehat{L}$ is injective and strict; if furthermore we assume that L is of finite rank, then α_L is an isomorphism in $\widehat{R}$-filt.

Proof. (1). Consider the following graded morphisms

$$G(M) \overset{canonical}{\underset{\cong}{\longrightarrow}} G(R) \otimes_{G(R)} G(M) \overset{G(\varphi_R) \otimes 1_{G(M)}}{\underset{\cong}{\longrightarrow}} G(\widehat{R}) \otimes_{G(R)} G(M) \overset{\varphi(\widehat{R}, M)}{\longrightarrow}$$

$$\longrightarrow G(\widehat{R} \otimes_R M) \overset{G(\alpha_M)}{\longrightarrow} G(\widehat{M}) \overset{G(\varphi_M)^{-1}}{\underset{\cong}{\longrightarrow}} G(M),$$

where φ_R resp. φ_M is the canonical morphism $R \to \widehat{R}$ resp. the canonical morphism $M \to \widehat{M}$ (it follows from Ch.I. Proposition 4.2.2. that $G(\varphi_R)$ resp. $G(\varphi_M)$ is an isomorphism), then one easily checks that $1_{G(M)} = G(\alpha_M) \circ \varphi(\widehat{R}, M)$. Thus $\varphi(\widehat{R}, M)$ is a monomorphism and hence an isomorphism. Take the composite of the first three morphisms in the above sequence and denote it by ϑ, we obtain the following commutative diagram

$$
\begin{array}{ccc}
 & & G(M) \\
 & \overset{\vartheta}{\nearrow} & \downarrow G(\varphi_M) \\
G(\widehat{R} \otimes_R M) & \underset{G(\alpha_M)}{\longrightarrow} & G(\widehat{M})
\end{array}
$$

Since ϑ and $G(\varphi_M)$ are isomorphisms it follows that $G(\alpha_M)$ is an isomorphism.

(2). If L is a filt-free R-module then one easily sees that $\widehat{R} \otimes_R L$ is a filt-free $\widehat{R}$-module with respect to the tensor product filtration on it by Ch.I. Lemma 6.13.. It again follows from Ch.I. Lemma 6.2. that $F(\widehat{R} \otimes_R L)$ is separated. Hence (1) and Ch.I. Corollary 4.5. entail that α_L is injective and strict. If furthermore L is of finite rank then, as a filtered R-module, $\widehat{R} \otimes_R L$ is complete with respect to $F(\widehat{R} \otimes_R L)$. Again by using Ch.I. Corollary 4.2.5. we may conclude that α_L is an isomorphism.   □

7. Corollary. In the commutative diagram (I) given before if FM is a good filtration then α_M is surjective and strict, hence $\widehat{R} \cdot \varphi_M(M) = \widehat{M}$ and $\widehat{M}$ is a finitely generated $\widehat{R}$-module.

Proof. Since FM is good we may choose a strict exact sequence in R-filt: $L \overset{\varepsilon}{\to} M \to 0$, where L is filt-free of finite rank. It follows from Ch.I. Theorem 3.4.13., Lemma 1.2.6. given above and the foregoing commutative diagram (II) we have an exact commutative diagram in $\widehat{R}$-filt:

$$
\begin{array}{ccc}
& 0 & \\
& \downarrow & \\
\hat{R} \otimes_R L & \xrightarrow{1 \otimes \epsilon} \hat{R} \otimes_R M & \to 0 \\
\alpha_L \downarrow & \quad\downarrow \alpha_M & \\
L & \xrightarrow{\;\hat{\epsilon}\;} M & \to 0 \\
\downarrow & & \\
0 & &
\end{array}
$$

Consequently α_M is surjective. The strictness of α_M follows from the strictness of $\hat{\epsilon}$, α_L and $1 \otimes \epsilon$. □

8. Proposition. Let $M \in R$-filt with filtration FM and consider the foregoing commutative diagram (I).

(1). If φ_M is injective then i is injective.

(2). If φ_M and α_M are injective then $i(M) \cap F_n(\hat{R} \otimes_R M) = i(F_n M)$ for all $n \in \mathbb{Z}$, i.e., the map i is strict.

(3). If $\hat{R}$ satisfies the condition that good filtrations in $\hat{R}$-filt are separated, then α_M is a strict filtered isomorphism of $\hat{R}$-modules in case FM is a good filtration.

Proof. (1). Trivial.

(2). by (1), i is injective. From $i(F_n M) \subseteq i(M) \cap F_n(\hat{R} \otimes_R M)$ we derive the following inclusion relations:

$$
\begin{aligned}
\varphi_M(F_n M) = \alpha_M i(F_n M) & \subseteq \alpha_M(i(M) \cap F_n(\hat{R} \otimes_R M)) \\
& \subseteq \varphi_M(M) \cap \alpha_M(F_n(\hat{R} \otimes_R M)) \\
& \subseteq \varphi_M(M) \cap F_n \widehat{M} \\
& = \varphi_M(F_n M) = \alpha_M i(F_n M).
\end{aligned}
$$

Hence $\alpha_M i(F_n M) = \alpha_M(i(M) \cap F_n(\hat{R} \otimes_R M))$. The injectivity of α_M then entails the equality $i(F_n M) = i(M) \cap F_n(\hat{R} \otimes_R M)$. This shows that i is strict.

(3). Suppose that good filtrations in $\hat{R}$-filt are separated and that the filtration FM on M is a good filtration, then the filtration $F(\hat{R} \otimes_R M)$ on $\hat{R} \otimes_R M$ is good by Ch.I. Lemma 6.15. and hence separated. Since $G(\alpha_M)$ is an isomorphism by Lemma 1.2.6. it follows from Ch.I. Corollary 4.2.5. that α_M is injective and strict. Finally, the surjectivity of α_M follows from Corollary 1.2.7. above. □

9. Lemma. Let R be a filtered ring with filtration FR. Taking the notations from Ch.I. §4., the following statements are equivalent:

(1). FR is faithful, i.e., $F_{-1}R \subseteq J(F_0R)$;

(2). $X \in J^g(\tilde{R})$;

(3). For a finitely generated $\widetilde{M} \in \tilde{R}$-gr, $X\widetilde{M} = \widetilde{M}$ if and only if $\widetilde{M} = 0$;

(4). For a finitely generated $\widetilde{M} \in \mathcal{F}_X$, $X\widetilde{M} = \widetilde{M}$ if and only if $\widetilde{M} = 0$;

(5). If $M \in R$-filt with good filtration FM then $G(M) = 0$ if and only if $M = 0$.

Proof. (1) $\Leftrightarrow$ (2). This is just Ch.I. Corollary 4.4.12..

(2) $\Leftrightarrow$ (3). The graded version of Nakayama's lemma (Ch.I. §4.1.).

(3) $\Rightarrow$ (4). Obvious.

(4) $\Rightarrow$ (3). Let $\widetilde{M}$ be any finitely generated graded $\tilde{R}$-module such that $X\widetilde{M} = \widetilde{M}$. Put $t(\widetilde{M}) = \{m \in \widetilde{M}, X^n m = 0, \text{ for some integer } n > 0\}$, then $t(\widetilde{M})$ is a graded submodule of $\widetilde{M}$. If $\widetilde{M} \neq t(\widetilde{M})$ then $\widetilde{M}/t(\widetilde{M})$ is a finitely generated graded X-torsionfree $\tilde{R}$-module and by the assumption $X(\widetilde{M}/t(\widetilde{M})) = \widetilde{M}/t(\widetilde{M})$. Hence (4) entails $\widetilde{M} = t(\widetilde{M})$, i.e., $\widetilde{M}$ must be an X-torsion $\tilde{R}$-module. But as $\widetilde{M}$ is finitely generated, this gives us $X^n\widetilde{M} = 0$ for a suitable positive integer n. Therefore, it follows from $\widetilde{M} = X\widetilde{M} = \cdots = X^n\widetilde{M}$ that $\widetilde{M} = 0$.

(4) $\Leftrightarrow$ (5). Just the translation for the category equivalence: R-filt $\leftrightarrow$ $\mathcal{F}_X$ (see Ch.I. Proposition 4.3.7.). $\square$

10. Theorem. Let R be a filtered ring with filtration FR. Suppose that $G(R)$ is left Noetherian and $\hat{R}$ is faithfully flat as a right R-module, then

(1). good filtrations in R-filt are separated, hence $F_{-1}R \subseteq J(F_0R)$ (Lemma 1.2.9. above);

(2). R is left Noetherian;

(3). every good filtration FM on $M \in R$-filt has the left Artin-Rees property;

(4). good filtrations induce good filtrations on R-submodules;

(5). the morphism $\alpha_M \colon \hat{R} \otimes_R M \to \widehat{M}$ (as given in the foregoing diagram (I)) is a strict filtered isomorphism in $\hat{R}$-filt for every $M \in R$-filt with good filtration FM, where $\widehat{M}$ is made with respect to FM.

Proof. (1). Let $M \in R$-filt have a good filtration FM. Put $N = \cap_{n\in\mathbb{Z}} F_n M$. Clearly N is an R-submodule of M. Since $\hat{R} \otimes_R N \subseteq \hat{R} \otimes_R M$ and $F(\hat{R} \otimes_R M)$ being a good filtration (Ch.I. Lemma 6.15.) it follows from Proposition 1.2.3. that $\hat{R} \otimes_r N \subseteq \cap_{n\in\mathbb{Z}} F_n(\hat{R} \otimes_R M) = 0$. But then the faithful flatness of $\hat{R}$ over R yields that $N = 0$, i.e., FM is separated.

(2). Consider any finitely generated R-module M, then M may be endowed with some good filtration FM say. Taking an arbitrary R-submodule of M, then since $\hat{R}$ is Noetherian (by Proposition 1.2.3.) the $\hat{R}$-submodule $\hat{R} \otimes_R N$ of the finitely generated $\hat{R}$-module $\hat{R} \otimes_R M$ is also finitely generated. Pick $x_1, \cdots, x_t \in N$ such that $1 \otimes x_1, \cdots, 1 \otimes x_t$ generate $\hat{R} \otimes_R N$ as an $\hat{R}$-module, then the finitely generated R-module $N' = Rx_1 + \cdots + Rx_t$ satisfies $\hat{R} \otimes_R (N/N') = 0$. It follows from the faithful flatness of $\hat{R}$ over R that $N = N'$. Hence M is Noetherian.

(3). Let FM be a good filtration on $M \in R$-filt and N a finitely generated R-submodule of M, say $N = \sum_{i=1}^{s} Ru_i$. Choosing a good filtration FN on N, it follows from Ch.I. Lemma 6.15. and the foregoing Proposition 1.2.3. that there exists an integer $c \in \mathbb{Z}$ such that for all $n \in \mathbb{Z}$

$$F_n(\hat{R} \otimes_R M) \cap (\hat{R} \otimes_R N) \subseteq F_{n+c}(\hat{R} \otimes_R N).$$

By the flatness assumption we may consider the map $i: N \to \hat{R} \otimes_R N$ as the restriction of $i: M \to \hat{R} \otimes_R M$ to N, where i is defined as in the foregoing diagram (I). Now, Proposition 1.2.8. yields that

$$i(F_n M) \cap i(N) \subseteq F_n(\hat{R} \otimes_R M) \cap (\hat{R} \otimes_R N) \subseteq F_{n+c}(\hat{R} \otimes_R N)$$

$$i(F_n M) \cap i(N) \subseteq F_{n+c}(\hat{R} \otimes_R N) \cap i(N) = i(F_{n+c}N).$$

Identifying N and $i(N)$ as R-submodules of $\hat{R} \otimes_R N$ we have for all $n \in \mathbb{Z}$, $F_n M \cap N \subseteq F_{n+c}N$. Therefore, FM has the left Artin-Rees property.

(4). Indeed, from Theorem 1.1.5. we derive that $\tilde{R}$ is left Noetherian and consequently good filtrations in R-filt induce good filtrations on R-submodules. Nevertheless, we include a direct proof here.

Let FM be a good filtration on $M \in R$-filt and N an arbitrary R-submodule of M. Take the filtration FN induced by FM on N, then since $G(R)$ is left Noetherian $G(N)$ is finitely generated in $G(R)$-gr (note that $G(M)$ is finitely generated in $G(R)$-gr), say $G(N) = \sum_{i=1}^{t} G(R)\sigma(u_i)$ with $u_i \in N$ and $\deg\sigma(u_i) = k_i$, $i = 1, \cdots, t$. It is clear that $F_n N \subseteq \cap_{m \in \mathbb{Z}}(\sum_{i=1}^{t} F_{n-k_i} Ru_i + F_m N)$ for all $n \in \mathbb{Z}$. Endow $N' = \sum_{i=1}^{t} Ru_i$ with the filtration FN' defined by putting $F_n N' = \sum_{i=1}^{t} F_{n-k_i} Ru_i$, for all $n \in \mathbb{Z}$, then FN' is a good filtration by definition. Using (1) and the fact that FN is induced by FM we arrive at for all $n \in \mathbb{Z}$

$$F_n N \subseteq \bigcap_{m \in \mathbb{Z}} (F_n N' + F_m N) \subseteq \bigcap_{m \in \mathbb{Z}} (N' + F_m M) = N'$$

Consequently, $N = N'$ and $F_n N = N \cap F_n M = N' \cap F_n M$, $n \in \mathbb{Z}$.

Finally, since $\hat{R} \otimes_R N' \subseteq \hat{R} \otimes_R M$ have good filtrations $F(\hat{R} \otimes_R N')$, $F(\hat{R} \otimes_R M)$ respectively, in view of Proposition 1.2.3. we have an integer $c \in \mathbb{Z}$ such that for all $n \in \mathbb{Z}$

$$F_n(\hat{R} \otimes_R M) \bigcap (\hat{R} \otimes_R N') \subseteq F_{n+c}(\hat{R} \otimes_R N').$$

Again by Proposition 1.2.3. we have for each $n \in \mathbb{Z}$

$$L = \bigcap_{p \in \mathbb{Z}} \left(F_n(\widehat{R} \otimes_R N') + F_p(\widehat{R} \otimes_R M) \right) \subseteq \widehat{R} \otimes_R N'.$$

Take $y \in L$, then $y \in F_n(\widehat{R} \otimes_R N') + F_{n-c}(\widehat{R} \otimes_R M)$, say $y = v + u$ with $v \in F_n(\widehat{R} \otimes_R N')$ and $u \in F_{n-c}(\widehat{R} \otimes_R M)$. Hence, $y - v = u \in F_{n-c}(\widehat{R} \otimes_R M) \cap (\widehat{R} \otimes_R N')$ yields $y - v \in F_n(\widehat{R} \otimes_R N')$ and thus $y \in F_n(\widehat{R} \otimes_R N')$. This shows that for all $n \in \mathbb{Z}$

$$\bigcap_{p \in \mathbb{Z}} \left(F_n(\widehat{R} \otimes_R N') + F_p(\widehat{R} \otimes_R M) \right) = F_n(\widehat{R} \otimes_R N').$$

Identifying N, N' and M with $i(N)$, $i(N')$ and $i(M)$ respectively, then the following inclusion relations

$$
\begin{aligned}
i(F_n N) &\subseteq \bigcap_{m \in \mathbb{Z}} \left(i(F_n N') + i(F_m N) \right) \\
&\subseteq \bigcap_{m \in \mathbb{Z}} \left(F_n(\widehat{R} \otimes_R N') + F_m(\widehat{R} \otimes_R M) \right) \\
&= F_n(\widehat{R} \otimes_R N')
\end{aligned}
$$

combined with Proposition 1.2.8. yield that

$$i(F_n N) = i(F_n N) \cap i(N') \subseteq F_n(\widehat{R} \otimes_R N') \cap i(N') = i(F_n N')$$

i.e., $F_n N = F_n N'$ for all $n \in \mathbb{Z}$. This establishes the assertion that the filtration induced by FM on N is good.

(5). This follows from Proposition 1.2.3. and Proposition 1.2.8..   □

11. Theorem. Let R be a filtered ring with filtration FR. The following statements are equivalent:

(1). FR is separated, $G(R)$ is left Noetherian, FR has the left Artin-Rees property and $F_{-1}R \subseteq J(F_0 R)$;

(2). $G(R)$ is left Noetherian and $\widehat{R}$ is a faithful flat right R-module.

Proof. This immediately follows from Theorem 1.2.4. and Theorem 1.2.10..   □

§2. Zariskian Filtrations

2.1. Definition and various characterizations

1. Definition. Let R be a filered ring with filtration FR. R is said to be a **left Zariski ring**, or FR is said to be a **left Zariskian filtration**, if the Rees ring $\tilde{R}$ of R associated with FR is left Noetherian and $F_{-1}R$ is contained in the Jacobson radical $J(F_0R)$ of F_0R (i.e., FR is faithful).

Similarly we may define **right Zariski ring**. If R is a left and right Zariski ring we will simplify terminology by referring to it as being a **Zariski ring**.

2. Theorem. (Characterizations of the Zariski property) For a filtered ring R with filtration FR, the following statements are equivalent:

(1). R is a left Zariski ring;

(2). FR is separated, $G(R)$ is left Noetherian, $F_{-1}R \subseteq J(F_0R)$ and every good filtration FM on $M \in R$-filt has the left Artin-Rees property;

(3). FR is separated, $G(R)$ is left Noetherian, $F_{-1}R \subseteq J(F_0R)$ and FR has the left Artin-Rees property;

(4). $G(R)$ is left Noetherian and the completion $\hat{R}$ of R with respect to the FR-topology on R is a faithful flat right R module;

(5). $G(R)$ is left Noetherian, good filtrations in R-filt induce good filtrations on R-submodules and good filtrations are separated;

(6). $G(R)$ is left Noetherian and for any $M \in R$-filt with good filtration FM, if N is any R-submodule of M with an arbitrary filtration FN (which is not necessarily good!) then each part F_nN, $n \in \mathbb{Z}$, of FN is closed in the FM-topology of M, or in other woeds, for each $n \in \mathbb{Z}$

$$F_nN = \bigcap_{p \in \mathbb{Z}}(F_nN + F_pM);$$

(7). $G(R)$ is left Noetherian, $F_{-1}R \subseteq J(F_0R)$ and for every left ideal L of R with good filtration FL we have: for each $n \in \mathbb{Z}$

$$F_nL = \bigcap_{p \in \mathbb{Z}}(F_nL + F_pR).$$

Proof. Let $\tilde{R}$ be the Rees ring of R associated with FR and X the canonical (homogeneous) element of degree 1 in $\tilde{R}_1$.

(1) $\Rightarrow$ (2). Note that $G(R) \cong \tilde{R}/X\tilde{R}$, (2) follows from Lemma 1.2.9., Remark 1.1.2. and Ch.I. Corollary 5.5..

(2) $\Rightarrow$ (3). Clear.

(3) $\Rightarrow$ (4). This follows from Theorem 1.2.4..

(4) $\Rightarrow$ (5). This follows from Theorem 1.2.10..

(5) $\Rightarrow$ (6). Since good filtrations yield good quotient filtrations on quotient modules we see that good filtrations in R-filt are separated if and only if for any good filtration FM and any R-submodule $N \subseteq M$ it follows that $N = \cap_{p \in \mathbf{Z}}(N + F_p M)$, namely, every R-submodule of M is closed in the FM-topology of M.

Let M have a good filtration FM and N an R-submodule of M with an arbitrary filtration FN. In view of (5) we have

$$(*) \qquad\qquad\qquad F_n M \cap N \subseteq F_{n+c} N$$

for all $n \in \mathbf{Z}$ and some $c \in \mathbf{Z}$. Clearly $L = \cap_{p \in \mathbf{Z}}(F_n N + F_p M) \subseteq N$ since N is closed. Take $m \in L$, then $m \in F_n N + F_{n-c} M$, say $m = y + z$ with $y \in F_n N$ and $z \in F_{n-c} M$. Hence, $m - y \in N \cap F_{n-c} M$ combined with $(*)$ yields $m - y \in F_n N$, hence $m \in F_n N$. This shows that the equality $\cap_{p \in \mathbf{Z}}(F_n N + F_p M) = F_n N$ holds for all $n \in \mathbf{Z}$.

(6) $\Rightarrow$ (7). Taking $N = 0$ in (6) yields that good filtrations are separated, hence FR is faithful by Lemma 1.2.9.. Since FR is itself a good filtration on R, (7) follows at once from (6).

(7) $\Rightarrow$ (1). From (7) and the definition of a left Zariski ring it remains to show that $\tilde{R}$ is left Noetherian. But in view of Remark 1.1.2. and Theorem 1.1.5. we only need to check that FR induces good filtrations on left ideals of R. To this end, let L be an arbitrary left ideal of R with the filtration FL induced by FR. Then by Ch.I. Theorem 4.2.4. $G(L)$ is a finitely generated left ideal of $G(R)$, say

$$(*) \qquad\qquad\qquad G(L) = \sum_{i=1}^{s} G(R)\sigma(u_i)$$

with $u_i \in L$ and $\deg\sigma(u_i) = k_i$, $i = 1, \cdots, s$. Putting the good filtration

$$F_n L' = \sum_{i=1}^{s} F_{n-k_i} R u_i, \; n \in \mathbf{Z},$$

on the left ideal $L' = \sum_{i=1}^{s} R u_i$ of R, then $(*)$ and (7) entail that for all $n \in \mathbf{Z}$ we have:

$$
\begin{aligned}
F_n L \;&\subseteq\; \bigcap_{p \in \mathbf{Z}}(F_n L' + F_p L) \\
&\subseteq\; \bigcap_{p \in \mathbf{Z}}(F_n L' + F_p R) \\
&=\; F_n L' \subseteq F_n L.
\end{aligned}
$$

Hence $F_n L = F_n L'$, $n \in \mathbb{Z}$, i.e., the induced filtration FL on L is good, as desired. $\square$

3. Remark. By an argumentation similar to the one given in the proof of Theorem 2.1.2. above it is not hard to check the following assertions:

(a). The conditions mentioned in (5) of Theorem 2.1.2. are also equivalent to say that $G(R)$ is left Noetherian and for every filt-free R-module L of finite rank, if M is an arbitrary R-submodule of L with the filtration FM induced by FL then $G(M) = \sum_{i=1}^{s} G(R)\sigma(m_i)$ implies that $M = \sum_{i=1}^{s} Rm_i$ and $F_n M = \sum_{i=1}^{s} F_{n-k_i} Rm_i$ for all $n \in \mathbb{Z}$, where $m_i \in M$ with $k_i = \deg\sigma(m_i)$, $i = 1, \cdots, s$. We also refer to [VE1] for a detail proof of this equivalence.

(b). The conditions mentioned in (7) of Theorem 2.1.2. are also equivalent to say that $\tilde{R}$ is left Noetherian and for every left ideal I of R we have $I = \cap_{m \in \mathbb{Z}} (I + F_m R)$. A complete proof of this equivalence may be found in [Sao].

The commutative Zariski rings as they appear in commutative algebra or algebraic geometry provide a first example (cf. [Z-S]): A commutative Zariski ring is a commutative Noetherian ring R with an I-adic filtration where I is an ideal of R contained in the Jacobson radical of R. Some interesting examples of noncommutative I-adic Zariskian filtrations will be given in subsection 2.2. later. To end this subsection, here we stress that the link between I-adic filtration and Zariskian filtration is in fact deeper, i.e., for a Zariskian filtration FR on the given ring R, the filtration $F(F_0R)$ induced on the subring F_0R by FR is "almost" $F_{-1}R$-adic. To be precise, let R be a left Zariski ring with filtration FR. Consider the filtration

$$F(F_0R) = \{F_0R \cap F_n R, \ n \in \mathbb{Z}\}$$

on F_0R induced by FR, then we have

4. Lemma. The subring F_0R of R is a left Zariski ring with respect to $F(F_0R)$.

Proof. $F_{-1}(F_0R) = F_{-1}R \subseteq J(F_0R) = J(F_0(F_0R))$ is clear. Since $\tilde{R}$ is left Noetherian it follows from [NVO1] that $F_0R = \tilde{R}_0$ is left Noetherian. Moreover, we see that $\widetilde{F_0R}^- = \oplus_{n \leq 0} F_n(F_0R) = \tilde{R}^-$ and $\widetilde{F_0R}^+ = \oplus_{n \geq 0} F_n(F_0R) = \oplus F_0R \cong F_0R[t]$. Hence from ([NVO1] Proposition II.3.4.) it follows that $\widetilde{F_0R}$ is left Noetherian. This shows that F_0R is a left Zariski ring. $\square$

5. Corollary. With notations as in the lemma, then every left ideal of F_0R is closed in the $F(F_0R)$-topology of F_0R; In particular, every $(F_{-1}R)^m$, $m > 0$, is closed in the $F(F_0R)$-topology.

Proof. This follows from Theorem 2.1.2.. $\square$

Now, if we look at the structure of $G(F_0R)^- = \oplus_{n\leq 0}G(R)_n$, it is left Noetherian by ([NVO1] Proposition II.3.4.) since $G(R)$ is left Noetherian. Thus the ideal $I = \oplus_{n<0}G(R)_n$ of $G(F_0R)^-$ is finitely generated. It follows that $G(F_0R)^-$ is of the form

$$G(R)_0[\bar{a}_1, \cdots, \bar{a}_p]$$

for some finite set of homogeneous elements $\bar{a}_i$, say of degree $d_i < 0$, $i = 1, \cdots, p$. If $x \in F_wR - F_{w-1}R$, where $w < 0$, then $\sigma(x) = x + F_{w-1}R = \sum \bar{r}_0^\gamma \bar{a}_1^{\gamma_1} \cdots \bar{a}_p^{\gamma_p}$ with $w = \gamma_1 d_1 + \cdots + \gamma_p d_p$ and $r_0 \in F_0R$. So we have $x - \sum r_0^\gamma a_1^{\gamma_1} \cdots a_p^{\gamma_p} \in F_{w-1}R$ with $x_1 = \sum r_0^\gamma a_1^{\gamma_1} \cdots a_p^{\gamma_p} \in (F_{-1}R)^{\mu_x}$, where $\mu_x = \gamma_1 + \cdots + \gamma_p$ is determined by $\sum \bar{r}_0^\gamma \bar{a}_1^{\gamma_1} \cdots \bar{a}_p^{\gamma_p}$. Put $\mu_w = \min\{\mu_x, \ x \in F_wR - F_{w-1}R\}$, then $x_1 \in (F_{-1}R)^{\mu_w}$. Noting that $w < 0$, $d_i < 0$, it is not hard to see that

$$(*) \qquad\qquad \mu_w < \mu_{w-1} < \mu_{w-2} < \cdots$$

If we repeat the same argumentation for $y = x - x_1 \in F_{w-1}R$ we obtain that $y_1 = y - x_2 \in F_{w-2}R$ where $x_2 \in (F_{-1}R)^{\mu_{w-1}} \subseteq (F_{-1}R)^{\mu_w}$. A repetition of this process for y_1 (and so on) yields

$$x \in \bigcap_{t\geq 1}\left((F_{-1}R)^{\mu_w} + F_{w-t}R\right) = (F_{-1}R)^{\mu_w}$$

by Corollary 2.1.5.. This shows that $F_wR \subseteq (F_{-1}R)^{\mu_w}$. Now choose $w < \min\{d_1, \cdots, d_p\}$, then $\mu_w > 2$. It follows from $(*)$ that for any $n > 0$ we have $\mu_{w-(n-2)} \geq n$ and hence

$$F_{w-(n-2)}R \subseteq (F_{-1}R)^{\mu_{w-(n-2)}} \subseteq (F_{-1}R)^n.$$

more clearly we have

$$
\begin{array}{ccccccccc}
 & \subseteq & (F_{-1}R)^n & \subseteq & \cdots & \subseteq & (F_{-1}R)^3 & \subseteq & (F_{-1}R)^2 & \subseteq & F_{-1}R \\
 & & \cup & & \cdots & & \cup & & \cup & & \cup \\
 & \subseteq & F_{w-(n-2)}R & \subseteq & \cdots & \subseteq & F_{w-1}R & \subseteq & F_wR & \subseteq & F_{w+1}R
\end{array}
$$

Therefore we have proved the following result:

6. Proposition. With notations as above, then the filtration $F(F_0R)$ on F_0R induced by FR is always topologically equivalent to the $F_{-1}R$-adic filtration on F_0R. In particular, if all a_i, $i = 1, \cdots, p$, are in $G(R)_{-1}$ then $F_wR = (F_{-1}R)^{-w}$ for all $w < 0$, i.e., in this case $F(F_0R)$ is the real $F_{-1}R$-adic filtration.

2.2. Examples of Zariskian filtration

a. Complete filtered rings with Noetherian associated graded ring

1. Proposition. Let R be a filtered ring with filtration FR. Suppose that R is complete with respect to FR and that $G(R)$ is left Noetherian, then R is a left Zariski ring.

Proof. From Proposition 1.2.3. we see that R satisfies the condition (5) of Theorem 2.1.2..
□

Since a filtered ring with discrete filtration is always complete we have the following

2. Corollary. Each one of the following filtered rings is left Zariskian:

(1). The skew polynomial ring $R[x, \varphi, \delta]$ over a left Noetherian ring R with the standard filtration as defined in Ch.I. Example 2.3.F., where φ is an automorphism of R and δ is a φ-derivation of R, since in this case $G(R[x, \varphi, \delta]) \cong R[\overline{x}, \varphi]$;

(2). The universal enveloping algebra $U(\mathbf{g})$ of a finite dimensional k-Lie algebra (where k is a field) with the standard filtration as defined in Ch.I. Example 2.3.F., since it is well known that $G(U(\mathbf{g}))$ is a polynomial ring over k in finitely many commuting indeterminates;

(3). The derivation ring $A[\mathbf{d}]$ of a commutative k-algebra over a commutative ring k with the standard filtration as defined in Ch.I. Example 2.3.H.. From [M-R] we know that if A is Noetherian then $G(A[\mathbf{d}])$ is Noetherian. In particular, if we consider the n-th Weyl algebra $A_n(k)$ over a field k with the Bernstein filtration ($\sum$-filtration) as defined in Ch.I. Example 2.3.H. (Ch.I. Example 4.3.9.), then the associated graded rings of $A_n(k)$ with respect to both filtrations are the same (in the ungraded sense), namely, the polynomial ring over k in $2n$ commuting variables;

(4). The ring of differential operators $\mathcal{D}(A)$ of a commutative k-algebra A (k is a commutative ring) with the filtration as defined in Ch.I. Example 2.3.I. such that $G(\mathcal{D}(A))$ is left Noetherian.

The above corollary also includes the well known rings of $\mathbb{C}$-linear differential operators studied in [Bj1]:

(1). The ring of $\mathbb{C}$-linear differential operators on V, denoted $\mathcal{D}(V)$ (i.e., $\mathcal{D}(V) = A(V)_{\mathbb{C}}[\mathrm{Der}_{\mathbb{C}} A(V)]$) with the standard (positive) filtration over $A(V)$ based on $\mathrm{Der}_{\mathbb{C}} A(V)$ as a generating set, where V is an irreducible smooth subvariety of the affine n-space $\mathbb{C}^n$ and $A(V)$ is the coordinate ring of V.

(2). The ring of $\mathbb{C}$-linear differential operators $\mathcal{D}_n = \mathcal{O}_n\langle \partial/\partial z_1, \cdots, \partial/\partial z_n \rangle$ on $\mathcal{O}_n$, with the positive filtration defined by putting $F_k \mathcal{D}_n = \{\sum_{\alpha \leq k} f_\alpha(z)\partial^\alpha\}$, $k \in \mathbb{Z}$, where $f_\alpha(z) \in \mathcal{O}_n$ whereas $\mathcal{O}_n = \mathbb{C}\{z_1, \cdots, z_n\}$ is the local ring of convergent power series in n variables with coefficients in the complex field $\mathbb{C}$.

b. Stalks $\mathcal{E}_p$ of the sheaf of microlocal differential operators

In order to include a non-trivial example such that the ring considered has a **nonzero negative part** in its filtration , and is also **non-complete** with respect to the given filtration, we refer to the ring $\mathcal{E}_p$ of germs of microlocal differential operators with its filtration $F\mathcal{E}_p$ as defined in [Shar] and [Bj1] (and we refer to loc.cit. for a precise definition and more detail concerning the properties of this ring).

3. Proposition. $\mathcal{E}_p$ is a left and right Zariski ring.

Proof. From ([Bj1] Ch.4. Proposition 1.4.) it follows that $G(\mathcal{E}_p)$ is a commutative Noetherian domain. More precisely, $G(\mathcal{E}_p) \cong \mathcal{O}_{2n-1}[t, t^{-1}]$ with $G(\mathcal{E}_p)_n \cong \mathcal{O}_{2n-1}t^n$, where $\mathcal{O}_{2n-1}[t, t^{-1}]$ is the ring of finite Laurent series $\sum \varphi_\nu t^\nu$ in the variable t and with coefficients φ_ν in the local ring $\mathcal{O}_{2n-1} = G(\mathcal{E}_p)_0$.

If $u \in F_{-1}\mathcal{E}_p$ then $\sigma(1 - u)(p) = \sigma(1)(p) \neq 0$. Since $\sigma(xy) = \sigma(x)\sigma(y)$ holds for every $x, y \in \mathcal{E}_p$ (as $G(\mathcal{E}_p)$ is a domain!) it follows that $(1-u)^{-1} \in \mathcal{E}_p$ and we claim: $(1-u)^{-1} \in F_0\mathcal{E}_p$. Indeed, if $\vartheta = (1 - u)^{-1} \in F_n\mathcal{E}_p - F_{n-1}\mathcal{E}_p$ with $n \geq 1$ then $1 = (1 - u)\vartheta = \vartheta - u\vartheta$ and $\vartheta = 1 + u\vartheta \in F_{n-1}\mathcal{E}_p$ follows from $u \in F_{-1}\mathcal{E}_p$, a contradiction. Thus $F_{-1}\mathcal{E}_p \subseteq J(F_0\mathcal{E}_p)$. Next, note that $G(\mathcal{E}_p)$ is a strongly $\mathbb{Z}$-graded ring, by using the relation $\widetilde{\mathcal{E}_p}/X\widetilde{\mathcal{E}_p} \cong G(\mathcal{E}_p)$ one easily derives for all $m, n \in \mathbb{Z}$, $F_n\mathcal{E}_p F_m\mathcal{E}_p = F_{n+m}\mathcal{E}_p$ (a filtered ring satisfying this condition is called strongly filtered, see §4. later). But this means that the Rees ring $\widetilde{\mathcal{E}_p}$ of $\mathcal{E}_p$ is strongly $\mathbb{Z}$-graded. Therefore it will suffice to establish that $F_0\mathcal{E}_p$ is left and right Noetherian because then $\widetilde{\mathcal{E}_p}$ will be Noetherian. That $F_0\mathcal{E}_p$ is Noetherian follows from the definition of the filtration on $\mathcal{E}_p$ and a boundness argument similar to ([Bj1] Theorem 2.1., the part of the proof on p.141).   □

Even if $\mathcal{E}_p$ is one of the main examples that motivated us to study the (noncommutative) filtered Zariski ring which has a filtration with non-trivial negative and non-trivial positive parts (in a purely algebraic way) we do not dig deeper into the analytic theory of rings of (micro-) differential operators in this book. It is the "micro" aspect that introduces filtrations with nonzero negative tails in our lives; we will study microlocalization of filtered rings in some generality in Ch.IV..

c. Discrete valuation rings

Let Δ be a skewfield and Λ a discrete valuation ring of Δ in the sense of [Schi]. If P is the unique maximal ideal of Λ then it is not hard to check that the filtration $F\Delta$ defined by the following increasing chain

$$\cdots \subseteq P^2 \subseteq P \subseteq \Lambda \subseteq P^{-1} \subseteq P^{-2} \subseteq \cdots$$

makes Δ into a left and right Zariski ring with with $F_0\Delta = \Lambda$ and $G(\Delta) = \Lambda/P[t, t^{-1}]$.
At this stage, connections between discrete valuations on skewfields and filtrations with suitable associated graded rings may be studied in some detail. The noncommutative theory will be considered later, here we first discuss the commutative case in this example.

Let R be an arbitrary ring. Put $\Gamma = \mathbb{Z} \cap \{\infty\}$ and $\infty + \infty = \infty$, $\infty > n$ and $\infty + n = \infty$ for all $n \in \mathbb{Z}$. Recall that a **discrete valuation function** $v(x)$ of R is a map defined on a subring R' of R, $v: R' \to \Gamma$, satisfying:
V1. $v(0) = \infty$,
V2. $v(ab) = v(a) + v(b)$, for all $a, b \in R'$,
V3. $v(a + b) \geq \min(v(a), v(b))$, for all nonzero $a, b \in R'$.
Then one checks that the family

$$F_nR = \{a \in R, v(a) = \infty \text{ or else } -v(a) \leq n\}, n \in \mathbb{Z}$$

is a filtration of R and $\cup_{n \in \mathbb{Z}} F_nR = R'$. Therefore FR is exhaustive if and only if $R' = R$; FR is separated if and only if: $v(a) = \infty$ if and only if $a = 0$. In case FR is separated, then $G(R)$ is a domain (because it has no homogeneous zero divisors).

4. Proposition. Let R be a commutative domain with field of fractions Q. If v is a Γ-valuation on R then there exist exhaustive and separated filtrations FR resp. FQ on R resp. on Q, such that
(1). $G(R)$ resp. $G(Q)$ is a domain;
(2). for all $n \in \mathbb{Z}$, $F_nR = F_nQ \cap R$ and F_0Q is a discrete valuation ring of Q with the unique maximal ideal $F_{-1}Q$.

Proof. Since any valuation function on R can be extended to Q, (1) and (2) are easy to check. $\qquad \square$

Now, let R be an arbitrary filtered ring with filtration FR. Then one may define a map $v: R \to \Gamma$ such that

$$v(a) = \begin{cases} \infty, & \text{if } a \in \cap_{n \in \mathbb{Z}} F_nR; \\ -n, & \text{if } a \in F_nR - F_{n-1}R. \end{cases}$$

From the definition it is easy to see that the map v satisfies the following properties:

FV1. $v(ab) \geq v(a) + v(b)$, for all a, $b \in R$,

FV2. $v(a+b) \geq \min(v(a), v(b))$, for all a, $b \in R$,

FV3. $v(a) = \infty$ if and only if $a \in \cap_{n \in \mathbb{Z}} F_n R$.

5. Lemma. Let R be an arbitrary filtered ring with separated filtration FR, then
(1). $\sigma(a) \in h(G(R))$ is a (right) regular homogeneous element of $G(R)$ if and only if $v(ab) = v(a) + v(b)$ for all $b \in R$, where $a \in F_n R - F_{n-1} R$ for some $n \in \mathbb{Z}$. Hence $G(R)$ has no (homogeneous) (right) zero divisors if and only if the equality in FV1 holds for all a, $b \in R$;
(2). if $G(R)$ has no non-trivial zero divisors then R has no non-trivial zero divisors.

Proof. Straightforward.   □

From the definition of v we may construct a filtration $F'R$ on R by putting: $F_n' R = \{a \in R, v(a) = \infty$ or else $-v(a) \leq n\}$, $n \in \mathbb{Z}$. Then obviously $F_n R = F_n' R$ for all $n \in \mathbb{Z}$.

Recombining Proposition 2.2.4. and Lemma 2.2.5. we easily obtain

6. Proposition. Let R be a commutative filtered ring with separated filtration FR. If $G(R)$ is a domain then
(1). R is a domain and $v: R \to \Gamma$ defined above is a Γ-valuation on R;
(2). there is an exhaustive and separated filtration FQ on the field Q of fractions of R such that: $F_n R = F_n Q \cap R$ for all $n \in \mathbb{Z}$, $G(Q)$ is a domain and $F_0 Q$ is a discrete valuation ring with the unique maximal ideal $F_{-1}Q$, $F_n Q = (F_{-1}Q)^{-n}$ for all $n \in \mathbb{Z}$.

The relation between discrete valuations on fields and filtrations having a domain as the associated graded ring is now completely expressed in

7. Theorem. Let Q be a commutative field, R a subring of Q. The following statements are equivalent:
(1). R is a discrete valuation ring of Q;
(2). There is an exhaustive and separated filtration FQ on Q such that $F_0 Q = R$ and $G(R)$ is a domain.

8. Corollary. Let R be a Noetherian local domain with the unique maximal ideal $P \neq 0$ and let Q be the field of fractions of R. Put $P^{-1} = \{x \in Q, \, xP \subseteq R\}$ and $F_n Q = P^{-n}$ for all $n \in \mathbb{Z}$, then the following statements are equivalent:
(1). R is a regular local ring of dimension 1;
(2). R is a discrete valuation ring of Q;
(3). FQ is an exhaustive and separated filtration on Q and $G(Q)$ is a domain.

d. Dehomogenization of gradings to Zariskian filtrations

One may also construct Zariskian filtrations on $\mathbb{Z}$-graded rings by using the dehomogenization technique for gradings in the sense of Ch.I. 4.3.. As we will see from now on, this has
some particular interest in the study of graded rings.

The first result given below is a generalization of Hilbert basis theorem.

9. Theorem. Let $S = \oplus_{n \in \mathbb{Z}} S_n$ be a $\mathbb{Z}$-graded ring and T a regular central homogeneous
element of degree 1 in S. Considering the dehomogenization $R = S/(1-T)S$ of S (with
respect to T) with filtration $F_n R = S_n + (1-T)S/(1-T)S$, $n \in \mathbb{Z}$, the following statements
are equivalent:

(1). S is a left Noetherian ring and $T \in J^g(S)$, where $J^g(S)$ is the graded Jacobson radical
of S;

(2). S/TS is left Noetherian and the completion $\hat{R}$ of R with respect to FR is a faithful flat
right R-module, or equivalently, R is a left Zariski ring.

Proof. From Lemma 4.4.11. of Ch.I. and Ch.I. 4.3. we know that $\tilde{R} \cong S$, $G(R) \cong S/TS$
and $T \in J^g(S)$ if and only if $F_{-1}R \subseteq J(F_0R)$, hence the equivalence of (1) and (2) now
follows from Theorem 2.1.2.. □

When the given graded ring S has a **positive** or **left limited** gradation, i.e., there exists
an integer $c \leq 0$ such that $S_n = 0$ for all $n < c$, Theorem 2.2.9. above may be reduced to

10. Theorem. Let S have a positive or left limited gradation, let T be a regular central
homogeneous element of degree 1 in S. With notations as in Theorem 2.2.9., the folllowing
statements are equivalent:

(1). S is left Noetherian;

(2). S/TS is left Noetherian.

Proof. Note that under the assumption in the theorem, FR is discrete and hence R is
complete with respect to FR, i.e., $\hat{R} = R$. Hence by the foregoing example a. R will be
a left Zariski ring and hence S ($\cong \tilde{R}$) will be left Noetherian if S/TS ($\cong G(R)$) is left
Noetherian. The implication (1) $\Rightarrow$ (2) is trivial. □

Let $S = \oplus_{n \in \mathbb{Z}} S_n$ be a $\mathbb{Z}$-graded ring and I a graded ideal of S with a centralizing sequence
of homogeneous generators $\{T_1, \cdots, T_n\}$ of degree 1 (in the sense of Ch.I. Definition 4.4.2.),
i.e., for each $j \in \{0, \cdots, n-1\}$ the image $\overline{T}_{j+1}$ of T_{j+1} in $\overline{S}_j = S/\sum_i^j T_i S$ is a central
element. Furthermore, let us introduce two extra conditions on I:

(RN). $\overline{T}_{j+1}$ is regular;

(JN). $\overline{T}_{j+1} \in J^g(\overline{S}_j)$.

Now, by using the foregoing results an easy induction yields the following

11. Corollary. Let S be a $\mathbb{Z}$-graded ring and I an ideal of S with a centralizing sequence of homogeneous generators $\{T_1, \cdots, T_n\}$ of degree 1. Suppose that I satisfies the condition (RN), then the following statements are equivalent:

(1). S is left Noetherian and I satisfies the condition (JN);

(2). For each $j \in \{0, \cdots, n-1\}$ the dehomogenization $\overline{S}_j/(1-\overline{T}_{j+1})\overline{S}_j$ of $\overline{S}_j$ $(= S/\sum_i^j T_i S)$ is a left Zariski filtered ring.

12. Corollary. Let S have a positive or left limited $\mathbb{Z}$-gradation and I an ideal of S with a centralizing sequence of homogeneous generators $\{T_1, \cdots, T_n\}$ of degree 1. Suppose that I satisfies the condition (RN), then the following statements are equivalent:

(1). S is left Noetherian;

(2). S/I is left Noetherian.

Another class of Zariskian filtrations is obtained by applying the dehomogenization trick to the generalized Rees rings determined by invertible ideals.

Let R be a ring and S be an overring of R. Let I be an invertible ideal of R. Consider the dehomogenization $A = \check{R}(I)/(1 - y)\check{R}(I)$ of $\check{R}(I)$ with respect to y, where $\check{R}(I) = \oplus_{n\in\mathbb{Z}} I^n t^n \subseteq S[t, t^{-1}]$ is the generalized Rees ring of R with respect to I (see the definition given in Ch.I. Example 4.3.9.(c).) and $y = t^{-1}$, then it is obvious that $\check{R}(I)$ and hence $G(A)$ are strongly $\mathbb{Z}$-graded. By the observations given in Ch.I. Example 4.3.9.(c). and Ch.I. Lemma 4.4.11. one easily derives the following theorem.

13. Theorem. Let R, I, and A be as described above, then the following statements are equivalent:

(1). R is left Noetherian (or equivalently $\check{R}(I)$ is left Noetherian since $\check{R}(I)_0 = R$) and I is contained in the Jacobson radical $J(R)$ of R;

(2). The I-adic filtration on R is left Zariskian;

(3). R/I is left Noetherian and $\hat{A}$ is faithfully flat over A as a right A-module (or equivalently, A is a left Zariski ring).

Note. To prove the above theorem one needs the following results of [NVO1]:

(1). A strongly $\mathbb{Z}$-graded ring $A = \oplus_{n\in\mathbb{Z}} A_n$ is left Noetherian if and only if A_0 is left

Noetherian.

(2). If a $\mathbb{Z}$-graded ring $A = \oplus_{n \in \mathbb{Z}} A_n$ is left Noetherian then the graded subrings $A^- = \oplus_{n \leq 0} A_n$ and $A^+ = \oplus_{n \geq 0} A_n$ of A are also left Noetherian.

e. Some related examples

In the foregoing examples we have seen a number of Zariskian filtrations, in particular, we have Zariski filtered rings which have a nonzero negative part in their filtrations and are not always complete. However, some interrelations between properties characterizing Zariskian filtrations remain unsettled.

Look at the following conditions:

  a. $G(R)$ is left Noetherian;
  b. Good filtrations in R-filt induce good filtrations on R-submodules;
  c. $F_{-1}R \subseteq J(F_0R)$,

as we have seen that a., b. and c. characterize a left Zariski filtered ring R with filtration FR. But if R satisfies a. and b. then c. may fail to hold, e.g., let R be a commutative Noetherian ring and I an ideal of R which is not contained in the Jacobson radical of R, then consider the I-adic filtration on R. It turns out that if P is a maximal ideal of R such that I is not contained in P and put $M = R/P$ with the good filtration $F_n M = I^n M, n \in \mathbb{Z}$, then $G(M) = 0$ but $M \neq 0$. This means that FR is not faithful in the sense of Lemma 1.2.9.. There also exists a commutative local ring with unique maximal ideal P that is finitely generated and satisfying $\cap_{n \geq 1} P^n = 0$ (hence the conditions a. and c. above do hold here) but R fails to be Noetherian and hence b. fails (we refer to [Naga] for such an example).

We are left with an intriguing question here:

14. Question. If the conditions b. and c. hold does it follow that the condition a. hold?

Finally, let us point out that there are examples of filtered rings R showing that the conditions
  (i). $G(R)$ is left Noetherian, and
  (ii). FR has the left Artin-Rees property

are satisfied (hence $\hat{R}$ is flat as a right R-module by Theorem 1.2.4.) but $\hat{R}$ is not faithfully flat as a right R-module. This may be used to illustrate that the important Fuchsian filtration on a Weyl algebra (see Ch.I. §2. Example L.) is not Zariskian.

Indeed, from Proposition 1.1.3. we know that any Noetherian $\mathbb{Z}$-graded ring $A = \oplus_{n \in \mathbb{Z}} A_n$ with the grading filtration (as defined in Ch.I. §2.) $F_n^{(1)} A = \sum_{k \leq n} A_k, n \in \mathbb{Z}$, satisfies the condition (i) and (ii) above, but $\hat{A}$ is not necessarily faithfully flat as a right A-module because in this case $F_{-1}A$ need not be in $J(F_0A)$. For instance, take the ring $A = R[t, t^{-1}]$ of finite Laurent series over a Noetherian ring R in variable t, then A has the natural gradation given by the power of t: $A_n = Rt^n$, $n \in \mathbb{Z}$, and the grading filtration $F^{(1)}A$ on it, but we see that $t^{-1} \in F_{-1}A$, $t^{-1} \notin J(F_0A)$.

Another interesting example is the ring of skew differential polynomials $A = R[X][Y, D]$ with the identity automorphism over $R[X]$, where $R[X]$ is the polynomial ring over a left Noetherian ring R in X and $D = \partial/\partial X$ on $R[X]$. So we have the following relations: $YX - XY = 1$, $Yf(X) = f(X)Y + D(f(X))$ for every $f(X) \in R[X]$. It is well known that A is a left Noetherian ring. On the other hand, A has a $\mathbb{Z}$-gradation defined by giving X-degree -1 and Y-degree $+1$; in other words, for $n \in \mathbb{Z}$ we have $A_n = \sum_{j-i=n} RX^iY^j$ where $i \geq 0$, $j \geq 0$. From what we have pointed out above the grading filtration $F^{(1)}A$ on A has the left (and right) Artin-Rees property but $F_{-1}A$ fails to be in $J(F_0A)$ since $X^2Y \in F_{-1}A$, $X^2Y \notin J(F_0A)$.

Now, let $A_{n+1}(k)$ be the $n + 1$-st Weyl algebra over a field k of characteristic zero, then the Fuchsian filtration (see Ch.I. §2. Example 2.3.L.) on $A_{n+1}(k)$ is not Zariskian because $1 + x_{n+1}^2 y_{n+1}$ is not invertible even if $x_{n+1}^2 y_{n+1}$ is in $F_{-1}A_{n+1}(k)$. The fact that $FA_{n+1}(k)$ has left and (right) Artin-Rees property (or satisfies the comparison condition in the sense of [Bj2]) follows from the foregoing argument.

Remark. The Fuchsian filtration on $A_{n+1}(k)$, as pointed out by J-E. Björk, has a natural geometric significance based upon Deligne's constructions of vanishing cycles and has been used to study the holonomic modules over $A_{n+1}(k)$ (cf. [Bj2]).

§3. Lifting Structures of Zariski Rings

The idea of this section is to provide some examples where certain ring structural properties are lifted from the associated graded ring to the given left Zariski ring. For some theory on gr-simple, gr-Artinian, gr-hereditary, Von Neuman gr-regular rings and gr-orders we refer to [NVO1] and [LeVV].

3.1. $G(R)$ is gr-Artinian, gr-regular or Von Neuman gr-regular

In what follows R is always a left **Zariski ring** with filtration FR.
Let us start by expliciting a property already inherent in the characterization of Zariskian filtrations.

1. Lemma. Let $M \in R$-filt with a good filtration FM. If N is an R-submodule of M with filtration FN induced by FM such that $G(N) = G(M)$, then $N = M$.

Proof. Since N has filtration FN induced by FM, $G(N) = \oplus_{n \in \mathbf{z}}(F_n M \cap N + F_{n-1}M)/F_{n-1}M$ is a graded submodule of $G(M)$. Let $0 \neq m \in M$. Since R is Zariskian and FM is good we may assume that $m \in F_n M - F_{n-1}M$ for some $n \in \mathbb{Z}$. It follows from $G(N) = G(M)$ that $m = y + m'$, where $y \in F_n M \cap N$ and $m' \in F_{n-1}M$. Hence $m \in N + F_{n-1}M$. Repeating this for $m - y = m'$ and so on we get that $m \in \cap_{p \in \mathbf{z}}(N + F_p M) = N$ by Theorem 2.1.2.. Since m is arbitrary it follows that $M = N$. □

2. Corollary. Let $M \in R$-filt with a good filtration FM.
(1). If $G(M)$ is gr-Artinian then M is Artinian. In particular, if $G(R)$ is a gr-skewfield then R is a skewfield.
(2). If the Krull dimension of $G(M)$ is well defined then the Krull dimension of M is defined and $K.\dim_R M \leq K.\dim_{G(R)}G(M)$. In particular, $K.\dim R \leq K.\dim G(R)$.
(3). If $K.\dim_R M = K.\dim_{G(R)}G(M) = \alpha$ and if $G(M)$ is α-critical then M is an α-critical module.

3. Corollary. If $G(R)$ is gr-Artinian semisimple (simple), then R is Artinian semisimple (simple).

Proof. If $G(R)$ is gr-Artinian simple then by Corollary 2. we know that R is Artinian simple. If R is Artinian then it is well known that the Jacobson radical $J(R)$ of R is nilpotent. Since $G(R)$ is gr-semisimple it has no nonzero graded nilpotent ideal, hence the fact $G(J(R)) = 0$ yields $J(R) = 0$ since R is Zariskian. Therefore R is semisimple. □

From Ch.I. Theorem 7.2.13. we easily derive the following theorem.

4. Theorem. Let R be as before.

(1). If $G(R)$ is left gr-regular (see the definition in Ch.I. §7.2.) then R is left regular.

(2). gl.dim$R \leq$ gr.gl.dim$G(R) \leq$ gl.dim$G(R)$.

The result stated in the next proposition may be viewed as special cases of Theorem 3.1.4. above. But in order to clarify how graded ring theoretical properties of $G(R)$ can be lifted to R we also include a detailed proof below.

5. Proposition. Let R be as before.

(1). If $G(R)$ is left gr-hereditary then R is left hereditary.

(2). If $G(R)$ is Von Neuman gr-regular then R is Von Neuman regular.

Proof. (1). Let $G(R)$ be left gr-hereditary. Then every graded left ideal of $G(R)$ is gr-projrctive in $G(R)$-gr (hence a projective $G(R)$-module). Taking a left ideal L of R, then L is finitely generated since R is left Zariskian. Consider the good filtration FL induced by FR, it follows that $G(L)$ is a finitely generated graded left ideal of $G(R)$ and hence gr-projective. Now Ch.I. Proposition 7.2.10. yields that L is filt-projective in R-filt and L is certainly projective as a left R-module.

(2). Let $G(R)$ be Von Neuman gr-regular. Then every graded left $G(R)$-module is gr-flat (hence flat). Since $G(R)$ is left Noetherian every finitely generated graded $G(R)$-module is finitely presented and hence gr-projective. Let M be any finitely generated R-module. Choose a good filtration FM on M, then $G(M)$ is gr-projective in $G(R)$-gr since $G(M)$ is finitely generated. It follows from Ch.I. Proposition 7.2.10. that M is filt-projective in R-filt. Now, take an arbitrary R-module M, then $M = \varinjlim M_i$, where M_i ranges over all finitely generated R-submodules of M and the index set is directed by usual inclusion. Hence the well known formula (cf. [Rot] Theorem 8.11.) $\mathrm{Tor}_n^R(\varinjlim M_i, N) = \varinjlim \mathrm{Tor}_n^R(M_i, N)$ (where N is any R-module and $n \geq 0$) shows that M is a flat R-module. Consequently R is Von Neuman regular. □

3.2. $G(R)$ is a gr-maximal order

The question we raise here is the following: if R is a filtered ring with filtration FR such that $G(R)$ is a (gr-) maximal order in a (gr-) simple Artinian ring is then R a maximal order in a simple Artinian ring? In [CH] a positive answer was given to this question in the particular case where the rings considered are Noetherian domains. In turn, the latter result has been used by [Ma] in order to prove the following result: If I is an ideal of a Noetherian ring R such that I is generated by a regular normalizing set of generators and I being contained in the Jacobson radical of R then R is a maximal order in a division ring when R/I is a maximal order in a division ring. In fact one may establish a more general result without using filtrations at all (cf. [PS2]): if P is an invertible ideal contained in the Jacobson radical of a Noetherian ring R then R is a maximal order in a simple Artinian ring if R/P is a maximal order in a simple Artinian ring. We now will present the graded version of a slight generalization of the forementioned result and apply it to obtain a positive answer to our original question in full generality. We make the section self-contained by introducing some terminology and classical facts as well as a few perhaps less well-known subtleties of graded nature.

Recall that a ring Q such that every $q \in Q$ which is a non-zero divisor is necessarily an invertible element of Q is said to be a **quotient ring**. A subring A of a quotient ring Q is said to be an **order in** Q if every regular element of A is invertible in Q and every $q \in Q$ is a left and right fraction with respect to regular elements of A. Let A and B be orders in Q; we say that A is **equivalent to** B in Q if there exist $a, b, c, d \in Q$ such that $aAb \subseteq B$ and $cBd \subseteq A$. An order in Q is said to be a **maximal order in** Q whenever for every nonzero ideal I of A we have: $A = \{q \in Q, \, qI \subseteq I\} = \{q' \in Q, \, Iq' \subseteq I\}$. When A is an order in Q the A-bimodules contained in Q may be thought of as being a kind of generalized fractional ideals. We say that an A-bimodule X contained in Q is **invertible (in** Q) if there exists an A-bimodule Y contained in Q such that $XY = YX = A$. Of course the ideals of A may be viewed as A-bimodules contained in Q. If I is an ideal of A that is invertible in Q then we have the properties:

(i). $A \subseteq I^{-1} = \{q \in Q, \, qI \subseteq A\} = \{q' \in Q, \, Iq' \subseteq\}$;

(ii). $I^{-1}I = II^{-1} = A$.

In [NVO1] the class of the gr-Goldie rings has been introduced and studied in detail. It has been observed in loc. cit. that the condition of being a gr-Goldie ring is not equivalent to being a $\mathbb{Z}$-graded ring and also a Goldie ring. On one hand a $\mathbb{Z}$-graded prime Goldie ring A does possess a graded quotient ring in the torsion theoretic sense, Q^g say. The ring Q^g is gr-simple gr-Artinian but it is not a graded ring of fractions of A (i.e., Q^g is not a gr-quotient ring as defined above) because not every graded left ideal of A that is an essential left ideal of A does contain a **regular homogeneous** element. We summarize some results of [NVO1], [NNVO], [VO3]:

1. Lemma. Let A be a $\mathbb{Z}$-graded (semiprime) prime Goldie ring. In each of the cases listed below A will have a graded ring of fractions Q^g that is gr-simple gr-Artinian (in other words A is a **graded order** in Q^g).

(1). A has a regular central homogeneous element of positive degree.

(2). A is positively graded.

(3). Every homogeneous element of nonzero degree in A is a nilpotent element.

(4). The ring A is a P.I. ring.

(5). The gradation of A satisfies the condition (E): if for $n \in \mathbb{Z}$, $x_n \neq 0$ in A_n then $A_{-n}x_n \neq 0$ (this condition is left-right symmetric if A is semiprime!) and A_0 is a prime ring.

In the filtered situation we will want to apply the lemma to $A = \tilde{R}$ and therefore the presence of $X \in \tilde{R}_1$ allows to use Lemma 1.(1).. So the only problem remains in applying the lemma to $G(R)$ but here the properties (2), (3), (4), (5) listed in the lemma will provide interesting criteria.

In a general theory of graded orders in gr-simple gr-Artinian rings the failure of a general graded version of Goldie's theorems only creates superficial problems; we may refer to the book on graded orders, [LeVV], for a detailed study of these rings.

We now assume that A has a graded ring of fractions Q^g. Such an A is said to be a **gr-maximal order** in Q^g if for any nonzero graded ideal I of A we have that $A = \{q \in Q^g, Iq \subseteq I\} = \{q' \in Q^g, q'I \subseteq I\}$, or equivalently: for any graded subring T of Q^g containing A such that $aTb \subseteq A$ for some regular homogeneous elements $a, b \in Q^g$ we have that $A = T$.

2. Lemma. (Lemma 3.1. of [LeVV] and see also [VO3]) Assume that the $\mathbb{Z}$-graded ring A has a graded ring of fractions Q^g which is a gr-simple gr-Artinian ring. The ring A is a gr-maximal order in Q^g if and only if A is a maximal order in Q, where Q is the quotient ring of Q^g, i.e., $Q = Q_{cl}(Q^g) = Q_{cl}(A)$ whereas Q_{cl} denotes the classical ring of quotients.

In order to obtain the main result in this section we first prove a general lemma concerning the **graded Jacobson radical**.

3. Lemma. Let $A = \oplus_{g \in G} A_g$ be an arbitrary (group) G-graded ring and I a graded ideal of A. Suppose that A is left Noetherian and I is an invertible ideal contained in the graded Jacobson radical $J^g(A)$ of A. Consider the I-adic filtration FA on A, then we have

(1). any good filtration on a graded A-module is separated;

(2). The Rees ring $\tilde{A}$ of A associated to FA is left Noetherian;

(3). good filtrations in A-filt induce good filtrations on A-submodules.

Proof. From the assumptions and from Ch.I. Proposition 4.4.7. it follows that I has the Artin-Rees property and hence one easily sees that $I \subseteq J^g(A)$ if and only if for any finitely generated graded A-module M and any graded A-submodule N of M, $N = \cap_{n \geq 1}(N + I^n M)$. Consequently (1) follows because any two good filtrations on the same module are equivalent. Next, since A is left Noetherian the generalized Rees ring $\check{A} = \oplus_{n \in \mathbb{Z}} I^n$ of A is left Noetherian and hence $\tilde{A}$ is left Noetherian (see the note after Theorem 2.2.13.). Finally, (3) follows immediately from (2). $\square$

4. Proposition. ([NVO1] Proposition II.1.4.) Let R be a G-graded ring where G is an ordered group. Then

(1). A graded ideal P of R is (semiprime) prime if and only if for any two homogeneous elements $a, b \in R$ such that $(aRa \subseteq P)$ $aRb \subseteq P$ it follows that $(a \in P)$ a or b is in P;

(2). R is a domain if and only if R has no homogeneous zero divisors.

One of the immediate consequences of the above proposition is the following

5. Corollary. If P is a graded ideal of R with the property: for any two graded ideals I and J of R, $(I^2 \subseteq P)$ $IJ \subseteq P$ implies $(I \subseteq P)$ $I \subseteq P$ or $J \subseteq P$, then P is a (semiprime) prime ideal (in other words, for a graded ideal there is no essential difference between the notions of (semiprime) prime ring and gr-(semiprime) prime ring).

6. Lemma. Let R be a strongly (ordered group) G-graded ring. Suppose that R_e is a (semiprime) prime ring, where e is the neutral element of G, then

(1). R is a (semiprime) prime ring;

(2). $R^+ = \oplus_{g \geq e} R_g$ and $R^- = \oplus_{g \leq e} R_g$ are (semiprime) prime rings.

Proof. In view of Corollary 5. we only need to consider graded ideals. But since R is strongly graded any graded ideal I of R is of the form RI_e (I_eR), where $I_e = I \cap R_e$. Hence (1) follows immediately from our assumption. Now, let I and J be two graded ideal of R^+ such that $IJ = 0$. Suppose $I_g = I \cap R_g \neq 0$ for some $g \in G$. Then $R_{g^{-1}}I_g J_\rho R_{\rho^{-1}} = 0$ for all $\rho \in G$. Note that $R_{g^{-1}}I_g$ and $J_\rho R_{\rho^{-1}}$ are ideals of R_e, it follows that $J_\rho R_{\rho^{-1}} = 0$ and hence $J_\rho = 0$ because R is strongly graded and $I_g \neq 0$. This shows that $J = 0$, hence R^+ is a prime ring. In a similar way one may prove that R^- is a prime ring. $\square$

7. Lemma. Let R be a filtered ring with separated filtration FR. If $G(R)$ is a (semiprime) prime ring then R is a (semiprime) prime ring. Consequently $\tilde{R}$ is a (semiprime) prime ring.

Proof. If I and J are two ideals of R such that $IJ = 0$ then $G(I)G(J) = 0$, where I and J have their respective filtrations induced by FR. Thus $G(I) = 0$ or $G(J) = 0$, and therefore $I = 0$ or $J = 0$ since FR is separated. That $\tilde{R}$ is a (semiprime) prime ring follows from the fact that $R = \tilde{R}/(1 - X)\tilde{R}$ and the ideal $(1 - X)$ does not contain nonzero homogeneous elments of $\tilde{R}$. $\square$

8. Corollary. Let R be a ring and I an invertible ideal of R with inverse I^{-1} in some overring S of R. Suppose that R/I is a (semiprime) prime ring and $\cap_{n\geq 1} I^n = 0$. With notation as given in Ch.I. Example 4.3.9., then

(1). $G(A) \cong \check{R}(I)/Y\check{R}(I)$ is a (semiprime) prime ring, and hence A is a (semiprime) prime ring, where $A = \check{R}(I)/(1-Y)\check{R}(I) \cong S(I) = \cup_{n\in\mathbb{Z}} I^n \subseteq S$;

(2). $G_I(R) = \oplus_{n\geq 0} I^n/I^{n+1} \cong \oplus_{n\leq 0} G(A)_n = G(A)^-$ is a (semiprime) prime ring, and hence R is a (semiprime) prime ring;

(3). the generalized Rees ring $\check{R}(I)$ of R with respect to $S(I)$ and the Rees ring $\widetilde{R}$ of R with respect to the I-adic filtration on R are (semiprime) prime rings.

9. Theorem. ([VVO]) Let $A = \oplus_{n\in\mathbb{Z}} A_n$ be a $\mathbb{Z}$-graded left Noetherian ring and assume that X is a central regular homogeneous element of positive degree contained in the graded Jacobson radical $J^g(A)$ of A. If A/XA is a gr-maximal order in a gr-simple gr-Artinian ring then A is a gr-maximal order in a gr-simple gr-Artinian ring.

Proof. By the assumptions AX is an invertible prime ideal of A, it follows from Corollary 3.2.8. above that A is a prime ring. In view of Lemma 3.2.1. the ring A has a gr-simple gr-Artinian ring of (homogeneous) fractions, say Q^g, and A is a graded order in Q^g. Consider a graded ideal P of A and let $q \in Q^g$ be an homogeneous element such that $qP \subseteq P$. We may assume that P is chosen to be maximal with respect to the latter property. Then we claim that $P \cap AX = PX$ since $P \subseteq \{a \in A, aX \in P\} = (P : X)_A$ and $q(P : X)_A \subseteq (P : X)_A$. Moreover, note that $X \in J^g(A)$ and P is a finitely generated graded ideal of A the graded Nakayama's lemma yields $P \not\subseteq AX$.

Next, the homogeneous regular elements of A/AX form an Ore set because A/AX is a gr-maximal order. We write $C^g(X)$ for the homogeneous multiplicatively closed set consisting of pre-images of homogeneous regular elements in A/AX. Then from $AX \subseteq J^g(A)$ it is easily deduced that A satisfies the Ore conditions with respect to $C^g(X)$ (this is the obvious graded version of Theorem 1.3. in [CG]). If for some $d \in C^g(X)$ and $y \in A$ we have $dy = 0$ then $y \in AX$, say $y = y'X$, follows. By applying similar argumentation to $dy' = 0$, etc $\cdots$, we arrive at the conclusion $y \in \cap_{n\geq 1} AX^n = 0$ by Lemma 3.2.3.. This shows that $C^g(X)$ consists of regular elements of A.

Now, observe that $\overline{P} = P \bmod AX$ is a nonzero homogeneous ideal and thus there exists a nonzero homogeneous element $d \in P \cap C^g(X)$. But then $qd = a \in P$. Applying the second Ore condition to d and a we have $cqd = ca = rd$ for some $c \in C^g(X)$, $r \in A$. The regularity of d entails $cq = r$ and thus $rP = cqP \subseteq cP$. Let $\pi \colon A \to A/AX$ be the canonical epimorphism. From $\pi(r)\pi(P) \subseteq \pi(c)\pi(P)$ and the regularity of $\pi(c)$ we may derive that $\pi(c)^{-1}\pi(r)\pi(P) \subseteq \pi(P)$ where $\pi(c)^{-1}\pi(r) \in Q^g(A/AX)$ and the latter one is the gr-simple gr-Artinian ring of (homogeneous) fractions of A/AX. Consequently $\pi(r) \in \pi(c)\pi(A)$ (note that A/AX is a maximal order in $Q^g(A/AX)$) or $r \in cA + AX$. If $r \in cA + AX^k$ for some $k \geq 1$, say $r = cb + sX^k$, then for each $a \in P$ we obtain: $ra - saX^k \in cP$ and $ra = cqa \in cP$ implies $saX^k \in cP$. Hence $saX^k \in cP \cap AX^k$. We claim that $cP \cap AX^k = cPX^k$. Indeed if $cp \in cP \cap AX^k$, say $cp = wX^k$ then $c \in C^g(X)$ entails that $p = p_1X$ for some $p_1 \in P$ (note that $P \cap AX = PX$). From $cp_1 = wX^{k-1}$ we go on to $cp_2 = wX^{k-2}$ for some $p_2 \in P$ and finally we will arrive at $cp_k = w$, hence $cp = cp_kX^k \in cPX^k$ as claimed. But then $saX^k \in cPX^k$ yields $sP \subseteq cP$. Again from $\pi(s)\pi(P) \subseteq \pi(c)\pi(P)$ we deduce $s \in cA + AX$ and thus $r \in cA + s'X^{k+1}$ for some $s' \in A$. We can now repeat the argument untill we decide that $r \in \cap_{n \geq 1}(cA + AX^n) = cA$, by Lemma 3.2.3.. Finally, $cq = r \in cA$ entails $q \in A$ because c is a regular element of A. This settles the problem in case q was homogeneous. In case q is not homogeneous and $qP \subseteq P$ we may decompose $q = q_{\alpha_1} + \cdots + q_{\alpha_n}$ with $q_{\alpha_i} \in Q^g_{\alpha_i}$, $\alpha_i \in \mathbb{Z}$ and $\alpha_1 < \cdots < \alpha_n$. Since P is a graded ideal $q_{\alpha_i}P \subseteq P$ holds for $i = 1, \cdots, n$ and we apply the foregoing part of the proof to the homogeneous components q_{α_i} to conclude that $q_{\alpha_i} \in A$ for $i = 1, \cdots, n$. Therefore $q \in A$ also follows in this case and the theorem has been proved. □

10. Proposition. Let $A = \oplus_{n \in \mathbb{Z}} A_n$ be a strongly $\mathbb{Z}$-graded ring. Suppose that A_0 is a Goldie prime ring. Then

(1). A_0 is a maximal order in a simple Artinian ring if and only if A is a gr-maximal order in a gr-simple gr-Artinian ring;

(2). if A_0 is a Noetherian maximal order in a simple Artinian ring then the graded subring $A^+ = \oplus_{n \geq 0} A_n$ resp. $A^- = \oplus_{n \leq 0} A_n$ of A is a gr-maximal order in a gr-simple gr-Artinian ring.

Proof. (1). Let $Q(A_0)$ resp. $Q^g(A)$ be the ring of fractions of A_0 resp. the graded ring of fractions of A, then by ([LeVV] Proposition II.1.4., Proposition II.1.5. or [NVVO]) $Q^g(A)$ is strongly $\mathbb{Z}$-graded and $Q^g(A)_0 = Q(A_0)$. Moreover, since any graded A-module M is generated by its part of degree zero it follows that for any graded ideal I of A we have $(I : I)_{Q^g(A)} = A((I : I)_{Q^g(A)})_0 = A(I_0 : I_0)_{Q(A_0)}$, where $(I : I)_{Q^g(A)} = \{q \in Q^g(A), qI \subseteq I\}$, $(I_0 : I_0)_{Q(A_0)} = \{q \in Q(A_0), qI_0 \subseteq I_0\}$. Hence (1) is clear now.

(2). From (1) we know that A is a gr-maximal order and hence a maximal order by Lemma 3.2.2.. Consider the polynomial ring $A[t]$ over A in t, then $A[t]$ has the natural positive gradation defined by the power of t. It follows from Theorem 3.2.9. and Lemma 3.2.2. that $A[t]$ is a maximal order. Now take the +-gradation on $A[t]$ as defined in Ch.I. Example 4.3.9.(d). then $A[t]$ is strongly $\mathbb{Z}$-graded and $A[t]_0 = A^+$. Therefore A^+ is a maximal order by Lemma 3.2.2., Lemma 3.2.6. and (1). Similarly one may prove that A^- is a maximal order. □

At this point we are able to turn back to the filtered situation.

11. Theorem. Let R be a left Zariski ring with filtration FR such that $G(R)$ is a gr-maximal order in a gr-simple gr-Artinian ring, then R is a maximal order in a simple Artinian ring.

Proof. The Zariskian condition yields that $\tilde{R}$ is left Noetherian and the canonical central regular homogeneous element X of $\tilde{R}$ is contained in $J^g(\tilde{R})$, the graded Jacobson radical of $\tilde{R}$. We may put $A = \tilde{R}$ in the foregoing theorem and so we obtain that $\tilde{R}$ is a gr-maximal order in a gr-simple gr-Artinian ring $Q^g(\tilde{R})$. Consequently $\tilde{R}_{(X)}$ is a gr-maximal order in $Q^g(\tilde{R})$ too. But $\tilde{R}_{(X)} \cong R[t, t^{-1}]$ as graded rings, where the latter one has the gradation defined by the power $t^n, n \in \mathbb{Z}$, of t. Hence R is a maximal order in its ring of fractions by Proposition 3.2.10.. $\square$

12. Corollary. Let A be a ring, σ an automorphism of A and δ a σ-derivation of A. Let R be one of the following rings: the ring of skew polynomials $A[t, \sigma]$, the t-adic completion $A[[t, \sigma]]$ of $A[t, \sigma]$ and the ring of differential skew polynomials $A[t, \sigma, \delta]$. If A is a Noetherian maximal order then R is a Noetherian maximal order.

Proof. With respect to the standard filtration on R we always have $A[t, \sigma]$ as the associated graded ring. It follows from Theorem 3.2.11. that we only need to show that $A[t, \sigma]$ is a maximal order. But this follows immediately from Proposition 3.2.10. because $A[t, \sigma]$ is the positive part of the strongly $\mathbb{Z}$-graded ring $A[t, t^{-1}, \sigma]$. $\square$

13. Theorem. ([PS2]) Let A be a Noetherian ring and P an invertible ideal of A contained in the Jacobson radical $J(A)$ of A. If A/P is a maximal order then A is a maximal order.

Proof. Since $P \subseteq J(A)$ and A is Noetherian the P-adic filtration FA on A is Zariskian by Theorem 2.2.13.. Moreover, $G_I(A)$ is the positive part of the strongly $\mathbb{Z}$-graded ring $\oplus_{n \in \mathbb{Z}} P^n/P^{n+1}$, it follows from Proposition 3.2.10. that $G_I(A)$ is a maximal order. Hence A is a maximal order by Theorem 3.2.11.. $\square$

§4. Zariskian Filtrations on Simple Artinian Rings

This section is devoted to a full generalization of Example 2.2.c. in the non-commutative case.

Let $A = \oplus_{g \in G} A_g$ be a (group) G-graded ring. Recall from [NVO1] that the notions gr-semisimple, gr-simple and gr-Artinian may be defined just as in the ungraded case but now in terms of graded modules, graded ideals and homogeneous elements, for example, a gr-skewfield is a graded ring such that each nonzero homogeneous element is invertible. A gr-field is necessarily of the form $k[t, t^{-1}]$ where k is a field and t is a variable, say $\deg(t) = s$. A gr-skewfield Δ is of the form $\Delta = \Delta_0[y, y^{-1}, \varphi]$ where Δ_0 is a skewfield, y is a variable, of degree ρ say, and φ is an automorphism of Δ_0 such that $y\lambda = \varphi(\lambda)y$ holds for all $\lambda \in \Delta_0$. If we assume moreover that Δ satisfies a polynomial identity then φ has finite order modulo the inner automorphisms of Δ_0 and then $Z(\Delta) = k[y^n, y^{-n}]$, where $Z(\Delta)$ is the centre of Δ and n is a positive integer.

4.1. Proposition. (cf. [NVO1]) If A is a gr-simple gr-Artinian ring then $A \cong M_n(\Delta)(\underline{d})$ as $\mathbb{Z}$-graded rings, where Δ is a graded skewfield and $\underline{d} \in \mathbb{Z}^n$ determines the gradation on A by the rule: $A_n = (\Delta_{n+d_i-d_j})_{ij}$. If A is uniformly simple (see the latter definition) then $A_0 \cong M_n(\Delta_0)$ is also simple (not just semisimple as it general is!) Artinian.

We call a filtered ring R with filtration FR **strongly filtered** if $F_n R F_m R = F_{n+m} R$ for all $n, m \in \mathbb{Z}$. By using the formula $\tilde{R}/X\tilde{R} \cong G(R)$ one easily derives: R is strongly filtered if and only if $G(R)$ is strongly graded. Then it turns out that for some integer e the condition $F_{ne} R F_{me} R = F_{n+m} eR$ is equivalent to $G(R)_{ne} G(R)_{me} = G(R)_{(m+n)e}$. This follows by defining $F'_n R = F_{ne} R$, $\tilde{R}' = \tilde{R}^{(e)} = \oplus_{n \in \mathbb{Z}} \tilde{R}_{ne}$, $G(R)' = \oplus_{n \in \mathbb{Z}} G(R)_{ne}$. If $G(R)_m = 0$ for $m \notin e\mathbb{Z}$ then we have the situation where $G(R)$ is strongly $e\mathbb{Z}$-graded; the filtration FR then satisfies $F_{ne} R = F_{ne+1} R = \cdots = F_{(n+1)e-1} R$ and $F_{(n+1)e} R \neq F_{ne} R$ for all $n \in \mathbb{Z}$, i.e., the filtration is stepwise with length of the step equal to e. In the ring theoretical sense it will make no difference to replace FR by $F'R$ obtained by $F'_n R = F_{ne} R$ if we are only interested in properties determined by properties of $G(R)$ that are insensitive for this "inflation" of the gradation.

Now, let us see how the results of Example 2.2.c. may be extended to the case of non-commutative rings.

4.2. Proposition. Let R be a left and right Artinian ring with a separated filtration FR. The following statements are equivalent:

(1). $G(R)$ has no (homogeneous) zero divisors;

(2). R is a skewfield and $G(R)$ is a gr-skewfield;

(3). R is a skewfield and $F_0 R$ is a discrete valuation ring of R with the unique maximal ideal $F_{-1} R$; Moreover, there is a positive integer $e \in \mathbb{Z}$ such that $F_{ne} R = (F_{-1} R)^{-n}$, i.e., the filtration is stepwise with step e, as in foregoing remark.

Proof. (1) $\Rightarrow$ (2). By Lemma 2.2.5. R is a domain and hence a skewfield because it is an Artinian ring. If $\sigma(a_n) \in G(R)_n$ then $a_n \in F_n R - F_{n-1} R$ and a_n has an inverse, say a_t with $a_t \in F_t R - F_{t-1} R$. From $a_t a_n = a_n a_t = 1$ it follows that $n + t \geq 0$ because $1 \notin F_{-1} R$ (since FR is separated). If $n + t > 0$ then $\sigma(a_t)\sigma(a_n) = 0$ contradicts the assumptions, hence $t = -n$ and thus $\sigma(a_n)$ has an inverse $\sigma(a_t) \in G(R)_{-n}$ or $G(R)$ is a gr-skewfield.

(2) $\Rightarrow$ (3). From Lemma 2.2.5. it follos that $v: R \to \Gamma = \mathbb{Z} \cap \{\infty\}$ with $v(a) = -n$, whenever $a \in F_n R - F_{n-1} R$, is a valuation function on R. From the definition it is obvious that $F_0 R$ is the corresponding discrete valuation ring of R. If e is the smallest positive number such that $G(R)_e \neq 0$ then it is clear that $G(R)$ is $e\mathbb{Z}$-strongly graded and it follows that $F_{ne} R = (F_{-1} R)^{-n}$ and FR is stepwise with step e.

(3) $\Rightarrow$ (1). Since $F_0 R$ is a discrete valuation ring of R, its unique maximal ideal $M = \pi F_0 R = F_0 R \pi$ for some $\pi \in M$. From $F_n R = (M)^{-n} = \pi^{-n} F_0 R$ it follows that $F_n R F_{-n} R = F_{-n} R F_n R = F_0 R$. Hence $G(R)$ is strongly graded by the previous remark, and then it has no (homogeneous)zero divisors because $G(R)_0 = F_0 R / M$ is a skewfield. $\qquad \square$

4.3. Remark. (1). In case $G(R)$ is trivially graded (we could say $e = \infty$ in this case) then FR is trivial, i.e., $F_{-n} R = 0$, $F_0 R = F_n R$ for all $n > 0$.

(2). Any subring R of a skewfield Δ is a discrete valuation ring if and only if there is an exhaustive and separated filtration $F\Delta$ on Δ with $F_0 \Delta = R$ and $G(\Delta)$ being without homogeneous zero divisors (a consequence of the foregoing proposition).

(3). If R is a filtered ring with separated filtration FR then FR satisfies:

i). $F_{-1} R \subseteq J(F_0 R)$ and

ii). $G(R)$ is a gr-skewfield,

if and only if R is a skewfield and $F_0 R$ is a discrete valuation ring with unique maximal ideal $F_{-1} R$.

The proof of this statement is not difficult, we omit it here.

In the sequel we will turn to the case where $G(R)$ is allowed to have some zero divisors. It is necessary to recall a few notions about gr-semisimple rings; Although many results may be phrased for arbitrary grading groups we restrict to the $\mathbb{Z}$-graded case. A $\mathbb{Z}$-graded ring A is said to be **gr-semisimple** if and only if $A = L_1 \oplus \cdots \oplus L_n$, where each L_i, $i = 1, \cdots, n$, is a minimal graded left ideal of A. We say that A is **gr-simple** if it is gr-semisimple and $\text{Hom}_A(L_i, L_j) \neq 0$ for every $i, j = 1, \cdots, n$, or equivalently: there exist $n_{ij} \in G$ such that

$L_j \cong T(n_{ij})L_i$ (where $T(n_{ij})$ is the shift functor in A-gr associated to $n_{ij} \in \mathbb{Z}$). A gr-semisimple ring A is said to be **uniformly gr-simple** if $A = L_1 \oplus \cdots \oplus L_n$ as before but with $L_i \cong L_j$ in A-gr for any $i, j = 1, \cdots, n$. Note that if A is gr-semisimple (gr-simple) then A_0 is generally (only) semisimple; however if A is gr-uniformly simple then A_0 is a simple ring (see [NVO1] Remarks I.5.9.).

Recall that a $\mathbb{Z}$-graded ring A satisfies condition (E) (see Lemma 3.2.1.) if for every $a_n \neq 0$ in A_n we have $A_{-n}a_n \neq 0$, or equivalently: if every nonzero graded left ideal of A intersects A_0 nontrivially. In case A_0 is a semiprime ring then property (E) is left-right symmetric. It is also clear that a minimal graded left ideal L of A is necessarily the form $L = Av$ for some idempotent element v of A_0. Put $L_i = Av_i$, $i = 1, \cdots, n$, then it follows from $A = Av_1 \oplus \cdots \oplus Av_n$ that $A_0 = A_0 v_1 \oplus \cdots \oplus A_0 v_n$ and every $A_0 v_i$ is a minimal left ideal of A_0. Observing that a nonzero graded left ideal in a gr-Artinian ring necessarily contains a minimal graded left ideal, so it is straightforward to establish the following proposition.

4.4. Proposition. For a $\mathbb{Z}$-graded ring A the following statements are equivalent:

(1). The garadation satisfies condition (E), A is gr-Artinian and A_0 is a semiprime ring;

(2). A is gr-semisimple gr-Artinian.

Proof. (1) $\Rightarrow$ (2). cf. [NNVO]; (2) $\Rightarrow$ (1). Easy enough. □

Let A be a gr-semisimple gr-Artinian ring, say $A = Av_1 \oplus \cdots \oplus Av_n$, then for every $i, j \in \{1, \cdots, n\}$ we have: $v_i Av_i \cong \mathrm{Hom}_A(Av_i, Av_i)$, $v_i Av_j \cong \mathrm{Hom}_A(Av_i, Av_j)$.

Let us conclude these general remarks concerning gr-semisimplicity by providing a list of conditions equivalent to the uniform gr-simplisity.

4.5. Proposition. With notations as before, the following statements are equivalent for a gr-semisimple gr-Artinian ring A:

(1). A_0 is simple;

(2). A_0 is prime;

(3). If $A_m \neq 0$ for some $m \in \mathbb{Z}$ then $v_i A_m v_j \neq 0$. In particular for every $i, j \in \{1, \cdots, n\}$ we have $v_i A_0 v_j \neq 0$;

(4). The ring A is uniformly gr-simple;

(5). The ring A is gr-simple and if $A_m \neq 0$ for some $m \in \mathbb{Z}$ then there exists a minimal graded left ideal L of A such that $(L_m)^2 \neq 0$ (assuming that A has nontrivial gradaion);

(6). By the graded version of the Wedderburn theorem (cf. [NVO1]) we know that $A = M_n(\Delta)(\underline{d})$ where Δ is a gr-skewfield and $\underline{d} \in \mathbb{Z}^n$. Now, if $A_m \neq 0$ for some $m \in \mathbb{Z}$ then $\Delta_m \neq 0$.

Proof. Most of the statements appear in [NVO1]. For the implications $(4) \Rightarrow (5) \Rightarrow (6)$ we may refer to [LVO2]. □

We now return to the setting of filtered rings: first with a general lemma preserving the taste of the foregoing statements.

4.6. Lemma. Let R be a filtered ring with separated filtration FR, then the following properties hold:

(1). If $G(R)$ satisfies condition (E) then $a_n \in F_n R - F_{n-1} R$ implies $F_{-n} R a_n \neq 0$.

(2). If $G(R)$ is $e\mathbb{Z}$-strongly graded for some $e \in \mathbb{N}$ and if $G(R)_0$ is a (semi-)prime ring then $F_0 R$ is a (semi-)prime ring.

Proof. (1). Suppose that $\bar{c} \in G(R)_{-n}$ is such that $\bar{c}\sigma(a_n) \neq 0$. Then $ca_n \neq 0$ for some $c \in F_{-n} R - F_{-n-1} R$ such that $\sigma(c) = \bar{c}$, hence $F_{-n} R a_n \neq 0$.

(2). Consider $F_0 R$ as a filtered ring with the filtration $F(F_0 R)$ induced by FR, then $F(F_0 R)$ is also separated. Moreover, since $G(F_0 R) = G(R)^-$ it follows from Lemma 3.2.6. and Lemma 3.2.7. that $F_0 R$ is a (semi-)prime ring. □

At this point we need the P.I. assumption in order to strengthen our case.

4.7. Proposition. Let R be a filtered ring with separated filtration FR. The following statements are equivalent:

(1). R is left and right Artinian, $F_0 R$ is a prime P.I. ring, $G(R)$ satisfies condition (E) and $G(R)_0$ is a prime ring;

(2). R is a central simple algebra and $G(R)$ is a uniformly graded central simple algebra.

Proof. $(1) \Rightarrow (2)$. By Posner's theorem (cf. [P]) the classical ring of fractions of the prime P.I. ring $F_0 R$ is obtained by central localization, i.e., $F_0 R Q(Z(F_0 R)) = \Delta_1$ is a central simple algebra. If a_0 is a regular element of $F_0 R$ then it is also a regular element of R (if $a_0 b = 0$ with $b \in R$, say $b \in F_t R - F_{t-1} R$, then $a_0 b F_{-t} R = 0$ with $b F_{-t} R \neq 0$ in $F_0 R$ would yield a contradiction in view of Lemma 4.6. above) and therefore $\Delta_1 \in R$ because R is (left and right) Artinian. If $F_{-n} R \neq 0$ with $n \geq 0$ then $F_n R F_{-n} R$ is an ideal of $F_0 R$. Hence $\Delta_1 = F_n R F_{-n} R \Delta_1 = F_n R (F_{-n} R \Delta_1) = F_n R \Delta_1$, thus $F_n R \subseteq \Delta_1$ for all $n \geq 0$ and $R = \Delta_1$ follows. It remains to establish that $G(R)$ is a uniformly graded central simple algebra. First, since $G(R)_0$ is prime and $G(R)$ satisfies condition (E) we know that $G(R)$ is a prime ring. Since $\tilde{R} \subseteq \tilde{R}_{(X)} \cong R[t, t^{-1}]$ and $G(R) \cong \tilde{R}/X\tilde{R}$ it follows that $G(R)$ is a prime P.I. ring so it will be a graded central simple algebra if $Z(G(R))$ is a gr-field (cf. [NVO1]). If $\bar{c} \in h(Z(G(R)))$, the set of homogeneous elements of $Z(G(R))$, then let $\bar{c} = \sigma(c)$ with

$c \in F_t R - F_{t-1} R$; Clearly c is regular in R because $\bar{c}$ is regular in $G(R)$. Hence $c^{-1} \in R$ and as $\bar{c}$ is regular in $G(R)$ we know that $c^{-1} \in F_{-t} R - F_{-t-1} R$ and $\sigma(c^{-1})$ is an inverse for $\sigma(c) = \bar{c}$. From Proposition 4.5. it follows that $G(R)$ is uniformly graded simple because $G(R)_0$ is prime.

$(2) \Rightarrow (1)$. A combination of Proposition 4.4., Proposition 4.5. and Lemma 4.6.. $\qquad \square$

4.8. Lemma. Let R and FR be as in Proposition 4.7.. Suppose that R is left and right Artinian, $F_0 R$ is a prime (left) Goldie ring and $G(R)$ satisfies condition (E) over the prime ring $G(R)_0$. Then $F_0 R$ is a maximal order in the simple Artinian ring R.

Proof. Again our assumptions imply that a regular element of $F_0 R$ is also a regular element in R and hence a unit of R. Therefore $F_0 R$ is a maximal order in a simple Artinian ring $Q \subseteq R$. If $x \in R - Q$, say $x \in F_n R - F_{n-1} R$, then $n > 0$ and $F_{-n} R x \subseteq F_0 R \subseteq Q$ and $Q F_{-n} R x \subseteq Q$. Since $F_{-n} R$ is a two-sided ideal of the prime Goldie ring $F_0 R$ it contains a regular element c of $F_0 R$, i.e., a unit of Q and thus $Q F_{-n} R = Q$ holds for all $n \geq 0$. From $Q F_{-n} R x \subseteq Q$ we then derive $x \in Q$ and consequently $R = Q$ as desired. $\qquad \square$

4.9. Theorem. Let R be a filtered ring with separated filtration FR. Suppose that R is left and right Artinian, $G(R)$ is $e\mathbb{Z}$-strongly graded over the Noetherian prime ring $G(R)_0$ and $F_0 R$ is a P.I. ring. Then the following properties hold:

(1). $F_0 R$ is a (local) classical maximal order over the discrete valuation ring $Z(F_0 R) \subseteq K = Z(R)$;

(2). R is a left and right Zariski ring.

Proof. By Lemma 4.6. $F_0 R$ is a prime ring. Since $G(R)$ is $e\mathbb{Z}$-strongly graded over the prime Noetherian ring $G(R)_0$ we obtain that $G(R)$ is Noetherian too. If $\bar{c}$ is a regular homogeneous element of $G(R)$ then $c \in F_n R - F_{n-1} R$ with $\sigma(c) = \bar{c}$ is regular too hence invertible in R. Moreover, one easily checks that $\deg \sigma(c^{-1}) = -n$ (see also the final part of the proof of $(1) \Rightarrow (2)$ in Proposition 4.7.) and that $\bar{c}$ is therefore invertible in $G(R)$. However, from [NNVO] we recall the following: if $G(R)$ satisfies condition (E) then all regular homogeneous elements of a semiprime Noetherian ring $G(R)$ are invertible if and only if $G(R)$ is gr-semisimple (indeed $G(R)$ coincides with its classical graded ring of fractions!). Clearly we are in this case because $G(R)$ is $e\mathbb{Z}$-strongly graded over the prime Noetherian ring $G(R)_0$. Hence $G(R)$ is certainly Noetherian semiprime, hence gr-semisimple by the preceding remark. From Proposition 4.5. it follows that $G(R)$ is in fact uniformly gr-simple (hence also prime) and this shows that we are in the situation where the equivalent conditions of Proposition 4.7. hold.

(1). In the proof of Proposition 4.7. we established that $F_0 R Q(Z(F_0 R)) = R$, hence

$Q(Z(F_0R)) = K = Z(R)$ and $Z(F_0R) \subseteq K = Z(R)$. For the filtration FK induced on K by FR it follows that the associated graded ring $G(K)$ is a subring of $Z(G(R))$ hence a domain. Theorem 2.2.7. now yields that $F_0K = F_0R \cap K$ is a discrete valuation ring of K and $G(K)$ is of course strongly $e_1\mathbb{Z}$-graded for some $e_1 \in \mathbb{N}$. Again from $R = KF_0R$ we may deduce that $Z(F_0R) = F_0R \cap K = F_0K$. Now by Lemma 4.6. F_0R is a prime P.I. ring with a Noetherian (even a discrete valuation ring) centre (cf. [M-R] P.467, Corollary 6.14.) hence by ([EF] Theorem 2.) F_0R is a finitely generated F_0K-module hence certainly an F_0K-order in R. Let $\mathcal{C}(F_{-1}R) = S = \{c \in F_0R, c$ is regular modulo $F_{-1}R$ i.e., its image $\bar{c}$ in $G(R)_0$ is regular$\}$. If $c \in S$ then $\bar{c}$ is regular in $G(R)$ too; Indeed if $\bar{c}\bar{y} = 0$ for some nonzero $\bar{y} \in G(R)_n$ (it suffices to check homogeneous elements) then $\bar{c}\bar{y}G(R)_{-n} = 0$ with $\bar{y}G(R)_{-n} \neq 0$, contradicts the regularity of $\bar{c}$ in $G(R)_0$ (similarly on the left). But then c is regular in R hence invertible in R, say $c^{-1} \in F_pR - F_{p-1}R$ for some $p \in \mathbb{Z}$. From $1 = cc^{-1}$ it follows that $p \geq 0$ (because $1 \notin F_{-1}R$) and then $p = 0$ follows from the fact that $p > 0$ leads to a contradiction $\bar{c}\sigma(c^{-1}) = 0$. Thus we have shown that the set S consists of invertible elements of F_0R and therefore F_0R equals $S^{-1}F_0R$. Consequently F_0R is "local" (it is the localization at the prime ideal $F_{-1}R$) and $F_{-1}R = J(F_0R)$, the Jacobson radical of F_0R.

To prove that F_0R is a maximal order over F_0K in R, let I be a nonzero ideal of F_0R, say $I \subseteq F_{-k}R$, $I \nsubseteq F_{-k-1}R$ for some $k \geq 0$. If $a_n \in F_nR - F_{n-1}R$ is such that $a_nI \subseteq I$ with $n \geq 0$ then we claim that $n = 0$. Indeed, write $I_{(k)}$ for $(I + F_{-k}R)/F_{-k-1}R$. Then we have $\sigma(a_n)I_{(k)} \neq 0$ (otherwise $G(R)_{-n}\sigma(a_n)I_{(k)}G(R)_k = 0$ with $G(R)_{-n}\sigma(a_n) \neq 0$ and $I_{(k)}G(R)_k \neq 0$ contradicts the fact that $G(R)_0$ is prime). Pick $i \in I - F_{-k-1}R$ such that $\sigma(a_n)\sigma(i) \neq 0$ then from $a_ni \in I \subseteq F_{-k}R$ we obtain that $\sigma(a_ni)$ has degree at most $-k$. But on the other hand $\sigma(a_n)\sigma(i) = \sigma(a_ni)$ because $\sigma(a_n)\sigma(i) \neq 0$ is in $G(R)_{n-k}$. Consequently, $0 \neq \sigma(a_n)\sigma(i) \in G(R)_{-k} \cap G(R)_{n-k}$, but that is only possible when $n = 0$. Hence we have shown for all ideals I of F_0R that $F_0R = (I : I)_R = \{a \in R, aI \subseteq I\}$ and it is then well known that F_0R is a maximal order in R.

(2). We have established that $F_{-1}R = J(F_0R)$ in (1), and also we have seen that F_0R is in particular Noetherian (as it is a finitely generated module over its centre F_0K). Since $G(R)$ is $e\mathbb{Z}$-strongly graded it follows that FR is $e\mathbb{Z}$-strongly filtered. But then $F_{-1}R = F_{-e}R$ is now an invertible ideal so we may conclude that $\tilde{R}$ is $e\mathbb{Z}$-strongly graded and hence Noetherian. Therefore R is a left and right Zariski ring by definition. $\square$

From the foregoing results we have seen that it is essentially the graded version of Goldies theorem that we needed. So we may obtain a similar result in the non-P.I. case replacing the P.I. condition by the condition that $G(R)$ is positively graded. However this restriction is somewhat unnatural in the context of our present work, so we look for a good condition on $G(R)_0$ to make all tricks work.

4.10. Theorem. Let R be a left and right Artinian ring with separated filtration FR. Assume that $G(R)$ satisfies condition (E) and $G(R)_0$ is a prime left Goldie ring, then we have:

(1). $G(R)$ is uniformly gr-simple and $G(R)_0$ is simple;

(2). $F_{-1}R = J(F_0R)$ and F_0R is "local" in the sense of [Go];

(3). If furthermore F_0R is Noetherian then FR is Zariskian.

Proof. Put $S = \{a \in R, \ \sigma(a) \text{ is regular in } G(R)\}$, $\sigma(S) = \{\sigma(s), \ s \in S\}$ and $\sigma(S)_0 = \{\sigma(a) \in G(R)_0, \ \sigma(a) \text{ is regular in } G(R)_0\}$.

(1). Recall from [NNVO] that $\sigma(S)^{-1}G(R)$ is gr-simple with $(\sigma(S)^{-1}G(R))_0 = \sigma(S)_0^{-1}G(R)_0$. From Proposition 4.5. we infer that $\sigma(S)^{-1}G(R)$ is uniformly gr-simple and hence it is strongly $e\mathbb{Z}$-graded for some $e \in \mathbb{N}$. In view of Lemma 2.2.5. S is a set of units in R since R is Artinian and consequently, as $\sigma(S)$ consists of regular elements in $G(R)$, $\sigma(S)$ will consist of units in $G(R)$. Hence $G(R) = \sigma(S)^{-1}G(R)$ is uniformly gr-simple and $G(R)_0$ is a simple ring.

(2). Let $a \in F_{-1}R$, then for all $c \in F_0R$, $\sigma(1 - ca) = (1 - ca) \mathrm{mod} F_{-1}R$ is in $\sigma(S)_0$. From the proof of (1) it follows that $1 - ca$ is invertible in F_0R. Hence $F_{-1}R \subseteq J(F_0R)$ is clear. But then $F_{-1}R = J(F_0R)$ is also clear because $G(R)_0 = F_0R/F_{-1}R$ is simple. From the fact that $G(R)$ is strongly $e\mathbb{Z}$-graded, for some $e \in \mathbb{N}$, it follows that $F_{-n}R = (F_{-1}R)^n$ (up to contracting the filtration as before) for all $n \geq 0$, hence $\cap_{n \geq 0}(F_{-1}R)^n = \cap_{n \geq 0}F_{-n}R = 0$ because FR is separated. Therefore we may say that F_0R is a "local" ring in the sense of [Go].

(3). By (2) it remains to prove that $\tilde{R}$ is Noetherian. The fact that $G(R)$ is strongly $e\mathbb{Z}$-graded implies that R is strongly filtered and hence $\tilde{R}$ is strongly $e\mathbb{Z}$-graded. Hence $\tilde{R}$ will be Noetherian if $F_0R = \tilde{R}_0$ is Noetherian. $\qquad\square$

§5. Global Dimension of Rees Rings Associated to Zariskian Filtrations

The aim of this section is to investigate the (homological) regularity and global dimension of the Rees ring associated to a left Zariskian filtration. At this stage certain graded homological properties will also be discussed.

5.1. Addendum to graded homological algebra

Although homological dimensions of graded rings have been investigated in some detail, e.g. in [NVO1] Ch.I., the results we need in the sequel seem to be missing in the literature so we took care to expound the theory in some detail here. For esthetical reasons only we allowed gradations by arbitrary groups.

Let $A = \oplus_{g \in G} A_g$ be a (group) G-graded ring. Then Lemma 4.1.2. from Ch.I. first yields:

1. Proposition. Let P be a graded A-module. Then P is gr-projective if and only if $\mathrm{HOM}_A(P, -)$ is an exact functor in A-gr, if and only if $\mathrm{Hom}_A(P, -)$ is an exact functor in A-mod.

In view of the fact that the category A-gr has enough projective objects we may provide graded versions of several results appearing in [No], i.e., Section 7.5., Proposition 3, Lemma 1., Theorem 11., 12. and 13., Section 7.6., Lemma 2., Theorem 15.. Most of these results may be obtained by straightforward "grading" of the proofs given in loc. cit. so we only provide the proofs for two statements, where a few more delicate points of graded nature do appear.

2. Proposition. (1). An object $P \in A$-gr is projective if and only if $\mathrm{Ext}^1_A(P, C) = 0$ for all graded left A-modules, if and only if $\mathrm{Ext}^1_A(P, C) = 0$ for all left A-modules.
(2). Let $M \in A$-gr be such that gr.p.dim$_A M < n$ for some $n \geq 0$, then if $s \geq n$ we have that $\mathrm{Ext}^s_A(M, C) = 0$ for all A-modules C.
(3). Let $M \in A$-gr and $n \in I\!N$ and let

$$0 \longrightarrow Q \longrightarrow P_{n-1} \longrightarrow \cdots \longrightarrow P_0 \longrightarrow M \longrightarrow 0$$

be an exact sequence in A-gr such that each P_j is graded projective. If gr.p.dim$_A M \leq n$ then Q is also graded projective.
(4). Let $M \in A$-gr, $n \in I\!N$. Then gr.p.dim$_A M < n$ if and only if $\mathrm{Ext}^n_A(M, C) = 0$ for all graded left A-modules, if and only if $\mathrm{Ext}^n_A(M, C) = 0$ for all left A-modules.
(5). Let

$$0 \longrightarrow M' \longrightarrow P \longrightarrow M \longrightarrow 0$$

be an exact sequence in A-gr, and assume that P is gr-projective. If M is not graded projective then gr.p.dim$_A M = 1+$ gr.p.dim$_A M'$, taking into account that both members in the equality may be infinite. On the other hand if M is gr-projective too, then so is M'. In either case we have: gr.p.dim$_A M \geq 1+$ gr.pdim$_A M'$.

(6). If

$$0 \longrightarrow M_1 \longrightarrow M \longrightarrow M_2 \longrightarrow 0$$

is an exact sequence of graded left A-modules then gr.p.dim$_A M_2 \leq 1+$ max$\{$gr.p.dim$_A M,$ gr.p.dim$_A M_1\}$.

Proof. Exercise, following the ungraded situation as explained in [No]. □

3. Proposition. (1). Let $\{A_i, \ i \in I\}$, $\{B_j, \ j \in I\}$ be families of graded left A-submodules of M, and assume that I is well ordered (by $\leq$ say) such that the following conditions hold:

a. $B_i \subseteq A_i$ for every $i \in I$;

b. If $i < j$, then $A_i \subseteq B_j$;

c. If $x \neq 0$ in B_j, then $x \in A_i$ for some $i < j$, in other words $B_j = \cup_{i<j} A_i$;

d. For all $i \in I$ we have: gr.p.dim$_A(A_i/B_i) \leq n$.

Then the smallest graded A-submodule of M containing all A_i has graded projective dimension at most equal to n.

(2). l.gr.gl.dim$A =$ sup$\{$l.p.dim$_A M, M$ a cyclic graded left A-module$\}$, i.e., in the latter set M ranges over the graded left A-modules of type Ax for some homogeneous $x \in M$.

Proof. First observe the following properties:

(i). If $i \leq j$ then $A_i \subseteq A_j$, $B_i \subseteq B_j$;

(ii). We may suppose without loss of generality that $M = \cup_{i \in I} A_i$;

(iii). From condition c. it follows that $B_j = \cup_{i<j} A_i$, so if i_0 is the first element of I then $B_{i_0} = 0$.

The proof now continues by induction on the integer n. First suppose that $n = 0$, then A_i/B_i is graded projective (hence projective in A-mod too) for each i; the exactness of the sequence

$$0 \longrightarrow B_i \longrightarrow A_i \longrightarrow A_i/B_i \longrightarrow 0$$

then yields $A_i = B_i \oplus C_i$ where $C_i \cong A_i/B_i$. The proof in the case $n = 0$ will be complete if we establish that $M = \oplus_{i \in I} C_i$. Put $M' = \sum_{i \in I} C_i$ and suppose that $M' \neq M$. Since we assumed $M = \cup i \in I A_i$ we can select a first i such that there is an $x \in A_i - M'$. Write $x = y + c_i$ with $y \in B_i$, $c_i \in C_i$. By condition c. we have that $y \neq 0$ and $y \in A_j$ for some $j < i$. But then $y \in M'$ and the same follows for x, a contradiction.

Next, assume that $\sum_{j \in I} c_j = 0$ with $c_j \in C_j$ and c_j being zero for almost all j. If some c_j are nonzero we may write $0 = c_{j_1} + c_{j_2} + \cdots + c_{j_s}$, with $j_1 \leq j_2 \leq \cdots \leq j_s$ and each $c_{j_v} \neq 0$. Using condition b. we have $c_{j_1} + \cdots + c_{j_{s-1}} \in B_{j_s}$, so from the fact that $B_{j_s} + C_{j_s}$ is a direct sum we obtain that $c_{j_s} = 0$, a contradiction. Thus we have established statement

(1). in the case $n = 0$. For the induction we now suppose $n > 0$ and the statement holds for all smaller values of the induction variable. Since A_i and B_i are graded we may construct a graded free A-module $L = \oplus\{Au_{m_\sigma}, m_\sigma \in M_\sigma, \sigma \in G\}$ with gradation given by: $L_\tau = \oplus\{A_{\tau\sigma^{-1}}u_{m_\sigma}, m_\sigma \in M_\sigma, \sigma \in G\}$ for $\tau \in G$ and the graded free A-modules: $F_i = \oplus\{Au_{m_\sigma}, m_\sigma \in (A_i)_\sigma, \sigma \in G\}$ with gradation given by: $(F_i)_\beta = \oplus\{A_{\beta\sigma^{-1}}u_{m_\sigma}, m_\sigma \in (A_i)_\sigma, \sigma \in G\}$ for $\beta \in G$, and $T_i = \oplus\{Au_{m_\sigma}, m_\sigma \in (B_i)_\sigma, \sigma \in G\}$ with gradation given by $(T_i)_\vartheta = \oplus\{A_{\vartheta\sigma^{-1}}u_{m_\sigma}, m_\sigma \in (B_i)_\sigma, \sigma \in G\}$ for $\vartheta \in G$. So we have $T_i \subseteq F_i \subseteq L$ as graded A-modules, and moreover we have the following exact sequence in A-gr:

$$L \xrightarrow{\alpha} M \longrightarrow 0 \qquad F_i \xrightarrow{\alpha'_i} A_i \longrightarrow 0 \qquad T_i \xrightarrow{\alpha''_i} B_i \longrightarrow 0$$

where α'_i and α''_i are restrictions of α to F_i and T_i respectively. Put $N = \mathrm{Ker}\alpha$, $K_i = \mathrm{Ker}\alpha'_i = N \cap F_i$, $E_i = \mathrm{Ker}\alpha''_i = N \cap T_i$. Using the 3×3-Lemma, we arrive at the following commutative diagram:

$$
\begin{array}{ccccccccc}
& & 0 & & 0 & & 0 & & \\
& & \downarrow & & \downarrow & & \downarrow & & \\
0 & \to & E_i & \to & T_i & \to & B_i & \to & 0 \\
& & \downarrow & & \downarrow & & \downarrow & & \\
0 & \to & K_i & \to & F_i & \to & A_i & \to & 0 \\
& & \downarrow & & \downarrow & & \downarrow & & \\
0 & \to & K_i/E_i & \to & F_i/T_i & \to & A_i/B_i & \to & 0 \\
& & \downarrow & & \downarrow & & \downarrow & & \\
& & 0 & & 0 & & 0 & &
\end{array}
$$

Furthermore, F_i/T_i is gr-free by construction and so it follows from Proposition 5.1.2. ((5). or (6).) that: $1+ \mathrm{gr.p.dim}_A(K_i/E_i) = \mathrm{gr.p.dim}_A(A_i/B_i) \leq n$. Consequently $\mathrm{gr.p.dim}_A(K_i/E_i) \leq n-1$.

Now consider the graded submodule N with A_i replaced by K_i and B_i replaced by E_i, $i \in I$, we obtain:

1. $E_i \subseteq K_i$, $i \in I$;
2. If $i < j$ then $K_i \subseteq E_j$;
3. If $x \in E_j$ and $x \neq 0$ then the nonzero coefficients in the expression for x as an element of T_j are associated to a finite number of nonzero elements corresponding to those of B_j considered as basis-elements of T_j. Therefore there exists an $i < j$ such that all the forementioned elements are A_i; hence it follows that $x \in N \cap F_i = K_i$ for this i.
4. $\mathrm{gr.p.dim}_A(K_i/E_i) \leq n-1$, $i \in I$.

Since by construction, $\cup_{i \in I} F_i = L$ (note that $\cup_{i \in I} A_i = M$!) we may conclude by the induction hypothesis that $\mathrm{gr.p.dim}_A N \leq n-1$. Now the proof is finished by the observation

that
$$0 \longrightarrow N \longrightarrow L \stackrel{\alpha}{\longrightarrow} M$$
is (graded) exact in A-gr and L is gr-free hence (again from Proposition 5.1.2.(5)) it follows that gr.p.dim$_A M \leq 1+$ gr.p.dim$_A N \leq n$.

(2). Assume that p.dim$_A M \leq n$ for some finite integer $n \geq 0$, where M varies over the graded cyclic A-modules, i.e., $M = Ax$ for some homogeneous $x \in M$. We have to establish for any graded A-module T that p.dim$_A T \leq \sup\{$p.dim$_A M$, M gr-cyclic$\}$. Let $\{x_i, \ i \in I\}$ be a set of homogeneous generators for T and put a well-ordering on I. For each $i \in I$ let A_i be the graded submodule of T generated by all the x_j with $j \leq i$ and let B_i be generated by x_j with $j < i$. Then we have: $T = \cup\{A_i, \ i \in I\}$, $A_i/B_i = A\overline{x}_i$, $\overline{x}_i = x_i \mathrm{mod} B_i$. Therefore p.dim$_A(A_i/B_i) \leq n$ and hence conditions a. b. c. d. of (1) hold and so p.dim$_A T \leq n$ follows from (1) above, finishing the proof. $\square$

Recall from ([Rot] Corollary 7.10. and Corollary 8.8.) that if
$$\cdots \longrightarrow P_1 \longrightarrow P_0 \longrightarrow M \longrightarrow 0$$
and
$$\cdots \longrightarrow Q_1 \longrightarrow Q_0 \longrightarrow N \longrightarrow 0$$
are flat resolutions of a left A-module N and a right A-module M respectively, then for all $n \geq 0$ we have:

$$H_n(\mathcal{P}_M \otimes_A N) \cong H_n(M \otimes_A \mathcal{Q}_N) = \mathrm{Tor}_n^A(M, N),$$

where $\mathcal{P}_M$ and $\mathcal{Q}_N$ are deleted flat resolutions. Hence Lemma 4.1.6. of Ch.I. and the foregoing results also allow to provide similar results with respect to the graded (global) weak dimension of A just as in the ungraded case.

4. Theorem. (Graded version of Auslander theorem) If A is left and right Noetherian then the left and right graded global dimension coincide.

Next, we also have the graded versions of "the change of rings" theorems. For the usual "change of rings" theorem we refer to [Kap].

5. Theorem. (First "change of rings" theorem) Let A be an arbitrary ring and $X \in A$ a regular non-invertible normalizing element. Put $A^* = A/XA$. If M is an A^*-module such that p.dim$_{A^*} M = n < \infty$ then p.dim$_A M = n + 1$. In particular gl.dim$A \geq 1+$ gl.dimA^* in case gl.dimA^* is finite.

In what follows we again let A be graded by an arbitrary group G. Let X be a regular noninvertible **homogeneous** normalizing element of A. Put $A^* = A/XA$.

6. Lemma. If M is an X-torsionfree A-module then any exact sequence

$$0 \longrightarrow K \longrightarrow L \overset{f}{\longrightarrow} M \longrightarrow 0$$

in A-mod yields an exact sequence in A^*-mod:

$$0 \longrightarrow K/XK \longrightarrow L/XL \overset{f^*}{\longrightarrow} M/XM \longrightarrow 0$$

7. Lemma. (Second "change of rings" theorem) Let A, X and M be as in the foregoing lemma, then we have: p.dim$_{A^*}(M/XM) \leq$ p.dim$_A M$.

Proof. If p.dim$_A M = \infty$ there is nothing to prove. So we may assume p.dim$_A M = n < \infty$. Consider an exact sequence in A-mod:

$$(*) \qquad\qquad 0 \longrightarrow K \longrightarrow L \longrightarrow M \longrightarrow 0$$

where L is a free A-module. If $n = 0$ then $(*)$ is split and then the foregoing lemma entails that the sequence

$$0 \longrightarrow K/XK \longrightarrow L/XL \longrightarrow M/XM \longrightarrow 0$$

splits too. Since L/XL is A^*-free it then follows that M/XM is A^*-projective, i.e., p.dim$_{A^*}(M/XM) = 0$. If $n > 0$ then we have p.dim$_A M = 1+$ p.dim$_A K$ by Proposition 5.1.2.. Hence p.dim$_A K = n - 1$. By Lemma 5.1.6. and Proposition 5.1.2. induction on n leads to the inequality p.dim$_{A^*}(M/XM) \leq n$. $\qquad\square$

8. Proposition. (Third "change of rings" theorem for graded rings) Let A and X be as before and suppose now that A is left Noetehrian and X is contained in the **graded Jacobson radical** $J^g(A)$ of A. If M is a finitely generated X-torsionfree graded A-module then p.dim$_{A^*}(M/XM) =$ p.dim$_A M$.

Proof. By Lemma 5.1.7. it remains to establish p.dim$_A M \leq$ p.dim$_{A^*}(M/XM)$. We may assume that p.dim$_{A^*}(M/XM) = n < \infty$. Consider an exact sequence in A-gr:

$$(*) \qquad\qquad 0 \longrightarrow K \longrightarrow L \overset{f}{\longrightarrow} M \longrightarrow 0$$

where L is a gr-free A-module of finite rank. If we have p.dim$_{A^*}(M/XM) = 0$ then the exact sequence

$$0 \longrightarrow K/XK \longrightarrow L/XL \overset{f^*}{\longrightarrow} M/XM \longrightarrow 0$$

is split. Put $T = K \oplus M$ in A-gr. Then T is obviously X-torsionfree and a finitely generated graded A-module. Thus $(K/XK) \oplus (M/XM) \cong (K \oplus M)/X(K \oplus M) = T/XT$ is gr-free of finite rank. By Proposition 4.4.7.of Ch.I. and Lemma 3.2.3. (or graded Nakayama's lemma) one easily sees that T is a gr-free A-module and hence M is a gr-projective A-module finishing the proof in case $n = 0$. In case p.dim$_{A^*}(M/XM) = n \geq 1$, Proposition 5.1.2. yields p.dim$_{A^*}(K/XK) = n - 1$ and so we may use induction on n in order to arrive at p.dim$_A M \leq n$ (again by applying Proposition 5.1.2.).   □

Now, we are able to provide further results dealing with the graded (homological) regularity and global dimension.

Let $A = \oplus_{n \in \mathbb{Z}} A_n$ be a $\mathbb{Z}$-graded ring. Consider the polynomial ring $A[t]$ over A in a commuting variable t. $A[t]$ has the "mixed" $\mathbb{Z}$-gradation defined by $A[t]_n = \{\sum_{i+j=n} a_i t^j, \ a_i \in A_i\}$, $n \in \mathbb{Z}$. Obviously, A is a graded subring of $A[t]$ with respect to the "mixed" gradation on $A[t]$.

9. Lemma. Let A and $A[t]$ be as above. Suppose that A is left Noetherian and left gr-regular, then for any finitely generated graded left $A[t]$-module M, p.dim$_A M < \infty$; and moreover $A[t]$ is left gr-regular with respect to the "mixed" gradation.

Proof. Let M be any finitely generated graded left $A[t]$-module, then $M \in A$-gr. Let $M^{(0)}$ be a finitely generated graded A-submodule of M such that $M = A[t]M^{(0)}$. Put $M^{(n)} = \sum_{i=0}^{n} t^i M^{(0)}$ for positive $n \in \mathbb{Z}$, then obviously $M^{(n)}$, and hence $M^{(n)}/M^{(n-1)}$ is finitely generated in A-gr. Moreover, since $M^{(n+1)} = tM^{(n)} + M^{(n)}$, left multiplication by t gives sequence of graded A-module surjections

$$M^{(0)} \longrightarrow M^{(1)}/M^{(0)} \longrightarrow M^{(2)}/M^{(1)} \longrightarrow \cdots .$$

Let K_n be the kernel of the resulting map $M^{(0)} \to M^{(n)}/M^{(n-1)}$, then K_n is a graded A-submodule of $M^{(0)}$ and $\{K_i\}$ is an incraesing chain. Hence $K_{n+l} = K_n$ for some n and all l. Consequently

$$M^{(n+l+1)}/M^{(n+l)} \cong M^{(n)}/M^{(n-1)}$$

as graded A-modules and

$$\text{p.dim}_A(M^{(n+l+1)}/M^{(n+l)}) = \text{p.dim}_A(M^{(n)}/M^{(n-1)}).$$

Since A is graded regular by assumption, it is then well known that

$$\text{p.dim}_A M^{(n+l)} \leq \text{Sup}\{\text{p.dim}_A M^{(0)}, \text{p.dim}_A(M^{(1)}/M^{(0)}), \cdots, \text{p.dim}_A(M^{(n)}/M^{(n-1)})\} = w,$$

say, hence p.dim$_A(\oplus_{n=0}^{\infty} M^{(n)}) = w$. However, there is an exact sequece

$$0 \longrightarrow \oplus M^{(n)} \xrightarrow{\ \varepsilon\ } \oplus M^{(n)} \xrightarrow{\ \pi\ } M \longrightarrow 0,$$

where ε: $(m^{(n)}) \mapsto (m'^{(n)})$ with $m'^{(n)} = m^{(n)} - m^{(n-1)}$ and π: $(m^{(n)}) \mapsto \sum m^{(n)}$. Therefore
p.dim$_A M \leq w + 1 < \infty$.
Now, consider the following exact sequence of $A[t]$-modules

$(*)$ $$0 \longrightarrow M[t] \longrightarrow M[t] \overset{e}{\longrightarrow} M \longrightarrow 0$$

where $M[t] = A[t] \otimes_A M$, e: $t^i \otimes m \mapsto t^i m$, then it is well known (cf. [Rot] Lemma 9.27.)
that p.dim$_A M =$ p.dim$_{A[t]} M[t]$ and the exactness of $(*)$ yields

$$\text{p.dim}_{A[t]} M \leq 1 + \text{p.dim}_{A[t]} M[t] \leq 1 + \text{p.dim}_A M < \infty.$$

This proves the graded regularity of $A[t]$. □

10. Theorem. Let A be a left Noetherian $\mathbb{Z}$-graded ring. If A is gr-regular then A is
regular.

Proof. Once again let $A[t]$ be the polynomial ring with the "mixed" gradation. Consider
the (graded) localization $A[t]_{(t)}$ of $A[t]$ at the multiplicatively closed subset $\{1, t, t^2, \cdots\}$,
then $A[t]_{(t)} \cong A[t, t^{-1}]$ as graded rings, where $A[t, t^{-1}]$ also has the "mixed" gradation:
$A[t, t^{-1}]_n = \{\sum_{i+j=n} a_i j^j, a_i \in A_i\}$, $n \in \mathbb{Z}$, in particular $A[t, t^{-1}]_0 = \sum_{i+j=0} A_i t^j \cong A$.
If M is any finitely generated graded $A[t, t^{-1}]$-module, say $M = \sum_{i=1}^s A[t, t^{-1}] \xi_i$ where all ξ_i
are homogeneous elements of M, then $M_0 = \sum_{i=1}^s A[t] \xi_i$ is a finitely generated graded $A[t]$-
module such that $A[t, t^{-1}] \otimes_{A[t]} M_0 = M$. It follows from Lemma 5.1.9. that p.dim$_{A[t,t^{-1}]} M <$
∞ and hence $A[t, t^{-1}]$ is gr-regular. Now the equivalence of categories: $A[t, t^{-1}]_0$-mod $\leftrightarrow$
$A[t, t^{-1}]$-gr, gives us the regularity of A. □

Question. Is it possible to drop the Noetherian condition in the theorem?

11. Theorem. Let A be a (group) G-graded ring and X a regular non-invertible homogeneous normalizing element of A. Put $A^* = A/XA$. Suppose that A is left Noetherian and
$X \in J^g(A)$.
(1). If A^* is left gr-regular then A is left gr-regular.
(2). gr.gl.dim$A \leq 1 +$ gr.gl.dimA^* and the equality holds in case A^* has finite graded global
dimension.
(3). If A is $\mathbb{Z}$-graded and A^* is left gr-regular then A is left regular.

Proof. Consider any finitely generated graded A-module M and put $t(M) = \{m \in$
M, $X^s m = 0$ for some $s \in \mathbb{N}\}$, then $t(M)$ is a graded submodule of M and the sequence

$(*)$ $$0 \longrightarrow t(M) \longrightarrow M \longrightarrow M/t(M) \longrightarrow 0$$

is exact in A-gr; Moreover note that $M/t(M)$ is X-torsionfree and $X^k t(M) = 0$ for some $k \in I\!N$ since $t(M)$ is finitely generated too. Hence an easy induction on k together with Theorem 5.1.5. yield: p.dim$_A t(M) \leq 1+$ gr.gl.dimA^*. Also Proposition 5.1.8. entails p.dim$_A(M/t(M)) \leq$ gr.gl.dimA^*. It follows from the exactness of $(*)$ that p.dim$_A M \leq 1+$ gr.gl.dimA^*. Now Proposition 5.1.3. yields that gr.gl.dim$A \leq 1+$ gr.gl.dimA^*. On the other hand, if gr.gl.dim$A^* < \infty$ then a graded version of Theorem 5.1.5. yields that gr.gl.dim$A \geq 1+$ gr.gl.dimA^*. Therefore we arrive at the equality gr.gl.dim$A = 1+$ gr.gl.dimA^*. From the proof we easily see that if every finitely generated graded A^*-module has finite projective dimensionthen every finitely generated graded A-module has finite projective dimension. Hence (1) and (2) are proved, and consequently (3) follows from Theorem 5.1.10..   □

Let us finish this addendum by giving two propositions concerning the regularity and global dimension of a strongly $Z\!\!\!Z$-graded ring.

12. Proposition. Let A be a strongly $Z\!\!\!Z$-graded ring. The following statements are equivalent:

(1). A has finite (graded) global dimension;

(2). A_0 has finite global dimension;

(3). $A^+ = \oplus_{n \geq 0} A_n$ has finite global dimension;

(4). $A^- = \oplus_{n \leq 0} A_n$ has finite global dimension.

Proof. (1) $\Leftrightarrow$ (2). Since A is strongly graded this is just ([NVO1] Lemma II.8.3.).

(1) $\Leftrightarrow$ (3). Consider the $+$-gradation (Ch.I. Example 4.3.9.(d).) on the polynomial ring $A[t]$, then $A[t]$ becomes a strongly $Z\!\!\!Z$-graded ring with $A[t]_0^+ = A^+$. Accordingly, if A has finite global dimension then it is well known that $A[t]$ has finite global dimension, and hence the equivalence (1) $\Leftrightarrow$ (2) finishes the proof.

(1) $\Leftrightarrow$ (4). By using the $-$-gradation on $A[t]$ the proof is similar to (1) $\Leftrightarrow$ (3).   □

13. Proposition. Let A be a strongly $Z\!\!\!Z$-graded ring. If A_0 is left Noetherian then the following are equivalent:

(1). A_0 is left regular;

(2). A is left (gr-)regular;

(3). $A^+ = \oplus_{n \geq 0} A_n$ (or $A^- = \oplus_{n \leq 0} A_n$) is left (gr-)regular.

Proof. (1) $\Leftrightarrow$ (2). In view of Theorem 5.1.10., this follows from the equivalence of categories: A-gr $\leftrightarrow$ A_0-mod.

(2) $\Leftrightarrow$ (3). Again consider the $+$-gradation on $A[t]$ as in the proof of Proposition 5.1.12., then the regularity of A^+ now yields the regularity of $A[t]$. Hence A is regular ([Rot] Lemma

9.27.). Conversely, if A is left Noetherian and (gr-)regular, then it follows from Lemma 5.1.9. and Theorem 5.1.10. that $A[t]$ is regular. Therefore $A^+ = A[t]_0^+$ is regular by (1).   □

5.2. Global dimension of Rees rings

An effect of subsection 5.1. is that we have been able to derive graded versions of "change of rings" type results under the condition $X \in J^g(A)$, that is considerably weaker than $X \in J(A)$. We had to pay for this extended generality by restricting to intrinsically graded notions, e.g., the "graded global dimension", which is unsatisfactory. What is still missing is a way to relate graded and ungraded homological dimensions under the same condition: $X \in J^g(A)$, we use a trick connected to the construction of a right adjoint for the forgetful functor A-gr $\to A$-mod (see [NVO1] Ch.I.) in order to obtain this type of relation.

Again let A be a (group) G-graded ring. Consider the groupring AG of G over A and view it as a G-graded ring by putting $(AG)_\sigma = \sum_{\gamma\tau=\sigma} A_\gamma\tau$ for $\sigma \in G$.

In AG, we may consider a subring $S = \sum_{\sigma\in G} A_\sigma \cdot \sigma$ which was introduced first time in [VDB1]. In fact S is isomorphic to A as a graded ring but it is only a subring of AG and not a graded subring of AG. Note that the part of degree e (e is the neutral element of G) in AG is equal to $\sum_{\sigma\in G} A_\sigma \cdot \sigma^{-1}$ and that this is isomorphic to A as an ungraded ring. An arbitrary A-module M may be viewed as an AG-module by restriction of scalars with respect to the canonical ring epimorphism $\pi\colon AG \to A$. The kernel Kerπ is the augmentation ideal of AG, i.e., Ker$\pi = \Delta_G$ is the ideal of AG generated by the elements $1 - g$, $g \in G$ (we write $1 = 1 \cdot e$ for the unit of AG). The AG-module structure of M may thus be defined by: $(\sum_{\sigma\in G} r_\sigma \cdot \sigma)m = (\sum_{\sigma\in G} r_\sigma)m$ for $m \in M$, $r_\sigma \in A$, $\sigma \in G$.

Note that Δ_G is a free A-module with basis $\{1 - g,\ g \in G,\ g \neq e\}$, indeed for $\tau \in G$ we have that $\tau - \tau\sigma = (1 - \tau\sigma) - (1 - \tau)$ for $\sigma \in G$. Although the results in the sequel remain true for arbitrary torsionfree grading groups G (in fact we only need that $(1 - g)MG \cong MG$ where $MG = AG \otimes_A M$) we restrict to the case $G = \mathbb{Z}$ now. In this case $AG \cong A[t, t^{-1}]$, $MG \cong M[t, t^{-1}]$, $S = \sum_{n\in\mathbb{Z}} A_n t^n$, $\Delta_G = \Delta = (t - 1)A[t, t^{-1}]$ and the $A[t, t^{-1}]$-linear map $\varepsilon\colon M[t, t^{-1}] \to M$, may be defined by $(\sum_{n\in\mathbb{Z}} t^n) \otimes m_n \mapsto \sum_{n\in\mathbb{Z}} m_n$. Obviously: Ker$\varepsilon = (t - 1)M[t, t^{-1}] \cong M[t, t^{-1}]$.

1. Lemma. Let $A = \oplus_{n\in\mathbb{Z}} A_n$ be a $\mathbb{Z}$-graded ring. With notation and conventions as above, we have:

(1). for any A-module M the $A[t, t^{-1}]$-module $M[t, t^{-1}]$ may be viewed as a graded A-module by putting $M[t, t^{-1}]_n = At^n \otimes M$ for $n \in \mathbb{Z}$ and $a_n(t^k \otimes m) = t^{k+n} \otimes a_n m$ for $a_n \in A_n$, $m \in M$ and $n \in \mathbb{Z}$.

(2). There is an $A[t, t^{-1}]$-linear exact sequence:

$$(*) \qquad 0 \longrightarrow (t-1)M[t, t^{-1}] \longrightarrow M[t, t^{-1}] \longrightarrow M \longrightarrow 0$$

that may also be viewed as an exact S-linear sequence:

$$(**) \qquad 0 \longrightarrow M[t, t^{-1}] \longrightarrow M[t, t^{-1}] \longrightarrow M \longrightarrow 0$$

(since the S-structure on M is just the A-structure we may identify A and S as graded rings and view $(**)$ as an exact sequence of A-modules where $M[t, t^{-1}]$ is a graded A-module).

Note. In the above lemma, the A-module $M[t, t^{-1}]$ does not contain M as a submodule nor is it a direct sum of copies of M in the newly defined A-module structure that makes $M[t, t^{-1}]$ into a graded A-module.

2. Proposition. Let A be a $\mathbb{Z}$-graded ring and let X be a regular non-invertible homogeneous normalizing element of A. Suppose that A is left Noetherian and that $X \in J^g(A)$. If M is any X-torsionfree graded A-module then f.dim$_A M \leq$ gr.gl.dim(A/XA), where f.dim$_A M$ denotes the flat dimension of M.

Proof. Note that $M = \varinjlim M_i$ where the M_i ranges over all finitely generated graded submodules of M and the index set is directed by usual inclusion. Recall ([Rot] Theorem 8.11): Tor$_n(\varinjlim M_i, N) \cong \varinjlim$Tor$_n(M_i, N)$, for all A-modules N and all $n \geq 0$. So we conclude that f.dim$_A M \leq$ gr.gl.dim(A/XA) since for each i, p.dim$_A M_i \leq$ gl.gl.dim(A/XA) because of Proposition 5.1.8.. $\qquad \square$

3. Theorem. Let A and X be as in Proposition 5.2.2. above. Then gl.dim$A \leq 1+$ gl.dim(A/XA) and the equality holds if A/XA has finite global dimension.

Proof. Consider a finitely generated A-module M and the A-linear exact sequence:

$$(*) \qquad 0 \longrightarrow t(M) \longrightarrow M \longrightarrow M/t(M) \longrightarrow 0$$

(as in Theorem 5.1.11.). By induction on the natural number k minimal such that $X^k t(M) = 0$ we arrive at: f.dim$_A t(M) \leq 1+$ gl.dimA/XA. Put $T = M/t(M)$. Then T is an X-torsionfree module. From Lemma 5.2.1. it follows that $T[t, t^{-1}]$ is an X-torsionfree graded A-module and f.dim$_A T \leq 1+$ f.dim$_A T[t, t^{-1}]$. Applying the foregoing proposition we obtain: f.dim$_A T \leq 1+$ gr.gl.dim$A/XA \leq 1+$ gl.dimA/XA. Exactness of the sequence $(*)$ now yields: f.dim$_A M \leq 1+$ gl.dimA/XA. Since A is left Noetherian the left global homological dimension equals the weak (flat) global dimension and the latter is completely determined by finitely generated A-modules. Therefore we finally arrive at the inequality: gl.dim$A \leq 1+$ gl.dimA/XA. On the other hand, if A/XA has finite global dimensoin then in view of Theorem 5.1.5. the inequality gl.dim$A \geq 1+$ gl.dimA/XA is clear enough. This completes the proof.   □

4. Corollary. Let A be a left Noetherian $\mathbb{Z}$-graded ring and I an invertible ideal of A generated by a finite normalizing system of homogeneous generators $\{x_1, \cdots, x_n\}$ say, such that x_i modulo $Ax_1 + \cdots + Ax_{i-1}$ satisfies the same conditions as X in A before. Then gl.dim$A \leq n+$ gl.dimA/I and the equality holds in case A/I has finite global dimension.

Concerning the Rees ring of a left Zariski ring, we now can mention the following

5. Theorem. Let R be a left Zariski ring with filtration FR.
(1). If $G(R)$ is left gr-regular (hence regular) then $\tilde{R}$ is left regular.
(2). gr.gl.dim$\tilde{R} \leq 1+$ gr.gl.dim$G(R)$; gl.dim$\tilde{R} \leq 1+$ gl.dim$G(R)$, and the equalities hold in case $G(R)$ has finite (gr-)global dimension.

Proof. Since R is a left Zariski ring it follows that $\tilde{R}$ is left Noetherian and the canonical regular central homogeneous element X of degree 1 is contained in the graded Jacobson radical $J^g(\tilde{R})$ of $\tilde{R}$. Noticing that $G(R) \cong \tilde{R}/X\tilde{R}$ we may apply Theorem 5.1.11. and Theorem 5.2.3. to $\tilde{R}$ and obtain the desired assertions.   □

6. Remark. Using the foregoing results, some classical results may be easily recaptured. For example, let A be a left Noetherian regular ring and σ an automorphism, then the skew polynomial ring $R = A[t, \sigma]$, regarded as a graded ring with gradation $R_n = At^n$, $n \in \mathbb{Z}$, satisfies the conditions of Theorem 11. (by putting $X = t$) and hence left regular. It follows from Ch.I. Theorem 7.2.12. that each of the following rings is left regular.
(1). $A[t, \sigma, \delta]$, where δ is a σ-derivation of A;
(2). $A[t, t^{-1}, \sigma]$;
(3). The crossed product $A * G$ of A by G, where G is a poly-infinite cyclic group;
(4). The crossed product $A * U(\mathbf{g})$ of A by $U(\mathbf{g})$, where R is a k-algebra over a commutative ring k and $\mathbf{g}$ a k-Lie algebra of finite dimension.
We refer to [M-R] for some detail about these rings.

§6. K_0 of Rings with Zariskian Filtration

In this section we develop some results concerning the K_0 of rings with Zariskian filtration. For some detail about algebraic K-theory we refer to [Bass]. We also refer to ([M-R] Ch.11. §4.) for a detail discussion of K_0 (in particular, Theorem 4.8.).

To start with, let A be any ring and $\mathcal{P}(A)$ the category of finitely generated projective left modules over A. By definition $K_0(A) = K_0(\mathcal{P}(A))$; Or more precisely, write $\mathcal{F} = \mathcal{F}(A)$ for the free abelian group whose free generators $< P >$ are the isomorphism classes of $P \in \mathcal{P}(A)$ (so $< P >=< P_1 >$ if and only if $P \cong P_1$) and let J be the subgroup of $\mathcal{F}$ generated by all expressions $< P > + < Q > - < P \oplus Q >$, then $K_0(A) = \mathcal{F}/J$. An element of $K_0(A)$ represented by $P \in \mathcal{P}(A)$ is denoted by $[P]$, then $[P]+[Q] = [P \oplus Q]$ for any $[P],[Q] \in K_0(A)$. Concerning $K_0(A)$, the following lemma is well known.

6.1. Lemma. With notations as above, each element of $K_0(A)$ can be written in the form $[P] - [Q]$; Moreover, let $P, Q \in \mathcal{P}(A)$, then $[P] = [Q]$ in $K_0(A)$ if and only if P, Q are stably isomorphic, i.e., there exists some $n \in I\!N$ such that $P \oplus A^n \cong Q \oplus A^n$.

If $A = \oplus_{n \in \mathbb{Z}} A_n$ is a $\mathbb{Z}$ graded ring, we put $K_{0g}(A) = K_0(\mathcal{P}_g(A))$, where $\mathcal{P}_g(A)$ is the category of graded finitely generated projective A-modules and morphisms in $\mathcal{P}_g(A)$ are graded morphisms of degree zero and $K_{0g}(A)$ is defined as in the ungraded case but intrinsically in terms of graded objects and graded morphisms, hence Lemma 6.1. holds for $K_{0g}(A)$.

6.2. Lemma. Let A be a $\mathbb{Z}$-graded ring and T a regular homogeneous normalizing element (i.e., $TA = AT$) of A. Suppose that $T \in J^g(A)$. Then the canonical map i: $K_{0g}(A) \to K_{0g}(A/TA)$ with $[P] \mapsto [A/TA \otimes_A P]$, which is induced by the natural map $A \to A/TA$, is an injection.

Proof. By Lemma 6.1. an element of Ker$i$ has the form $[P] - [Q]$, where $P,\ Q \in \mathcal{P}_g(A)$ and $[P/TP] = [Q/TQ]$. That implies that P/TP and Q/TQ are stably isomorphic. Now modify P and Q, by adding a suitable gr-free module of finite rank, to make a graded isomorphism $P/TP \cong Q/TQ$; this will not change $[P] - [Q]$. We want to prove that P and Q are graded isomorphic as A-modules, and consequently it will follow that $[P] = [Q]$. To this end, we put $\overline{P} = P/TP, \overline{Q} = Q/TQ$ and let $\vartheta \colon \overline{P} \to \overline{Q}$ be the given graded isomorphism of A/TA-modules (it is certainly an isomorphism of A-modules). Since P is graded projective, and the natural homomorphism $\pi_2 \colon Q \to \overline{Q}$ is an epimorphism, $\vartheta \pi_1 \colon P \to \overline{Q}$ induces a graded homomorphism $\tau \colon P \to Q$ so that the diagram

$$
\begin{array}{ccc}
 & & 0 \\
 & & \downarrow \\
P & \xrightarrow{\pi_1} & \overline{P} \to 0 \\
\tau \downarrow & & \downarrow \vartheta \\
Q & \xrightarrow{\pi_2} & \overline{Q} \to 0 \\
 & & \downarrow \\
 & & 0
\end{array}
$$

commutes in A-gr. We claim that τ is an isomorphism. Since π_1 and ϑ are epimorphisms, hence so is $\vartheta\pi_1 = \pi_2\tau$. Thus $Q = \tau(P) + \mathrm{Ker}\pi_2 = \tau(P) + TQ$. Since $T \in J^g(A)$ by assumption it follows from the graded version of Nakayama's lemma that $\tau(P) = Q$, that is, τ is an epimorphism. But Q is projective, so there exists a graded morphism $\alpha\colon Q \to P$ with $\tau\alpha = 1_Q$, and then $P = \alpha(Q) \oplus \mathrm{Ker}\tau$ in A-gr. Therefore,

$$\mathrm{Ker}\tau \subseteq \mathrm{Ker}\pi_2\tau = \mathrm{Ker}\vartheta\pi_1 = \mathrm{Ker}\pi_1 = TP,$$

and hence $P = \alpha(Q) + TP$. Again by the graded version of Nakayama's lemma we obtain $P = \alpha(Q)$, and so $\mathrm{Ker}\tau = 0$. This shows that τ is an isomorphism as desired. $\square$

6.3. Corollary. Let R be a filtered ring with filtration FR. Suppose that $F_{-1}R \subseteq J(F_0R)$, i.e., FR is faithful, then the canonical map $i\colon K_{0g}(\tilde{R}) \to K_{0g}(G(R))$ with $[P] \mapsto [\tilde{R}/X\tilde{R} \otimes_{\tilde{R}} P]$, which is induced by the natural map $\tilde{R} \to \tilde{R}/X\tilde{R} \cong G(R)$, is an injection, where X is the canonical central regular element of degree 1 of $\tilde{R}$.

Proof. Indeed, from Ch.I. Corollary 4.4.12. we have $X \in J^g(\tilde{R})$. Hence if we put $T = X$ and $A = \tilde{R}$ in Lemma 6.2. then the assertion follows. $\square$

6.4. Lemma. Let R be a complete filtered ring with respect to its filtration FR. Then $i\colon K_{0g}(\tilde{R}) \to K_{0g}(G(R))$ is an isomorphism.

Proof. By Corollary 6.2. above it remains to show that i is surjective. Let P be in $\mathcal{P}_g(G(R))$, Then $P \oplus Q = F$, where F is gr-free of finite rank in $\mathcal{P}_g(G(R))$. It follows from Ch.I. Lemma 6.2.(4). that there exists a finitely generated filt-free $L \in R$-filt with filtration FL such that $G(L) \cong F$. Let $F \xrightarrow{\pi} P$ be the (graded) projection of F onto P, then $\pi \in \mathrm{End}_{G(R)}(F)$ and $\pi^2 = \pi$. Again from Ch.I. Lemma 6.2.(6). there is a filtered morphism of degree zero, say $f\colon L \to L$ such that $G(f) = \pi$. Since L is of finite rank it is complete with respect to its filtration FL (Ch.I. Lemma 6.2.(7).). Hence by a classical result concerning topological rings

(Ch.I. Corollary 6.8.) there is a strict filtered morphism $g: L \to L$ such that $g^2 = g$ and $G(g) = G(f) = \pi$. One easily checks that $\mathrm{Im}(g)$ is a filt-projective R-module and by Ch.I. Remark 4.2.6. $\widetilde{g(L)}/X\widetilde{g(L)} = G(g(L)) = G(g)(G(L)) = \pi(G(L)) = P$. It follows from Ch.I. Lemma 6.4. that $i([\widetilde{g(L)}]) = [P]$. This completes the proof.   □

Let us now recall the following theorem.

6.5. Theorem. (K_0-part of Quillen's theorem, [Qui] Theorem 7.) Let R be a positively filtered ring with filtration FR such that the associated graded ring $G(R)$ is left Noetherian and each finitely generated left $G(R)$-module has a finite projective resolution (i.e., GR is left regular). Then the inclusion map $F_0R \hookrightarrow R$ induces an isomorphism $K_0(F_0R) \xrightarrow{\cong} K_0(R)$.

The next theorem generalizs this result.

6.6. Theorem. Let R be a left Zariski ring with filtration FR such that $G(R)$ is left gr-regular (hence left regular). There is an injection $\nu: K_0(R) \hookrightarrow K_0(G(R))$ with $\nu([R]) = [G(R)]$.

Proof. It follows from Theorem 5.2.5. that $\tilde{R}$ is left regular. Let us rewrite the (exact) localization sequence from [Qui] (or see [M-R] P.429)

$$K_{0g}(G(R)) \longrightarrow K_{0g}(\tilde{R}) \longrightarrow K_{0g}(\tilde{R}_{(X)}) \longrightarrow 0$$

as

$$K_{0g}(G(R)) \xrightarrow{\alpha} K_{0g}(\tilde{R}) \xrightarrow{\beta} K_0(R) \longrightarrow 0$$

using that $\tilde{R}_{(X)} \cong R[t, t^{-1}]$ and the fact that there is an equivalence of categories between $R[t, t^{-1}]$-gr and R-mod. The canonical map $i: K_{0g}(\tilde{R}) \to K_{0g}(G(R))$ is injective (Corollary 6.3.). We also have an exact sequence from [VDB2]:

$$K_{0g}(G(R)) \xrightarrow{\varphi} K_{0g}(G(R)) \xrightarrow{\gamma} K_0(G(R)) \longrightarrow 0$$

where γ is given by the forgetful functor on $\mathcal{P}_g(G(R))$ (that is, forgetting the gradation), $\varphi([P]) = [P] - [T(-1)P]$ and $T(-1)P$ is the image of P under the shift functor $T(-1)$ (see Ch.I. §4.1.). (The surjectivity of γ was not explicitly stated in [VDB2], but it follows easily from the proof.) We claim that the following diagram is commutative:

$$K_{0g}(\tilde{R})$$

$$\alpha \nearrow \qquad \downarrow i$$

$$K_{0g}(G(R))$$

$$\varphi \searrow$$

$$K_{0g}(G(R))$$

To this end, let P be in $\mathcal{P}_g(G(R))$. Then by Theorem 5.1.5. we have a projective resolution of P in $\tilde{R}$-gr:

$$0 \longrightarrow P_1 \longrightarrow P_0 \longrightarrow P \longrightarrow 0$$

where P_1, $P_0 \in \mathcal{P}_g(\tilde{R})$. Thus $\alpha([P]) = [P_0] - [P_1]$ and, hence, $i\alpha([P]) = [P_0/XP_0] - [P_1/XP_1]$. Now, regard XP_0 resp. XP_1 as a graded submodule of P_0 resp. P_1 (i.e., $(XP_0)_n = X(P_0)_{n-1}, (XP_1)_n = X(P_1)_{n-1}, n \in \mathbb{Z}$). Note that $XP_0 \subseteq P_1$ (since $P \in G(R)$-gr) and the sequence

$$0 \longrightarrow T(-1)P_0/P_1 \longrightarrow XP_0/XP_1 \longrightarrow 0$$

is exact in $\tilde{R}$-gr we then have the following (exact) commutative diagram in $\tilde{R}$-gr:

$$
\begin{array}{ccccccc}
 & & 0 & \to & 0 & \to & T(-1)P_0/P_1 \\
 & & \downarrow & & \downarrow & & \downarrow \\
0 & \to & XP_1 & \to & XP_0 & \to & XP_0/XP_1 & \to & 0 \\
 & & \downarrow & & \downarrow & & \downarrow \\
0 & \to & P_1 & \to & P_0 & \to & P_0/P_1 & \to & 0 \\
 & & \downarrow & & \downarrow & & \downarrow \\
 & & P_1/XP_1 & \to & P_0/XP_0 & \to & P & \to & 0 \\
 & & \downarrow & & \downarrow & & \downarrow \\
 & & 0 & & 0 & & 0
\end{array}
$$

It follows from the snake-lemma that the sequence

$$0 \longrightarrow T(-1)P \longrightarrow P_1/XP_1 \longrightarrow P_0/XP_0 \longrightarrow P \longrightarrow 0$$

is exact. Therefore in $K_{0g}(G(R))$ we have $[P_0/XP_0] - [P_1/XP_1] = [P] - [T(-1)P]$. Accordingly, $i\alpha([P]) = [P] - [T(-1)P] = \varphi([P])$.

Finally, since i is injective and β, γ are surjective, it follows that there is a commutative diagram

$$\begin{array}{ccc} & K_{0g}(\tilde{R}) & \xrightarrow{\beta} & K_0(R) & \to & 0 \\ {\scriptstyle\alpha}\nearrow & \downarrow{\scriptstyle i} & & \downarrow{\scriptstyle \nu} & & \\ K_{0g}(G(R)) & & & & & \\ {\scriptstyle\varphi}\searrow & & & & & \\ & K_{0g}(G(R)) & \xrightarrow{\gamma} & K_0(G(R)) & \to & 0 \end{array}$$

such that ν is injective and $\nu([R]) = [G(R)]$. This completes the proof. ☐

6.7. Corollary. In the situation of the theorem, if image $(K_{0g}(G(R)) \to K_0(G(R))) =< [G(R)] >$ (i.e, every finitely generated projective graded $G(R)$-module is stably free as a $G(R)$-module) then $K_0(R) =< [R] >$.

Proof. Use the fact that ν is injective and chase the diagram above. ☐

6.8. Corollary. Let R be a filtered ring with filtration FR. Suppose that R is complete with respect to its filtration topology and that $G(R)$ is left Noetherian and regular, then the morphism ν in the theorem is an isomorphism, i.e., $K_0(R) \cong K_0(G(R))$.

Proof. From the assumption we know that R is a left Zariski ring. Hence it follows from Lemma 6.4. and Theorem 6.6. that ν is an isomorphism. ☐

6.9. Corollary. If R is a left Zariski ring with filtration FR such that $G(R) \cong A[x_1, \cdots, x_m][y_1^{\pm}, \cdots, y_n^{\pm}]$, where A is a ring such that $K_0(A) =< [A] >$, then $K_0(R) =< [R] >$. In particular, if A, is a not necessarily commutative local ring, or if A is a field, or if A is a principal ideal domain then every finitely generated filt-projective R-module is free.

Proof. This follows from $K_0(A[x_1, \cdots, x_m][y_1^{\pm}, \cdots, y_n^{\pm}]) = K_0(A)$ ([Qui]), the famous Serre theorem and Quillen-Suslin theorem (cf. [Rot] Ch.4.). ☐

6.10. Proposition. Let A be a strongly $\mathbb{Z}$-graded ring and suppose that A_0 is left Noetherian and regular, then $K_{0g}(A) \cong K_0(A_0) \cong K_0(A^+) \cong K_0(A^-)$.

Proof. The isomorphism $K_{0g}(A) \cong K_0(A_0)$ follows from the category equivalence A-gr $\leftrightarrow$ A_0-mod. Since A_0 is left Noetherian and regular it follows from Proposition 5.1.13. that A^+ resp. A^- is left Noetherian and regular too. Therefore we may apply Theorem 6.5. to the positively graded ring A^+ resp. A^- and obtain the desired isomorphisms. ☐

6.11. Theorem. Let R be a ring and I an ideal of R. Let FR be the I-adic filtration on

R and $G(R)$ the associated graded ring of R. Suppose that $G(R)$ is isomorphic to A^+ or to A^- as graded rings for some strongly $\mathbb{Z}$-graded ring A (e.g., I is an invertible ideal).

(1). If R/I is left Noetherian and regular then $K_0(R/I) \cong K_0(G(R))$;

(2). If I is contained in the Jacobson radical of R and $\tilde{R}$ is left Noetherian (i.e., FR is Zariskian) then there is an embedding of K_0-groups: $K_0(R) \hookrightarrow K_0(R/I) = K_0(G(R))$ in case R/I is left regular.

Proof. Follows from Proposition 6.10. and Theorem 6.6..   □

CHAPTER III
Auslander Regular Filtered (Graded) Rings

Since the late sixties, various Auslander regularity conditions have been widely investigated in both commutative and noncommutative cases, e.g., [Ba2], [FGR], [Roos], [Lev], etc $\cdots$. In particular, the dimension and multiplicity theory for modules, the pure (holonomic) module theory over noncommutative Auslander regular rings have been developed in the study of rings of differential operators and enveloping algebras, e.g., [Ka], [Be], [B-B], [K-T], etc $\cdots$. A typical result concerning purity is the so-called **Gabber-Kashiwara theorem** (see [Gi] Theorem V8) which has been used to: (1). determine the existence of holonomic submodules in $\mathcal{D}$-modules and the existence of a b-function; (2). determine the equidimensionality of the characteristic variety of a simple module over the enveloping algebra of a Lie algebra. In this chapter we first present a general study of Auslander regularity of filtered rings and graded rings; Then we generalize the important Gabber-Kashiwara theorem and Roos theorem to Zariski rings with commutative associated graded rings including most of the examples given in Ch.II. §2; and consequently the pure module theory over Zariski ring is developed as well.

§1. Spectral Sequence Determined by a Filt-complex

The aim of this section is to provide a self-contained exposition of the theory of spectral sequences determined by filtered complexes in such a way that the reader need not go into the details of the general theory of spectral sequences, i.e., without using the usual exact couple, each object in usual consideration may be given by a concrete filtered or graded structure and hence may be separately treated according to our needs.

1.1. Spectral sequence determined by a filt-complex

Let

$$(\mathbf{M}, f, s) \qquad \cdots \longrightarrow M_{j-1} \xrightarrow{f_j} M_j \xrightarrow{f_{j+1}} M_{j+1} \longrightarrow \cdots$$

be a given **cochain complex of filtered abelian groups**, namely, every M_j has a increasing filtration consisting of subgroups, say $FM_j = \{F_nM_j,\ F_nM_j \subseteq F_{n+1}M_j,\ n \in \mathbb{Z}\}$, such that

    (i). all FM_j are exhaustive, i.e, $M_j = \cup_{n \in \mathbb{Z}} F_nM_j,\ j \in \mathbb{Z}$,

    (ii). $f_j(F_nM_{j-1}) \subseteq F_{n+s}M_j,\ n \in \mathbb{Z}$, for some integer $s \geq 0$ and all $j \in \mathbb{Z}$, i.e., all f_j are
       of degree s.

For every cohomology group $\mathbf{H}_{j-1}(\mathbf{M}) = \mathrm{Ker}f_j/\mathrm{Im}f_{j-1}$ of $(\mathbf{M}, f, s)$ we may define a filtration $F\mathbf{H}_{j-1}(\mathbf{M})$ on $\mathbf{H}_{j-1}(\mathbf{M})$ by putting

$$F_n\mathbf{H}_{j-1}(\mathbf{M}) = (\mathrm{Ker}f_j \cap F_nM_{j-1}) + \mathrm{Im}f_{j-1}/\mathrm{Im}f_{j-1},\ n \in \mathbb{Z},$$

such that the associated graded abelian group is

$$G(\mathbf{H}_{j-1}(\mathbf{M})) = \bigoplus_{n \in \mathbb{Z}} \frac{\mathrm{Ker}f_j \cap F_nM_{j-1} + \mathrm{Im}f_{j-1}}{\mathrm{Ker}f_j \cap F_{n-1}M_{j-1} + \mathrm{Im}f_{j-1}}$$

On the other hand, we have the associated cochain complex of graded abelian groups

$$(\mathbf{GM}, G(f), s) \qquad \cdots \longrightarrow G(M_{j-1}) \xrightarrow{G(f_j)} G(M_j) \xrightarrow{G(f_{j+1})} G(M_{j+1}) \longrightarrow \cdots$$

where $G(M_p)$ is the associated graded abelian group of M_p and $G(f_j)$ is of degree s. Hence we have the corresponding cohomology groups

$$\{\mathbf{H}_{gr}^{j-1}(\mathbf{GM}) = \mathrm{Ker}G(f_j)/\mathrm{Im}G(f_{j-1}),\ j \in \mathbb{Z}\}$$

and each $\mathbf{H}_{gr}^{j-1}(\mathbf{GM})$ is a graded abelian group.

Now, for each $j \in \mathbb{Z}$, any $n \in \mathbb{Z}$ and $k \geq 0$ we define

$$
\begin{aligned}
Z_n^k(j) &= \{x \in F_nM_{j-1},\ f_j(x) \in F_{n-k+s}M_j\} \\
B_n^k(j) &= F_nM_{j-1} \cap f_{j-1}(F_{n+k-s-1}M_{j-2}) \\
Z_n^\infty(j) &= \{x \in F_nM_{j-1},\ f_j(x) \in \cap_{p \in \mathbb{Z}} F_pM_j\} \\
B_n^\infty(j) &= F_nM_{j-1} \cap \mathrm{Im}f_{j-1}
\end{aligned}
$$

Note that $f_j f_{j-1} = 0$ (and all FM_p are exhaustive) so we have $B_n^k(j) \subseteq Z_n^k(j)$ and

$$
\begin{aligned}
F_nM_{j-1} &= Z_n^0(j) \supseteq Z_n^1(j) \supseteq \cdots \supseteq Z_n^\infty(j) = \cap_{k \geq 1} Z_n^k(j) \\
B_n^0(j) &\subseteq B_n^1(j) \subseteq \cdots \subseteq B_n^\infty(j) = \cup_{k \geq 0} B_n^k(j)
\end{aligned}
$$

Thus for each j we may construct a sequence of graded abelian groups $\{E_j^k\}_{k=1}^\infty$ by putting

$$E_j^k = \bigoplus_{n \in \mathbb{Z}} \frac{Z_n^k(j) + F_{n-1}M_{j-1}}{B_n^k(j) + F_{n-1}M_{j-1}}$$

1. Observations. (0). $E_j^0 = G(M_{j-1})$; $E_j^1 = \mathbf{H}_{gr}^{j-1}(\mathbf{GM})$.

(1). $Z_n^k(j) \cap F_{n-1}M_{j-1} = Z_{n-1}^{k-1}(j)$; and

$$E_j^k \cong \bigoplus_{n \in \mathbb{Z}} \frac{Z_n^k(j)}{B_n^k(j) + Z_{n-1}^{k-1}(j)}$$

where $k \geq 1$, in particular

$$E_j^\infty \cong \bigoplus_{n \in \mathbb{Z}} \frac{Z_n^\infty(j)}{B_n^\infty(j) + Z_{n-1}^\infty(j)}$$

(2). $f_j(Z_n^k(j)) = B_{n-k+s}^{k+1}(j+1)$, $n \in \mathbb{Z}$, $k \geq 1$.

(3). Since $f_j f_{j-1} = 0$, for each $k \geq 1$, there is a graded morphism $\chi_j(s-k) \colon E_{j-1}^k \to E_j^k$ such that for any

$$\overline{z_{n+k-s}^k} \in (E_{j-1}^k)_{n+k-s} = \frac{Z_{n+k-s}^k(j-1)}{B_{n+k-s}^k(j-1) + Z_{n+k-s-1}^{k-1}(j-1)},$$

we have

$$\chi_j(s-k)\left(\overline{z_{n+k-s}^k}\right) = \overline{f_{j-1}(z_{n+k-s}^k)} \in (E_j^k)_n = \frac{Z_n^k(j)}{B_n^k(j) + Z_{n-1}^{k-1}(j)},$$

where $n \in \mathbb{Z}$, i.e., $\chi_j(s-k)$ is of degree $s-k$, and $\chi_j(s-k)\chi_{j-1}(s-k) = 0$.

(4). It follows from (2) and the formula $f_j f_{j-1} = 0$ that

$$
\begin{aligned}
\mathrm{Im}\chi_j(s-k) &= \bigoplus_{n \in \mathbb{Z}} \frac{B_n^{k+1}(j) + Z_{n-1}^{k-1}(j)}{B_n^k(j) + Z_{n-1}^{k-1}(j)} \\
&\cong \bigoplus_{n \in \mathbb{Z}} \frac{B_n^{k+1}(j)}{B_n^{k+1}(j) \cap (B_n^k(j) + Z_{n-1}^{k-1}(j))} \\
&= \bigoplus_{n \in \mathbb{Z}} \frac{B_n^{k+1}(j)}{B_n^k(j) + B_n^{k+1}(j) \cap F_{n-1}M_{j-1}} \\
&= \bigoplus_{n \in \mathbb{Z}} \frac{B_n^{k+1}(j) + F_{n-1}M_{j-1}}{B_n^k(j) + F_{n-1}M_{j-1}}
\end{aligned}
$$

(5). Let us now compute $\mathrm{Ker}\chi_{j+1}(s-k)$. If $z_n^k \in Z_n^k(j)$ is such that

$$
\begin{aligned}
f_j(z_n^k) &\in B_{n+s-k}^k(j+1) + Z_{n+s-k-1}^{k-1}(j+1) \\
&\subseteq B_{n+s-k}^k(j+1) + F_{n+s-k-1}M_j,
\end{aligned}
$$

then since $B_{n+s-k}^k(j+1) = f_j(Z_{n-1}^{k-1}(j))$ we have $f_j(z_n^k) = f_j(y) + u$, for some $y \in Z_{n-1}^{k-1}(j) \subseteq F_{n-1}M_{j-1}$, $u \in F_{n+s-k-1}M_j$. Hence $f_j(z_n^k - y) = u \in F_{n+s-k-1}M_j$ and $z_n^k - y \in Z_n^{k+1}(j)$ because $z_n^k - y \in F_n M_{j-1}$. It follows that $z_n^k \in F_{n-1}M_{j-1} + Z_n^{k+1}(j) \subseteq F_{n-1}M_{j-1} + Z_n^k(j)$. On the other hand,

$$
\frac{F_{n-1}M_{j-1} + Z_n^{k+1}(j)}{F_{n-1}M_{j-1} + B_n^k(j)} \subseteq \frac{F_{n-1}M_{j-1} + Z_n^k(j)}{F_{n-1}M_{j-1} + B_n^k(j)}
$$

$$
\cong \frac{Z_n^k(j)}{B_n^k(j) + Z_n^k(j) \cap F_{n-1}M_{j-1}}
$$

$$
= \frac{Z_n^k(j)}{B_n^k(j) + Z_{n-1}^{k-1}(j)}
$$

$$
= (E_j^k)_n.
$$

This shows that

$$
\text{Ker}\chi_{j+1}(s-k) \subseteq \bigoplus_{n \in \mathbb{Z}} \frac{F_{n-1}M_{j-1} + Z_n^{k+1}(j)}{F_{n-1}M_{j-1} + B_n^k(j)}.
$$

Since $f_j(Z_n^{k+1}(j)) = B_{n+s-k-1}^{k+2}(j+1) \subseteq Z_{n+s-k-1}^{k-1}(j+1)$, it follows that

$$
\frac{F_{n-1}M_{j-1} + Z_n^{k+1}(j)}{F_{n-1}M_{j-1} + B_n^k(j)} \subseteq \text{Ker}\chi_{j+1}(s-k),
$$

and therefore

$$
\text{Ker}\chi_{j+1}(s-k) = \bigoplus_{n \in \mathbb{Z}} \frac{F_{n-1}M_{j-1} + Z_n^{k+1}(j)}{F_{n-1}M_{j-1} + B_n^k(j)}.
$$

Summing up, we have obtained the following equalities

$$
\text{Ker}\chi_{j+1}(s-k)/\text{Im}\chi_j(s-k) = E_j^{k+1}, \; k \geq 1.
$$

(6). The sequence $\{E_j^k, \; \chi_j(s-k); \; k \geq 1, \; j \in \mathbb{Z}\}$ of graded abelian groups E_j^k and graded morphisms of degree $s - k$ is called the **spectral sequence** determined by the filt-complex $(\mathbf{M}, f, s)$, and denoted (E, E^k). More precisely, from the spectral sequence constructed above we have the following table consisting of cochain complexes

$$
\begin{array}{ccccccc}
\cdots & \to & E_{j-1}^1 & \longrightarrow & E_j^1 & \longrightarrow & E_{j+1}^1 & \to & \cdots \\
& & \cdots & & \cdots & & \cdots \\
\cdots & \to & E_{j-1}^k & \xrightarrow{\chi_j(s-k)} & E_j^k & \xrightarrow{\chi_{j+1}(s-k)} & E_{j+1}^k & \to & \cdots \\
\cdots & \to & E_{j-1}^{k+1} & \longrightarrow & E_j^{k+1} & \longrightarrow & E_{j+1}^{k+1} & \to & \cdots \\
& & \cdots & & \cdots & & \cdots
\end{array}
$$

where $E_j^{k+1} \cong \mathrm{Ker}\chi_{j+1}(s-k)/\mathrm{Im}\chi_j(s-k)$, $k \geq 1$, $j \in \mathbb{Z}$, and the k-th complex is denoted by $(E^k, \chi(s-k))$.

(7). If $E_{j-1}^1 = 0$ then $E_{j-1}^k = 0$, where $k \geq 1$, and also $\mathrm{Ker}\chi_{j+1}(s-k) \cong E_j^{k+1}$, i.e., at this stage we may view E_j^{k+1} as a submodule of E_j^k. Hence we arrive at an exact sequence:

$$0 \longrightarrow E_j^{k+1} \longrightarrow E_j^k \longrightarrow S_{j+1}^k \longrightarrow 0$$

where $S_{j+1}^k = \mathrm{Im}\chi_{j+1}(s-k) \subseteq E_{j+1}^k$.

2. Definition. For any $j \in \mathbb{Z}$ the sequence $\{E_j^k\}_{k=1}^{\infty}$ is said to be **weakly convergent** if there is a filtered abelian group H with filtration FH such that $G(H) \cong E_j^{\infty}$ as graded groups. In this case we also say that the sequence $\{E_j^k\}_{k=1}^{\infty}$ weakly converges to H. If for every $j \in \mathbb{Z}$ the sequence $\{E_j^k\}_{k=1}^{\infty}$ is weakly convergent, then we say that the spectral sequence (E, E^k) is weakly convergent.

3. Remark. Suppose that the filtration FM_j on M_j is separated, then by the definition we see that $Z_n^{\infty}(j) = F_n M_{j-1} \cap \mathrm{Ker}f_j$. But on the other hand we have

$$
\begin{aligned}
G(\mathbf{H}_{j-1}(\mathbf{M})) &= \bigoplus_{n \in \mathbf{Z}} \frac{\mathrm{Ker}f_j \cap F_n M_{j-1} + \mathrm{Im}f_{j-1}}{\mathrm{Ker}f_j \cap F_{n-1} M_{j-1} + \mathrm{Im}f_{j-1}} \\
&= \bigoplus_{n \in \mathbf{Z}} \frac{Z_n^{\infty}(j) + \mathrm{Im}f_{j-1}}{Z_{n-1}^{\infty}(j) + \mathrm{Im}f_{j-1}} \\
&\cong \bigoplus_{n \in \mathbf{Z}} \frac{Z_n^{\infty}(j)}{Z_n^{\infty}(j) \cap (Z_{n-1}^{\infty}(j) + \mathrm{Im}f_{j-1})} \\
&\cong \bigoplus_{n \in \mathbf{Z}} \frac{Z_n^{\infty}(j)}{Z_{n-1}^{\infty}(j) + B_n^{\infty}(j)} \\
&\cong E_j^{\infty}
\end{aligned}
$$

This shows that the sequence $\{E_j^k\}_{k=1}^{\infty}$ is weakly convergent to $\mathbf{H}_{j-1}(\mathbf{M})$.

4. Definition. (1). Let M be a filtered abelian group with filtration FM. FM is said to be **left bounded** if there exists a $t \in \mathbb{Z}$ such that $F_n M = 0$ for $n \leq t$; FM is said to be **right bounded** if there exists an $s \in \mathbb{Z}$ such that $F_n M = M$ for all $n \geq s$.

(2). For any $j \in \mathbb{Z}$ the sequence $\{E_j^k\}_{k=1}^{\infty}$ is said to be **strongly convergent** if there is a filtered abelian group H with a left and right bounded filtration FH such that as graded groups $G(H) \cong E_j^{\infty}$. In this case we also say that the sequence $\{E_j^k\}_{k=1}^{\infty}$ strongly converges to H. If for every $j \in \mathbb{Z}$ the sequence $\{E_j^k\}_{k=1}^{\infty}$ strongly converges then we say that the spectral sequence (E, E^k) is strongly convergent.

The following proposition and its corollary show that under certain conditions one may get information for the cohomology group $\mathbf{H}_{j-1}(\mathbf{M})$ from the cohomology group $\mathbf{H}_{gr}^{j-1}(\mathbf{GM})$, although the sequence $\{E_j^k\}_{k=1}^{\infty}$ is now not necessarily convergent.

5. Proposition. With notation as before, for any $j \in \mathbb{Z}$ if there exists an $\omega \in I\!N$ such that for all $n \in \mathbb{Z}$ the following equalities hold

$$\operatorname{Ker} f_j \cap F_n M_{j-1} + F_{n-1} M_{j-1} = Z_n^{\omega}(j) + F_{n-1} M_{j-1}$$

then $G(\mathbf{H}_{j-1}(\mathbf{M}))$, is as a graded group isomorphic to a subquotient of $\mathbf{H}_{gr}^{j-1}(\mathbf{GM})$, i.e., there exist graded subgroups $T_1 \subseteq T_2 \subseteq \mathbf{H}_{gr}^{j-1}(\mathbf{GM})$ such that $G(\mathbf{H}_{j-1}(\mathbf{M})) \cong T_2/T_1$.

Proof. By definition we have

$$
\begin{aligned}
G(\mathbf{H}_{j-1}(\mathbf{M})) &= \bigoplus_{n \in \mathbb{Z}} \frac{\operatorname{Ker} f_j \cap F_n M_{j-1} + \operatorname{Im} f_{j-1}}{\operatorname{Ker} f_j \cap F_{n-1} M_{j-1} + \operatorname{Im} f_{j-1}} \\
&\cong \bigoplus_{n \in \mathbb{Z}} \frac{\operatorname{Ker} f_j \cap F_n M_{j-1}}{(\operatorname{Ker} f_j \cap F_n M_{j-1}) \cap (\operatorname{Ker} f_j \cap F_{n-1} M_{j-1} + \operatorname{Im} f_{j-1})} \\
&\cong \bigoplus_{n \in \mathbb{Z}} \frac{\operatorname{Ker} f_j \cap F_n M_{j-1}}{\operatorname{Ker} f_j \cap F_{n-1} M_{j-1} + B_n^{\infty}(j)};
\end{aligned}
$$

On the other hand, our assumption entails that

$$
\begin{aligned}
E_j^{\omega} &= \bigoplus_{n \in \mathbb{Z}} \frac{Z_n^{\omega}(j) + F_{n-1} M_{j-1}}{B_n^{\omega}(j) + F_{n-1} M_{j-1}} \\
&= \bigoplus_{n \in \mathbb{Z}} \frac{\operatorname{Ker} f_j \cap F_n M_{j-1} + F_{n-1} M_{j-1}}{B_n^{\omega}(j) + F_{n-1} M_{j-1}} \\
&\cong \bigoplus_{n \in \mathbb{Z}} \frac{\operatorname{Ker} f_j \cap F_n M_{j-1}}{(\operatorname{Ker} f_j \cap F_n M_{j-1}) \cap (B_n^{\omega}(j) + F_{n-1} M_{j-1})} \\
&\cong \bigoplus_{n \in \mathbb{Z}} \frac{\operatorname{Ker} f_j \cap F_n M_{j-1}}{B_n^{\omega}(j) + \operatorname{Ker} f_j \cap F_{n-1} M_{j-1}}.
\end{aligned}
$$

From $B_n^{\omega+1}(j) \subseteq B_n^{\infty}(j)$ we obtain a graded epimorphism from $E_j^{\omega+1}$ to $G(\mathbf{H}_{j-1}(\mathbf{M}))$ and thus $G(\mathbf{H}_{j-1}(\mathbf{M}))$ is a quotient group of $E_j^{\omega+1}$. In view of Observation 1.1.1.(6). we may conclude that $G(\mathbf{H}_{j-1}(\mathbf{M}))$ is a subquotient of $\mathbf{H}_{gr}^{j-1}(\mathbf{GM}) = E_j^1$. $\square$

6. Corollary. For any $j \in \mathbb{Z}$, if there exists an $\omega \in I\!N$ such that $E_j^\omega = 0$ then $G(\mathbf{H}_{j-1}(\mathbf{M})) = 0$.

Proof. By the definition of E_j^ω and the assumption we have for all $n \in \mathbb{Z}$:

$$Z_n^\omega(j) + F_{n-1}M_{j-1} = B_n^\omega(j) + F_{n-1}M_{j-1}$$
$$\subseteq F_n M_{j-1} \cap \operatorname{Ker} f_j + F_{n-1}M_{j-1}$$

and hence $Z_n^\omega(j) + F_{n-1}M_{j-1} = F_n M_{j-1} \cap \operatorname{Ker} f_j + F_{n-1}M_{j-1}$. Thus we see from the proof of Proposition 1.1.5. that $G(\mathbf{H}_{j-1}(\mathbf{M}))$ is an epimorphic image of E_j^ω and it follows from the assumption that $G(\mathbf{H}_{j-1}(\mathbf{M})) = 0$. □

From the proof of Proposition 1.1.5. we also derive the following

7. Corollary. For any $j \in \mathbb{Z}$, if there exists an $\omega \in I\!N$ such that for all $n \in \mathbb{Z}$ the following equalities hold

$$\operatorname{Ker} f_j \cap F_n M_{j-1} + F_{n-1}M_{j-1} = Z_n^\omega(j) + F_{n-1}M_{j-1}$$
$$\operatorname{Im} f_{j-1} \cap F_n M_{j-1} + F_{n-1}M_{j-1} = B_n^\omega(j) + F_{n-1}M_{j-1}$$

then for all $k \geq \omega$, $G(\mathbf{H}_{j-1}(\mathbf{M})) = E_j^k$.

8. Corollary. If the filtration FM_j on M_j is left bounded then the sequence $\{E_j^k\}_{k=1}^\infty$ is weakly convergent to $\mathbf{H}_{j-1}(\mathbf{M})$ and $G(\mathbf{H}_{j-1}(\mathbf{M}))$ is isomorphic to a subquotient of $\mathbf{H}_{gr}^{j-1}(\mathbf{GM})$; If FM_j is left bounded and FM_{j-2} is right bounded, then the sequence $\{E_j^k\}_{k=1}^\infty$ is strongly convergent to $\mathbf{H}_{j-1}(\mathbf{M})$ and there is some $\omega \geq 1$ such that $G(\mathbf{H}_{j-1}(\mathbf{M})) = E_j^k$ for all $k \geq \omega$.

9. Proposition. For any $j \in \mathbb{Z}$, if there exists an $\omega_1 \in I\!N$ such that

$$F_n M_j \cap f_j(M_{j-1}) \subseteq f_j(F_{n+\omega_1} M_{j-1})$$

for all $n \in \mathbb{Z}$, then the first equality of Corollary 1.1.7. holds; If there exists an $\omega_2 \in I\!N$ such that

$$F_n M_{j-1} \cap f_{j-1}(M_{j-2}) \subseteq f_{j-1}(F_{n+\omega_2} M_{j-2})$$

for all $n \in \mathbb{Z}$, then the second equality of Corollary 1.1.7. holds.

Proof. Suppose that there exists an $\omega_1 \in I\!N$ such that

$$F_n M_j \cap f_j(M_{j-1}) \subseteq f_j(F_{n+\omega_1} M_{j-1}), \ n \in \mathbb{Z}.$$

If $x \in Z_n^{\omega_1+1}(j)$, then $f_j(x) \in F_{n-\omega_1-1}M_j \cap f_j(M_{j-1}) \subseteq f_j(F_{n-1}M_{j-1})$. Therefore $f_j(x) = f_j(y)$ for some $y \in F_{n-1}M_{j-1}$ and hence $f_j(x-y) = 0$, i.e., $x - y \in F_nM_{j-1} \cap \mathrm{Ker} f_j$. It follows that $x \in \mathrm{Ker} f_j \cap F_nM_{j-1} + F_{n-1}M_{j-1}$ and consequently we arrive at $Z_n^{\omega_1+1}(j) \subseteq \mathrm{Ker} f_j \cap F_nM_{j-1} + F_{n-1}M_{j-1}$. Put $\omega' = \omega_1 + 1$, then we obtain

$$Z_n^{\omega'}(j) + F_{n-1}M_{j-1} = \mathrm{Ker} f_j \cap F_nM_{j-1} + F_{n-1}M_{j-1}, \ n \in \mathbb{Z}.$$

Next, since all filtrations in consideration are increasing filtrations we may suppose that there exists some $\omega_2 \in \mathbb{N}$ such that for all $n \in \mathbb{Z}$, $F_nM_{j-1} \cap f_{j-1}(M_{j-2}) \subseteq f_{j-1}(F_{n+\omega_2-1}M_{j-2})$, i.e., for all $n \in \mathbb{Z}$ we have

$$F_nM_{j-1} \cap f_{j-1}(M_{j-2}) \subseteq F_nM_{j-1} \cap f_{j-1}(F_{n+\omega_2-1}M_{j-2}) = B_n^{\omega_2}(j).$$

Hence the equality

$$\mathrm{Im} f_{j-1} \cap F_nM_{j-1} + F_{n-1}M_{j-1} = B_n^{\omega_2}(j) + F_{n-1}M_{j-1}$$

hods for all $n \in \mathbb{Z}$. $\square$

The reader is invited to formulate the foregoing results for any **filtered chain complex**.

1.2. Spectral sequence determined by a complete filt-complex

Now, Suppose that we are given two cochain complexes of filtered abelian groups $(\mathbf{M}, f, 0)$ and $(\mathbf{M}', f', 0)$, and a cochain mapping of degree zero $\varphi \colon \mathbf{M} \to \mathbf{M}'$. Then we have the following cochain mapping:

$$
\begin{array}{ccccccccc}
\cdots & \to & M_{j-1} & \xrightarrow{f_j} & M_j & \xrightarrow{f_{j+1}} & M_{j+1} & \to & \cdots \\
 & & \downarrow{\varphi_{j-1}} & & \downarrow{\varphi_j} & & \downarrow{\varphi_{j+1}} & & \\
\cdots & \to & M'_{j-1} & \xrightarrow{f'_j} & M'_j & \xrightarrow{f'_{j+1}} & M'_{j+1} & \to & \cdots
\end{array}
$$

where each φ_j is a filtered morphism of degree zero. It follows that φ induces a morphism of the associated spectral sequences, say $\varphi^* \colon E \to E'$, namely, for every $k \geq 1$ there is a cochain mapping $\varphi^*(k) \colon E^k \to E'^k$ as follows

$$
\begin{array}{ccccccccc}
\cdots & \to & E_{j-1}^k & \xrightarrow{\chi_j(-k)} & E_j^k & \xrightarrow{\chi_{j+1}(-k)} & E_{j+1}^k & \to & \cdots \\
 & & \downarrow{\varphi^*_{j-1}} & & \downarrow{\varphi^*_j} & & \downarrow{\varphi^*_{j+1}} & & \\
\cdots & \to & E_{j-1}'^k & \xrightarrow{\chi'_j(-k)} & E_j'^k & \xrightarrow{\chi'_{j+1}(-k)} & E_{j+1}'^k & \to & \cdots
\end{array}
$$

where each φ^*_j is a graded morphism of degree zero.

Note that $Z_n^\infty(j) = \cap_{n\in\mathbb{Z}} Z_n^k(j)$ and $B_n^\infty(j) = \cup_{n\in\mathbb{Z}} B_n^k(j)$, if we look at the table consisting of E^k-sequences in subsection 1.1. then the following lemma is obvious:

1. Lemma. With notation as above, if $\varphi^*(k)$: $E^k \xrightarrow{\cong} E'^k$ for some $k \geq 1$, then $\varphi^*(n)$: $E^n \xrightarrow{\cong} E'^n$ for $k \leq n \leq \infty$.

2. Remark. From Remark 1.1.3. we know that if FM_j on M_j is separated, then $G(\mathbf{H}_{j-1}(\mathbf{M})) \cong E_j^\infty$. Hence, if we assume that FM_j and FM_j' are separated then under the condition of Lemma 1.2.1. we have $\varphi^*(\infty)$: $G(\mathbf{H}_{j-1}(\mathbf{M})) \cong G(\mathbf{H}_{j-1}'(\mathbf{M}'))$ as graded abelian groups and $\varphi^*(\infty) = G(\varphi_{j-1}^H)$, where φ_{j-1}^H is the filtered morphism from $\mathbf{H}_{j-1}(\mathbf{M})$ to $\mathbf{H}_{j-1}'(\mathbf{M}')$ induced by φ. So the natural question is when one may conclude from the isomorphism $G(\varphi_{j-1}^H)$ that φ_{j-1}^H is an isomorphism.

3. Proposition. With notation as above, if $\mathbf{H}_{j-1}(M)$ is complete with respect to its filtration topology and $F\mathbf{H}_{j-1}'(\mathbf{M}')$ is separated, then φ_{j-1}^H is an isomorphism whenever $G(\varphi_{j-1}^H)$ is an isomorphism.

Proof. By the assumption this follows from Ch.I. Corollary 4.2.5.. □

4. Definition. A cochain complex $(\mathbf{M}, f, 0)$ of filtered abelian groups is said to be a **complete filt-complex** if every M_j is complete with respect to its given filtration topology.

5. Lemma. Suppose that $(\mathbf{M}, f, 0)$ is a complete filt-complex, then

(1). $\{E_j^k\}_{k=1}^\infty$ weakly converges to $\mathbf{H}_{j-1}(\mathbf{M})$, $j \in \mathbb{Z}$,

(2). $\mathrm{Ker} f_j$ is complete with respect to the $F\mathrm{Ker} f_j$-topology, $j \in \mathbb{Z}$, where $F\mathrm{Ker} f_j$ is the filtration on $\mathrm{Ker} f_j$ induced by FM_{j-1}.

Proof. (1). Since every FM_j is separated this follows from Remark 1.1.3..

(2). If $(x_i)_{i>0}$ is any Cauchy sequence in $\mathrm{Ker} f_j$ then it is also a Cauchy sequence in M_{j-1} because $F\mathrm{Ker} f_j$ is induced by FM_{j-1}. Hence $(x_i)_{i>0}$ converges in M_{j-1}, say $\lim_{i\to\infty} x_i = x$. Thus for any integer $p > 0$ there exists an integer $N(p) > 0$ such that for all $i \geq N(p)$ we have $x_i - x \in F_{-p}M_{j-1}$. Note that $x_i \in \mathrm{Ker} f_j$ we have $f_j(x) \in \cap_{k\in\mathbb{Z}} f_j(F_kM_{j-1}) \subseteq \cap_{k\in\mathbb{Z}} F_kM_j = 0$, hence $x \in \mathrm{Ker} f_j$. Moreover, the induced filtration $F\mathrm{Ker} f_j$ is also separated since FM_{j-1} is separated. Therefore $\mathrm{Ker} f_j$ is complete. □

6. Proposition. Let $(\mathbf{M}, f, 0)$ be a complete filt-complex as described before. For any

$j \in \mathbb{Z}$ if there exists an $\omega \in \mathbb{N}$ such that $E_j^\omega = 0$ and f_{j-1} is a strict filtered morphism, then $\mathbf{H}_{j-1}(\mathbf{M}) = 0$.

Proof. By Corollary 1.1.6. we have $G(\mathbf{H}_{j-1}(\mathbf{M})) = 0$. It follows from the definition of $G(\mathbf{H}_{j-1}(\mathbf{M}))$ that

$$\operatorname{Ker} f_j \cap F_n M_{j-1} + \operatorname{Im} f_{j-1} = \operatorname{Ker} f_j \cap F_{n-1} M_{j-1} + \operatorname{Im} f_{j-1}, \ n \in \mathbb{Z}.$$

Then for any $x_n \in \operatorname{Ker} f_j \cap F_n M_{j-1}$ we have $x_n = x_{n-1} + f_{j-1}(z_n)$ where $x_{n-1} \in \operatorname{Ker} f_j \cap F_{n-1} M_{j-1}$, $f_{j-1}(z_n) \in \operatorname{Im} f_{j-1}$. By an induction we have for any $k \geq 1$

$$x_n = x_{n-k} + \sum_{i=0}^{k-1} f_{j-1}(z_{n-i})$$

where $x_{n-k} \in F_{n-k} M_{j-1}$ and $f_{j-1}(z_{n-i}) \in F_{n-i} M_{j-1}$. By the assumption f_{j-1} is strict, so we may assume that $z_{n-i} \in F_{n-i} M_{j-2}$. Finally, since M_{j-1} and M_{j-2} are complete we have $\sum_{i \geq 0} z_{n-i} = y \in F_n M_{j-2}$ and $\lim\limits_{k \to \infty} x_{n-k} = 0$. Hence $x_n = \lim\limits_{k \to \infty} x_{n-k} + f_{j-1}(\sum_{i \geq 0} z_{n-i}) = f_{j-1}(y) \in \operatorname{Im} f_{j-1}$. This shows that $\operatorname{Ker} f_j \cap F_n M_{j-1} \in \operatorname{Im} f_{j-1}$ and consequently $F_n \mathbf{H}_{j-1}(\mathbf{M}) = 0$, $n \in \mathbb{Z}$, i.e., $\mathbf{H}_{j-1}(\mathbf{M}) = 0$. $\qquad\square$

7. Lemma. Let $(\mathbf{M}, f, 0)$ be a complete filt-complex as before. For any $j \in \mathbb{Z}$, $\operatorname{Im} f_{j-1}$ is complete with respect to the $F\operatorname{Im} f_{j-1}$-topology, where $F\operatorname{Im} f_{j-1}$ is defined by putting: $F_n \operatorname{Im} f_{j-1} = f_{j-1}(F_n M_{j-2})$, $n \in \mathbb{Z}$.

Proof. Since M_{j-2} is complete and by Lemma 1.2.5. $\operatorname{Ker} f_{j-1}$ is complete with respect to the filtration induced by FM_{j-2}, Ch.I. Theorem 3.4.13. yields an exact and commutative diagram

$$
\begin{array}{ccccccccc}
 & & 0 & & 0 & & & & \\
 & & \downarrow & & \downarrow & & & & \\
0 & \to & \operatorname{Ker} f_{j-1} & \longrightarrow & M_{j-2} & \longrightarrow & \operatorname{Im} f_{j-1} & \to & 0 \\
 & & \downarrow & & \downarrow & & \downarrow & & \\
0 & \to & \widehat{\operatorname{Ker} f_{j-1}} & \longrightarrow & \widehat{M_{j-2}} & \longrightarrow & \widehat{\operatorname{Im} f_{j-1}} & \to & 0 \\
 & & \downarrow & & \downarrow & & & & \\
 & & 0 & & 0 & & & &
\end{array}
$$

Hence $\operatorname{Im} f_{j-1} \cong \widehat{\operatorname{Im} f_{j-1}}$, i.e., $\operatorname{Im} f_{j-1}$ is complete. $\qquad\square$

8. Theorem. With notation as before, let $(M, f, 0)$ and $(M', f', 0)$ be two complete filt-complexes and φ: $M \to M'$ a cochain mapping of degree zero. Suppose that f and f' have the Artin-Rees property, i.e., all f_j and f'_j are filtered morphisms of degree zero having the Artin-Rees property in the sense of Ch.I. Defoinition 3.2.1.. If $\varphi^*(k)$: $E^k \overset{\cong}{\to} E'^k$ for some $k \geq 1$, then the induced filtered morphism φ^H_{j-1}: $\mathbf{H}_{j-1}(\mathbf{M}) \to \mathbf{H}'_{j-1}(\mathbf{M}')$ is an isomorphism for every $j \in \mathbb{Z}$.

Proof. Under the assumptions it follows from Lemma 1.2.5. that $\mathrm{Ker} f_j$ is complete; On the other hand, from Lemma 1.2.7. $\mathrm{Im} f_{j-1}$ is complete with respect to the filtration defined in Lemma 1.2.7.. Hence by Ch.I. Theorem 3.4.16. the Artin-Rees property now yields an exact and commutative diagram

$$
\begin{array}{ccccccc}
& & 0 & & 0 & & \\
& & \downarrow & & \downarrow & & \\
0 & \to & \mathrm{Im} f_{j-1} & \longrightarrow & \mathrm{Ker} f_j & \longrightarrow & \mathbf{H}_{j-1}(\mathbf{M}) \to 0 \\
& & \downarrow & & \downarrow & & \downarrow \\
0 & \to & \widehat{\mathrm{Im} f_{j-1}} & \longrightarrow & \widehat{\mathrm{Ker} f_j} & \longrightarrow & \mathbf{H}_{j-1}(\mathbf{M}) \to 0 \\
& & \downarrow & & \downarrow & & \\
& & 0 & & 0 & &
\end{array}
$$

Consequently $\mathbf{H}_{j-1}(\mathbf{M})$ is complete. Similarly $\mathbf{H}'_{j-1}(\mathbf{M}')$ is complete. Finally, it follows from Proposition 1.2.3. that φ^H_{j-1} is an isomorphism. $\qquad\square$

1.3. Double complexes and their spectral sequences

Recall that a **double complex** of abelian groups, denoted (A, δ, d), is a commutative diagram of abelian groups $A^{p,q}$:

$$
\begin{array}{ccc}
\uparrow & & \uparrow \\
\to \quad A^{p+1,q} & \longrightarrow & A^{p+1,q+1} \quad \to \\
\delta_{p,q} \uparrow & & \uparrow \\
\to \quad A^{p,q} & \xrightarrow{d_{p,q}} & A^{p,q+1} \quad \to \\
\uparrow & & \uparrow
\end{array}
$$

where the horizontal maps are denoted by d and the vertical maps by δ. We assume that $d^2 = \delta^2 = 0$ so that each and each column is a complex of abelian groups. We say that (A, δ, d) is a **double complex in the first quadrant** if $A^{vk} = 0$ whenever v or $k < 0$. Similarly one defines the double complex in other quadrants.

1. The first (second) filtration

For each $n \in \mathbb{Z}$, put

$$M_n = \oplus_{p+q=n} A^{p,q}$$

then we can define a filtration $F^I M_n$ on M_n as follows

$$F_l^I M_n = \oplus_{i \leq l} A^{n-i,i}, \; l \in \mathbb{Z}.$$

Obviously $G^I(M_n) \cong M_n$ and $G^I(M_n)_l \cong A^{n-l,l}$, where $G^I(M_n) = \oplus_{l \in \mathbf{z}} F_l^I M_n / F_{l-1}^I M_n$ denotes the associated graded abelian group of M_n. The filtration $F^I M_n$ defined above is called the **first filtration** (with respect to the given double complex). More clearly, the first filtration may be indicated by the following diagram:

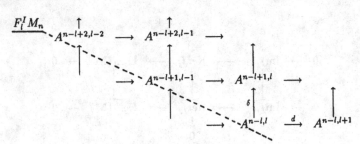

Moreover, let us define the following map Δ: $M_n \to M_{n+1}$ with

$$\Delta(\sum x_{p,q}) = \sum (-1)^p d(x_{p,q}) + \sum \delta(x_{p,q})$$

Then we have the following observations

(i). $\Delta^2 = 0$ (since $\delta d - d\delta = 0$).

(ii). $\Delta(F_l^I M_n) \subseteq F_{l+1}^I M_{n+1}$, $0 \leq l \leq n$.

(iii). The sequence

$$(\mathbf{M}, \Delta, 1): \qquad \cdots \longrightarrow M_{j-1} \xrightarrow{\Delta_j} M_j \xrightarrow{\Delta_{j+1}} M_{j+1} \longrightarrow \cdots$$

is a complex of filtered abelian groups with Δ being of **degree 1**.

(iV). In the associated complex of graded abelian groups

$$(\mathbf{GM}, G(\Delta), 1): \qquad \cdots \longrightarrow G(M_{j-1}) \xrightarrow{G(\Delta_j)} G(M_j) \xrightarrow{G(\Delta_{j+1})} G(M_{j+1}) \longrightarrow \cdots$$

each $G(\Delta_j)$ is of **degree 1**.

(V). For each $j \in \mathbb{Z}$, $\mathbf{H}_{j-1}(\mathbf{M}) = \mathrm{Ker}\Delta_j / \mathrm{Im}\Delta_{j-1}$ has a filtration

$$F_l \mathbf{H}_{j-1}(\mathbf{M}) = \frac{\mathrm{Ker}\Delta_j \cap F_l^I M_{j-1} + \mathrm{Im}\Delta_{j-1}}{\mathrm{Im}\Delta_{j-1}}, \; 0 \leq l \leq j-1,$$

and hence

$$G(\mathbf{H}_{j-1}(\mathbf{M})) = \bigoplus_{l=0}^{j-1} \frac{\mathrm{Ker}\Delta_j \cap F_l^I M_{j-1} + \mathrm{Im}\Delta_{j-1}}{\mathrm{Ker}\Delta_j \cap F_{l-1}^I M_{j-1} + \mathrm{Im}\Delta_{j-1}}$$

(Vi). Furthermore, look at the diagram

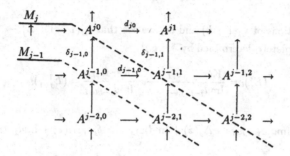

then by the definitions of Δ_j and the first filtration $F^I M_{j-1}$ on M_{j-1} we have the following chain isomorphism:

$$\cdots \to \quad A^{j-1,0} \quad \overset{d_{j-1,0}}{\Longrightarrow} \quad A^{j-1,1} \quad \overset{d_{j-1,1}}{\Longrightarrow} \quad A^{j-1,2} \quad \to \cdots$$
$$\cong \Big\downarrow \qquad\qquad \cong \Big\downarrow \qquad\qquad \cong \Big\downarrow$$
$$\cdots \to \quad G^I(M_{j-1})_0 \quad \overset{G(\Delta_j)_0}{\Longrightarrow} \quad G^I(M_j)_1 \quad \overset{G(\Delta_{j+1})_1}{\Longrightarrow} \quad G^I(M_{j+2})_2 \quad \to \cdots$$

Hence for $j \in \mathbb{Z}$ we have

$$\begin{aligned}
E_j^1 &= \mathbf{H}_{gr}^{j-1}(\mathbf{GM}) \\
&= \mathrm{Ker}G(\Delta_j)/\mathrm{Im}G(\Delta_{j-1}) \\
&\cong \bigoplus_{l\in\mathbb{Z}} \mathrm{Ker}d_{j-1-l,l}/\mathrm{Im}d_{j-1-l,l-1}
\end{aligned}$$

where E_j^1 is the first term in the spectral sequence determined by the complex $(\mathbf{M}, \Delta, 1)$. This means that $G(\Delta)$ is determined by the map d.

(Vii). Finally, we want to see how to calculate E_j^2 in the spectral sequence. If we look at the commutative diagram from the given double complex:

$$\begin{array}{ccccc}
A^{j-l,l-1} & \longrightarrow & A^{j-l,l} & \longrightarrow & A^{j-l,l+1} \\
\uparrow & & \uparrow{\scriptstyle\delta_{j-1-l,l}} & & \uparrow \\
A^{j-1-l,l-1} & \xrightarrow{d_{j-1-l,l-1}} & A^{j-1-l,l} & \longrightarrow & A^{j-1-l,l+1} \\
\uparrow & & \uparrow{\scriptstyle\delta_{j-2-l,l}} & & \uparrow \\
A^{j-2-l,l-1} & \xrightarrow{d_{j-2-l,l-1}} & A^{j-2-l,l} & \longrightarrow & A^{j-2-l,l+1}
\end{array}$$

then by the definitions of $\chi_j(1-k)$ and Δ_j we see that $\chi_j(1-1)$: $E^1_j \to E^1_{j+1}$ is of degree zero and it is completely determined by δ, i.e.,

$$(E^1_j)_l = \frac{\mathrm{Ker}d_{j-1-l,l}}{\mathrm{Im}d_{j-1-l,l-1}} \longrightarrow \frac{\mathrm{Ker}d_{j-l,l}}{\mathrm{Im}d_{j-l,l-1}} = (E^1_{j+1})_l$$

with

$$z + \mathrm{Im}d_{j-1-l,l-1} \mapsto \Delta_j(z) + \mathrm{Im}d_{j-l,l-1} = \delta_{j-1-l,l}(z) + \mathrm{Im}d_{j-l.l-1}.$$

Let M_n be as before. If we put

$$F^{II}_l M_n = \oplus_{i \leq l} A^{i,n-i},\ l \in \mathbb{Z},$$

then this is also a filtration on M_n; it is called the **second filtration** (with respect to the given double complex). Comparing with the first filtration we may indicate the second filtration by the following diagram:

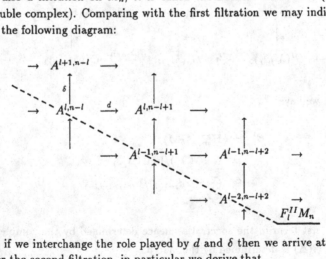

Accordingly, if we interchange the role played by d and δ then we arrive at a similar argumentation for the second filtration, in particular we derive that

$$(E^1_j)_l = \mathrm{Ker}\delta_{l,j-1-l}/\mathrm{Im}\delta_{l-1,j-1-l}$$

and χ_j is now completely determined by d.

Now, let

$$(\mathbf{M}, f, s) \qquad \cdots \longrightarrow M_{j-1} \xrightarrow{f_j} M_j \xrightarrow{f_{j+1}} M_{j+1} \longrightarrow \cdots$$

be a filt-complex of filtered abelian groups with f being of degree s for some $s \geq 0$ (i.e., all f_j are of degree s). We say that $(\mathbf{M}, f, s)$ is **bounded** if, for each $j \in \mathbb{Z}$, there exist integers $t = t(j)$, $v = v(j)$ such that $v(j) \geq v(j-1)$, $t(j) \leq t(j-1)$ and that $F_t M_j = 0$, $F_v M_j = M_j$.

2. Lemma. Let $(\mathbf{M}, f, s)$ be as before. If $(\mathbf{M}, f, s)$ is bounded then for each $j \in \mathbb{Z}$ there exists an $w \geq 0$ such that the two equalities in Corollary 1.1.7. hold and hence $G(\mathbf{H}_{j-1}(\mathbf{M})) \cong E_j^k$ for all $k \geq w$, where $\mathbf{H}_{j-1}(\mathbf{M}) = \operatorname{Ker} f_j / \operatorname{Im} f_{j-1}$ and E_j^k is a term of the spectral sequence (E, E^k) determined by $(\mathbf{M}, f, s)$.

Proof. By the definition:

$$Z_n^k(j) = \{x \in F_n M_{j-1}, \ f_j(x) \in F_{n-k+s} M_j\}$$
$$B_n^k(j) = F_n M_{j-1} \cap f_{j-1}(F_{n+k-s-1} M_{j-2}).$$

Taking $w = v(j) - t(j) + s$, then for $n - w + s \leq t(j)$ we obtain

$$F_{n-1} M_{j-1} + Z_n^w(j) \subseteq Z_n^\infty(j) + F_{n-1} M_{j-1}$$

and hence

$$F_{n-1} M_{j-1} + Z_n^w(j) = Z_n^\infty(j) + F_{n-1} M_{j-1}.$$

If $n - w + s > t(j)$, then since $n - w + s = n - v(j) + t(j)$ it follows that $n > v(j)$. Consequently $n - 1 \geq v(j) \geq v(j-1)$ and we also obtain

$$F_{n-1} M_{j-1} + Z_n^w(j) = M_{j-1}$$
$$= Z_n^\infty(j) + F_{n-1} M_{j-1}.$$

If $n \leq t(j-1)$, then it is easy to see that $B_n^w(j) = 0 = B_n^\infty(j)$. If $n > t(j-1)$, then

$$n + w - s - 1 = n + v(j) - t(j) - 1$$
$$> t(j-1) + v(j) - t(j) - 1$$
$$\geq t(j) + v(j) - t(j) - 1$$
$$= v(j) - 1,$$

i.e., $n + w - s - 1 \geq v(j) \geq v(j-2)$. Hence $B_n^w(j) = F_n M_{j-1} \cap f_{j-1} M_{j-2} = B_n^\infty(j)$. $\square$

3. Corollary. The first (second) filtration determined by a double complex in the first quadrant or third quadrant gives bounded filt-complex $(\mathbf{M}, \Delta, 1)$. Hence the conclusion of the last lemma holds for both filtrations.

Let us now pay some attention to the actual construction of a double complex.
Let R be a ring and

$$0 \longrightarrow S^0 \xrightarrow{f^0} S^1 \xrightarrow{f^1} S^2 \longrightarrow \cdots$$

a complex of R-modules. Let $H^j = Z^j/B^j$ be the j-th cohomology group of this complex, where $Z^j = \mathrm{Ker} f^j$, $B^j = \mathrm{Im} f^{j-1}$. If $j = 0$ there is no boundary and thus $H^0 = Z^0$. We also put $B^0 = 0$. For each $j \geq 0$ we may apply the well known **Horseshoe lemma** to the exact sequence

$$0 \longrightarrow B^j \longrightarrow Z^j \longrightarrow H^j \longrightarrow 0$$

This yields commutative diagrams

$$
\begin{array}{ccccc}
\downarrow & & \downarrow & & \downarrow \\
T^{1j} & \longrightarrow & K^{1j} & \longrightarrow & R^{1j} \\
\alpha \downarrow & & \sigma \downarrow & & \beta \downarrow \\
T^{0j} & \longrightarrow & K^{0j} & \longrightarrow & R^{0j} \\
\alpha' \downarrow & & \sigma' \downarrow & & \beta' \downarrow \\
0 \to B^j & \longrightarrow & Z^j & \longrightarrow & H^j & \to 0 \\
\downarrow & & \downarrow & & \downarrow \\
0 & & 0 & & 0
\end{array}
$$

where $K^{vj} = T^{vj} \oplus R^{vj}$ for all v and j, and each column gives a projective resolution of B^j, Z^j and H^j respectively.

Next, put

$$F^{vj} = T^{vj} \oplus R^{vj} \oplus T^{v,j+1} = K^{vj} \oplus T^{v,j+1}$$

we may obtain an R-linear map ϑ from F^{vj} to $F^{v-1,j}$ by putting $\vartheta(t_{vj}, r_{vj}, t_{v,j+1}) = (\alpha(t_{vj}), \beta(r_{vj}), \alpha(t_{v,j+1}))$. Since the diagram given above has exact columns this definition implies that each column in the diagram

$$
\begin{array}{ccc}
\downarrow & & \downarrow \\
F^{10} & \xrightarrow{\ \tau\ } & F^{11} & \rightarrow \\
\vartheta \downarrow & & \vartheta \downarrow \\
F^{00} & \xrightarrow{\ \tau\ } & F^{01} & \rightarrow
\end{array}
$$

is exact, where the map $\tau\colon F^{vj} \to F^{v,j+1}$, is defined as follows: $\tau(t_{vj}, r_{vj}, t_{v,j+1}) = (t_{v,j+1}, 0, 0)$. From the definition it is obvious that each row

$$0 \longrightarrow F^{v0} \longrightarrow F^{v1} \longrightarrow F^{v2} \longrightarrow \cdots$$

in the above diagram is a complex and $\tau\vartheta = \vartheta\tau$. Thus if we use the foregoing projective resolutions of Z^j and B^{j+1} and apply the Horseshoe lemma to the exact sequence

$$0 \longrightarrow Z^j \longrightarrow S^j \longrightarrow B^{j+1} \longrightarrow 0$$

then we obtain a projective resolution of S^j which is given by F^{vj}:

$$\cdots \longrightarrow F^{2j} \xrightarrow{\vartheta} F^{1j} \xrightarrow{\vartheta} F^{0j} \xrightarrow{\psi} S^j \longrightarrow 0$$

Therefore, we have obtained the following double complex in the first quadrant

$$
\begin{array}{ccccccc}
& \downarrow & & \downarrow & & \downarrow & \\
& F^{10} & \longrightarrow & F^{11} & \longrightarrow & F^{12} & \to \\
& {\scriptstyle\vartheta}\downarrow & & \downarrow & & \downarrow & \\
& F^{00} & \xrightarrow{\tau} & F^{01} & \longrightarrow & F^{02} & \to \\
& \downarrow & & \downarrow & & \downarrow & \\
0 \to & S^0 & \longrightarrow & S^1 & \longrightarrow & S^2 & \to \\
& \downarrow & & \downarrow & & \downarrow & \\
& 0 & & 0 & & 0 &
\end{array}
$$

By the construction of this double complex we easily see that

$$\operatorname{Ker}\tau_{vj}/\operatorname{Im}\tau_{v,j-1} = K^{vj}/T^{vj} = R^{vj},$$

for all v and j. Furthermore, since the sequence

$$0 \longrightarrow R^{vj} \longrightarrow K^{vj} \longrightarrow T^{vj} \longrightarrow 0$$

is splitting exact we check that the sequence

$$0 \longrightarrow \operatorname{Hom}_R(T^{vj}, R) \longrightarrow \operatorname{Hom}_R(K^{vj}, R) \longrightarrow \operatorname{Hom}_R(R^{vj}, R) \longrightarrow 0$$

is exact, too. Hence we get

$$(K^{vj})^*/(T^{vj})^* = (R^{vj})^* = (\operatorname{Ker}\tau_{vj}/\operatorname{Im}\tau_{v,j-1})^*.$$

On the other hand, from the construction of F^{vj} we have

$$\mathrm{Hom}_R(F^{vj},R) = (\mathrm{Hom}_R(T^{vj},R) \oplus \mathrm{Hom}_R(R^{vj},R)) \oplus \mathrm{Hom}_R(T^{v,j+1},R)$$

It follows from the definition of τ_{vj} that

$$\begin{aligned} \mathrm{Im}(\tau_{vj})^* &= \mathrm{Hom}_R(T^{v,j+1},R) = (T^{v,j+1})^*, \\ \mathrm{Ker}(\tau_{v,j-1})^* &= \mathrm{Hom}_R(R^{vj},R) \oplus \mathrm{Hom}_R(T^{v,j+1},R) \\ &= (R^{vj})^* \oplus (T^{v,j+1})^*. \end{aligned}$$

Therefore we arrive at the following formula

$$\mathrm{Ker}(\tau_{v,j-1})^*/\mathrm{Im}(\tau_{vj})^* = (R^{vj})^* = (\mathrm{Ker}\tau_{vj}/\mathrm{Im}\tau_{v,j-1})^*.$$

1.4. An application to Noetherian regular rings

Let R be a left and right Noetherian regular ring, i.e., every finitely generated left (right)R-module has finite projective dimension. Let M be a finitely generated left R-module.Choose a projective resolution of M as follows

$$0 \longrightarrow P_\mu \longrightarrow P_{\mu-1} \longrightarrow \cdots \longrightarrow P_0 \longrightarrow M \longrightarrow 0$$

where $\mu \geq \mathrm{p.dim}_R M$ and $P_0, \cdots, P_\mu$ are finitely generated projective R-modules, then $P_j^* = \mathrm{Hom}_R(P_j,R)$ are finitely generated projective right R-modules. In particular, the cohomology groups $\{\mathrm{Ext}^j_R((M,R)\}$ of the complex

$$(*) \qquad 0 \longrightarrow P_0^* \longrightarrow \cdots \longrightarrow P_\mu^* \longrightarrow 0$$

are finitely generated right R-modules.

Now, applying the foregoing argumentation, there is a resolution of the complex $(*)$ as follows

$$(\mathbf{Q}) \qquad \begin{array}{ccccccc} \downarrow & & \downarrow & & & & \downarrow \\ Q_{10} & \to & Q_{11} & \to & \cdots & \to & Q_{1\mu} \\ \downarrow & & \downarrow & & & & \downarrow \\ Q_{00} & \to & Q_{01} & \to & \cdots & \to & Q_{0\mu} \\ \downarrow & & \downarrow & & & & \downarrow \\ P_0^* & \to & P_1^* & \to & \cdots & \to & P_\mu^* \\ \downarrow & & \downarrow & & & & \downarrow \\ 0 & & 0 & & & & 0 \end{array}$$

In the resolution $(\mathbf{Q})$ given above we have

(1). $\{Q_{jv}\}$ are projective right R-modules and each column is exact;

(2). the cohomology groups $\{H_{v0},\ H_{v1},\ \cdots\}$ of the the row complexes

$$0 \longrightarrow Q_{v0} \longrightarrow Q_{v1} \longrightarrow \cdots \longrightarrow Q_{v\mu} \longrightarrow 0$$

are also projective R-modules;

(3). the complexes

$$\cdots \longrightarrow H_{2j} \longrightarrow H_{1j} \longrightarrow H_{0j} \longrightarrow \mathrm{Ext}_R^j(M, R) \longrightarrow 0$$

are projective resolutions of the right R-modules $\mathrm{Ext}_R^j(M, R)$ for all $j \geq 0$.

Looking at the dual left R-modules $Q_{vj}^* = \mathrm{Hom}_R(Q_{vj}, R)$, we obtain a double complex having the previous arrows reversed

$$
(\mathbf{Q}^*) \quad
\begin{array}{ccccccccc}
& \uparrow & & \uparrow & & & \uparrow & & \uparrow \\
0 & \to & Q_{1\mu}^* & \to & Q_{1,\mu-1}^* & \to & \cdots & \to & Q_{11}^* & \to & Q_{10}^* \\
& \uparrow & & \uparrow & & & \uparrow & & \uparrow \\
0 & \to & Q_{0\mu}^* & \to & Q_{0,\mu-1}^* & \to & \cdots & \to & Q_{01}^* & \to & Q_{00}^* \\
& \uparrow & & \uparrow & & & \uparrow & & \uparrow \\
& 0 & & 0 & & & 0 & & 0
\end{array}
$$

where the vertical map resp. the horizontal map is denoted by δ resp. by d (for example, $Q_{0\mu}^* \overset{\delta_{0\mu}}{\to} Q_{1\mu}^*$, $Q_{0\mu}^* \overset{d_{0\mu}}{\to} Q_{0,\mu-1}^*$). Note that

$$\cdots \longrightarrow Q_{1j} \longrightarrow Q_{0j} \longrightarrow P_j^* \longrightarrow 0$$

are exact sequences of projective modules and P_j are reflexive, we see that the dual sequences

$$0 \longrightarrow P_j^{**} \longrightarrow Q_{0j}^* \longrightarrow Q_{1j}^* \longrightarrow \cdots$$

are also exact and $P_j^{**} = P_j$.

Consider the bounded filt-complex $(\mathbf{M}^{II}, \Delta, 1)$ with the second filtration deriving from the double complex $(\mathbf{Q}^*)$, we find that

$$E_j^1 = P_{\mu+1-j} = \mathrm{Ker}\,\delta_{0,\mu+1-j} = (E_j^1)_0$$

for all $j \geq 1$. By the definition of $\chi(s-1)$ and the fact that the sequence

$$0 \longrightarrow P_\mu \longrightarrow \cdots \longrightarrow P_0 \longrightarrow M \longrightarrow 0$$

is a projective resolution we have the following table

$$0 \; \to \; \underset{0}{E_1^1 = P_\mu} \; \to \; \underset{0}{E_2^1 = P_{\mu-1}} \; \to \; \cdots \; \to \; \underset{M}{E_{\mu+1}^1 = P_0} \; \to \; 0$$

Hence $E_{\mu+1}^k = M = (E_{\mu+1}^k)_0$, i.e., it only has the part of degree zero, for all $k \geq 2$. By the foregoing lemma we have $G(\mathbf{H}_\mu) = E_{\mu+1}^k$ for large k. Recalling the definition of $G(\mathbf{H}_\mu)$ we can mention the following

1. Proposition. With notations as above, then $\mathbf{H}_v = 0$ for $v \neq \mu$ and $\mathbf{H}_\mu = M$, where $\mathbf{H}_v$ is the v-th cohomology group of the filt-complex $(\mathbf{M}^{II}, \Delta, 1)$.

On the other hand, if we consider the bounded filt-complex $(\mathbf{M}^I, \Delta, 1)$ with the first filtration which comes from the double complex $(\mathbf{Q}^\bullet)$, the by the foregoing argumentation

$$E_j^1 = \bigoplus_{l=0}^{j-1} \frac{\mathrm{Ker} d_{j-1-l,\mu-l}}{\mathrm{Im} d_{j-1-l,\mu-l-1}} = \bigoplus_{l=0}^{j-1} H_{j-1-l,\mu-l}^*$$

where $j \geq 1$ and $\{H_{vt}\}$ are the cohomology groups of the row complexes

$$0 \longrightarrow Q_{v0} \longrightarrow Q_{v1} \longrightarrow \cdots \longrightarrow Q_{v\mu} \longrightarrow 0$$

Note that the sequences

$$\cdots \longrightarrow H_{2j} \longrightarrow H_{1j} \longrightarrow H_{0j} \longrightarrow \mathrm{Ext}_R^j(M, R) \longrightarrow 0$$

are projective resolutions of the right R-modules $\mathrm{Ext}_R^j(M, R)$ for all $j \geq 0$, by the definition of $\chi(s-1)$ and the construction of double complex $(\mathbf{Q}^\bullet)$ we get

$$E_j^2 = \bigoplus_{l=0}^{j-1} \mathrm{Ext}_R^{j-1-l}(\mathrm{Ext}_R^{\mu-l}(M, R), R).$$

If we assume that $\mathrm{Ext}_R^p(\mathrm{Ext}_R^q(M, R), R) = 0$ for all $p < q$, then

$$E_\mu^2 = \bigoplus_{l=0}^{\mu-1} \mathrm{Ext}_R^{\mu-1-l}(\mathrm{Ext}_R^{\mu-l}(M, R), R) = 0.$$

Hence $E_\mu^k = 0$ for all $k \geq 2$. This shows that $E_{\mu+1}^k$ is a submodule of $E_{\mu+1}^2$ for all $k \geq 3$. Since the bounded filt-complexes $(\mathbf{M}^I, \Delta, 1)$ and $(\mathbf{M}^{II}, \Delta, 1)$ have the same cohomology groups $\{\mathbf{H}_{j-1}\}_{j\geq 1}$ with respect to both the first and the second filtrations and note that $\mathbf{H}_\mu = M$, it follows from the foregoing lemma 1.3.2. that we have the following

2. Theorem. Let R be a left and right Noetherian regular ring and M a finitely generated left (right) R-module. With notation as before, if $\text{Ext}_R^p(\text{Ext}_R^q(M, R), R) = 0$ for all $p < q$, then for any $\mu \geq \text{p.dim}_R M$ there is a canonical filtration FM on M consisting of submodules:

$$F_0 M \subseteq F_1 M \subseteq \cdots \subseteq F_\mu M = M$$

such that each factor $F_v M / F_{v-1} M$ is isomorphic to a submodule of $\text{Ext}_R^{\mu - v}(\text{Ext}_R^{\mu - v}(M, R), R)$.

§2. Auslander Regularity of Zariskian Filtered Rings

2.1. An introduction to Auslander regularity

Let R be a left (right) Noetherian ring and M an arbitrary finitely generated left (right) R-module. By ([Nor1] P.147) we know that $\text{p.dim}_R M = n < \infty$ if and only if $\text{Ext}_R^{n+1}(M, N) = 0$ for all finitely generated left (right) R-module N, and hence one easily sees that $\text{Ext}_R^n(M, F) \neq 0$ for some free R-module F. Since Ext behaves as Hom does on direct sums and products ([Rot] P.200) it follows that $\text{Ext}_R^n(M, R) \neq 0$.

1. Definition. For any R-module M the **grade number** of M, denoted $j_R(M)$, is the unique smallest integer such that $\text{Ext}_R^{j_R(M)}(M, R) \neq 0$. If such an integer does not exist we write $j_R(M) = \infty$.

2. Lemma. Let

$$0 \longrightarrow M_1 \longrightarrow M \longrightarrow M_2 \longrightarrow 0$$

be an exact sequence of R-modules. Then
(1). $j_R(M) \geq \inf\{j_R(M_1),\ j_R(M_2)\}$;
(2). $j_R(M_2) \geq \inf\{j_R(M),\ j_R(M_1)\}$.

Proof. A consequence of the long exact Ext-sequence.   □

3. Definition. Let M be an R-module. We say that M satisfies the **Auslander condition**. if for any $k \geq 0$ and any R-submodule N of $\text{Ext}_R^k(M, R)$ it follows that $j_R(N) \geq k$.

From Lemma 2.2. we immediately derive the following

4. Lemma. Let

$$0 \longrightarrow M_1 \longrightarrow M \longrightarrow M_2 \longrightarrow 0$$

be an exact sequence of R-modules. If M_1 and M_2 satisfy the Auslander condition then M does too.

5. Proposition. Let R be a left and right Noetherian regular ring and M a finitely generated R-module satisfying the Auslander condition. Suppose that $\text{p.dim}_R M \leq \mu$ for some integer $\mu \geq 0$ and that $\text{Ext}_R^k(M, R)$ satisfies the Auslander condition for every $k \geq 0$. If

$$0 \longrightarrow P_\mu \longrightarrow \cdots \longrightarrow P_1 \longrightarrow P_0 \longrightarrow M \longrightarrow 0$$

is a projective resolution of M then the following statements hold:
(1). If $\delta(M)$ is the smallest integer such that $M = F_{\delta(M)} M$, where FM is the canonical filtration on M as defined in Theorem 1.4.2., then $\mu = j_R(M) + \delta(M)$.

(2). If M' is a subquotient of M, i.e., there are submodules $T_1 \subset T_2 \subseteq M$ such that $M' = T_2/T_1$, then $j_R(M') \geq j_R(M)$.

proof. (1). Since each $\text{Ext}_R^k(M, R)$ satisfies the Auslander condition by assumption it follows from Theorem 1.4.2. and the definition of $\delta(M)$ that $\mu - \delta(M) \geq j_R(M)$. Using Lemma 2.2. and the assumption, an easy induction yields $\mu - \delta(M) \leq j_R(M)$. Hence $\mu = j_R(M) + \delta(M)$. (2). Because of Lemma 2.1.2. we may assume that M' is a submodule of M. Put $N = M' + F_{\delta(M)-1}M/F_{\delta(M)-1}M \subseteq F_{\delta(M)}M/F_{\delta(M)-1}M = M/F_{\delta(M)-1}M$. By (1) and the assumption we have $j_R(N) \geq \mu - \delta(M) = j_R(M)$. Now consider the exact sequence:

$$0 \longrightarrow F_{\delta(M)-1}M \cap M' \longrightarrow M' \longrightarrow N \longrightarrow 0$$

Lemma 2.2. entails that we only have to establish the inequality $j_R(F_{\delta(M)-1}M \cap M') \geq j_R(M)$. If $\delta(M) = 0$ then $M' \cong N$ and hence $j_R(M') = j_R(N) \geq j_R(M)$. If $\delta(M) = 1$, then $F_{\delta(M)-1}M \cap M' = F_0 M \cap M' \subseteq \text{Ext}_R^\mu(\text{Ext}_R^\mu(M, R), R)$ and so, our assumptions imply that $j_R(F_{\delta(M)-1}M \cap M') \geq \mu \geq j_R(M)$. From consideration of the following exact sequence

$$0 \to F_{\delta(M)-2}M \; \cap \; M' \to F_{\delta(M)-1}M \cap M' \to$$
$$\to (F_{\delta(M)-1}M \cap M')/(F_{\delta(M)-2}M \cap M') \to 0$$

and combined with

$$(F_{\delta(M)-1}M \cap M')/(F_{\delta(M)-2}M \cap M')$$
$$\cong (F_{\delta(M)-1}M \cap M' + F_{\delta(M)-2}M/F_{\delta(M)-2}M$$
$$\subseteq F_{\delta(M)-1}M/F_{\delta(M)-2}M$$

we may derive by an easy induction argumentation on $\delta(M) - 1$ that $j_R(F_{\delta(M)-1}M \cap M') \geq \mu - \delta(M) = j_R(M)$ as desired. $\qquad\square$

6. Corollary. Let R and M be as in Proposition 2.5.. If

$$0 \longrightarrow M' \longrightarrow M \longrightarrow M'' \longrightarrow 0$$

is an exact sequence of R-modules, then $j_R(M) = \inf\{j_R(M'), \; j_R(M'')\}$.

7. Definition. Let R be a left and right Noetherian ring with **finite global dimension**. If every finitely generated left and right R-module satisfies the Auslander condition then we say that R is an **Auslander regular ring**.

2.2. Auslander regularity of Zariskian filtered rings

In this subsection R is always a **left and right** Zariskian filtered ring with filtration FR. Our aim is to prove that if the associated graded ring $G(R)$ of R is Auslander regular then R is Auslander regular.

We start this subsection with a general lemma.

1. Lemma. Let

$$0 \longrightarrow M_0 \xrightarrow{f_1} M_1 \longrightarrow \cdots \longrightarrow M_{j-1} \xrightarrow{f_j} M_j \xrightarrow{f_{j+1}} \cdots$$

be any filt-complex in R-filt such that each FM_j is a good filtration on M_j. Then there exists an $\omega \in I\!N$ such that for all $n \in Z\!\!Z$ and $j \geq 1$ we have:

$$Z_n^\infty(j) + F_{n-1}M_{j-1} = Z_n^\omega(j) + F_{n-1}M_{j-1}$$
$$B_n^\infty(j) + F_{n-1}M_{j-1} = B_n^\omega(j) + F_{n-1}M_{j-1},$$

where $Z_n^\infty(j)$, $Z_n^\omega(j)$, $B_n^\infty(j)$, $B_n^\omega(j)$ are defined as in §1.

Proof. Since R is a Zariski ring, both filtrations $\{F_nM_j \cap f_j(M_{j-1}),\ n \in Z\!\!Z\}$ and $\{f_j(F_nM_{j-1}),\ n \in Z\!\!Z\}$ on $f_j(M_{j-1})$ are good filtrations. It follows from Ch.I. Lemma 5.3. that there exists an $\omega_1 \in I\!N$ such that $F_nM_j \cap f_j(M_{j-1}) \subseteq f_j(F_{n+\omega_1}M_{j-1})$, $n \in Z\!\!Z$. If $x \in Z_n^{\omega_1+1}(j)$, then $f_j(x) \in F_{n-\omega_1-1}M_j \cap f_j(M_{j-1}) \subseteq f_j(F_{n-1}M_{j-1})$. Therefore $f_j(x) = f_j(y)$ for some $y \in F_{n-1}M_{j-1}$ and so $f_j(x - y) = 0$, i.e., $x - y \in F_nM_{j-1} \cap \mathrm{Ker}f_j = Z_n^\infty(j)$. It follows that $x \in Z_n^\infty(j) + F_{n-1}M_{j-1}$ and hence we get that $Z_n^{\omega_1+1}(j) \subseteq Z_n^\infty(j) + F_{n-1}M_{j-1}$. Putting $\omega' = \omega_1 + 1$ we then obtain $Z_n^{\omega'}(j) + F_{n-1}M_{j-1} = Z_n^\infty(j) + F_{n-1}M_{j-1}$ for all $n \in Z\!\!Z$. Next, consider the filtrations $\{F_nM_{j-1} \cap f_{j-1}(M_{j-2}),\ n \in Z\!\!Z\}$ and $\{f_{j-1}(F_nM_{j-2}),\ n \in Z\!\!Z\}$ on $f_{j-1}(M_{j-2})$, then again both filtrations are good filtrations. Hence for some $\omega_2 \in I\!N$ we obtain $F_nM_{j-1} \cap f_{j-1}(M_{j-2}) \subseteq f_{j-1}(F_{n+\omega_2-1}M_{j-2})$, $n \in Z\!\!Z$. This means that for all $n \in Z\!\!Z$

$$B_n^\infty(j) = F_nM_{j-1} \cap f_{j-1}(M_{j-2}) \subseteq F_nM_{j-1} \cap f_{j-1}(F_{n+\omega_2-1}M_{j-2}) = B_n^{\omega_2}(j).$$

Consequently we have: $B_n^\infty(j) + F_{n-1}M_{j-1} = B_n^{\omega_2}(j) + F_{n-1}M_{j-1}$, $n \in Z\!\!Z$. Finally, it suffices to put $\omega = \max\{\omega', \omega_2\}$ in order to complete the proof. $\qquad\square$

Let M, N be in R-filt with filtrations FM and FN respectively. Recall from Ch.I. that the abelian group $\mathrm{HOM}_R(M, N) = \cup_{p \in Z\!\!Z}\mathrm{HOM}_R(M, N)_p$ may be filtered by putting: $F_p\mathrm{HOM}_R(M, N) = \mathrm{HOM}_R(M, N)_p$ for $p \in Z\!\!Z$.

Now, if $L = \oplus_{i=1}^t Re_i$ is a filt-free R-module of finite rank with good filtration $F_nL = \oplus_{i=1}^t F_{n-k_i}Re_i$ where n, $k_i \in Z\!\!Z$, then $L^* = \mathrm{Hom}_R(L, R) = \mathrm{HOM}_R(L, R)$ by Ch.I. Proposition 6.6.. Hence we may filter L^* as described above and we call this filtration FL^*.

2. Lemma. Let L be as above.

(1). FL^* is a good filtration on FL^*.

(2). $G(L^*) \cong \mathrm{Hom}_{G(R)}(G(L), G(R)) = G(L)^*$ in $G(R)$-gr.

Proof. (1). Since $L = \oplus_{i=1}^t Re_i$, $F_nL = \oplus_{i=1}^t F_{n-k_i}Re_i$ fo $n \in \mathbb{Z}$, it follows that $e_i \in F_{k_i}L$ and $L^* = \oplus_{i=1}^t e_i^*R$ where $\{e_i^*, i = 1, \cdots, t\}$ is the standard dual basis of L^* as a free right R-module. Consequently $e_i^*(F_nL) = F_{n-k_i}R$ and $F_pL^* = \sum_{i=1}^t e_i^* F_{p+k_i}R$, $p \in \mathbb{Z}$. This establishes that FL^* is good.

(2). Since L is filt-free, $G(L) = \oplus_{i=1}^t G(R)\sigma(e_i)$ with $\deg\sigma(e_i) = k_i$, it is easily seen that $G(L)^*$ is a graded free right $G(R)$-module. Thus the morphism $i: G(L^*) \to G(L)^*$ with $\sigma(f) \mapsto G(f)$, is obviously an injective graded morphism of degree zero. If $\varphi \in G(L)^*$ is of degree p then $\varphi(\sigma(e_i)) \in G(R)_{p+k_i}$, $i = 1, \cdots, t$. Choose $r_i \in F_{p+k_i}R$ such that $\varphi(\sigma(e_i)) = \sigma(r_i)$ and define $\psi: L \to R$, $e_i \to r_i$, $i = 1, \cdots, t$, then $i(\psi) = \varphi$. Hence this proves that i is an isomorphism. $\square$

Let $M \in R$-filt with good filtration FM. Since R is a Zariski ring it follows that good filtrations induce good filtrations on R-submodules, hence we have a filt-free resolution of M in R-filt:

$$\cdots \longrightarrow L_j \xrightarrow{f_j} L_{j-1} \longrightarrow \cdots \longrightarrow L_1 \xrightarrow{f_1} L_0 \xrightarrow{\varepsilon} M \longrightarrow 0$$

where each L_j is a filt-free object of finite rank. It follows from Theorem 4.2.4. and Lemma 6.2. of Ch.I. that we obtain a gr-free resolution of $G(M)$ in $G(R)$-gr:

$$\cdots \longrightarrow G(L_j) \xrightarrow{G(f_j)} G(L_{j-1}) \longrightarrow \cdots \longrightarrow G(L_1) \xrightarrow{G(f_1)} G(L_0) \xrightarrow{G(\varepsilon)} G(M) \longrightarrow 0$$

where each $G(L_j)$ is a gr-free object of finite rank. Consider the dualized sequences:

$$F(*) \qquad 0 \longrightarrow L_0^* \xrightarrow{f_1^*} L_1^* \longrightarrow \cdots \longrightarrow L_{j-1}^* \xrightarrow{f_j^*} L_j^* \longrightarrow \cdots$$

$$G(*) \qquad 0 \longrightarrow G(L_0)^* \xrightarrow{G(f_1)^*} G(L_1)^* \longrightarrow \cdots \longrightarrow G(L_{j-1})^* \xrightarrow{G(f_j)^*} G(L_j)^* \longrightarrow \cdots$$

then the Ext-groups corresponding to these are:

$$\{\mathrm{Ext}_R^{j-1}(M, R) = \mathrm{Ker} f_j^* / \mathrm{Im} f_{j-1}^*, \ j = 1, 2, \cdots\};$$
$$\{\mathrm{Ext}_{G(R)}^{j-1}(G(M), G(R)) = \mathrm{Ker} G(f_j)^* / \mathrm{Im} G(f_{j-1})^*, \ j = 1, 2, \cdots\}.$$

For every $j = 1, 2, \cdots$, define a filtration on the right R-module $\mathrm{Ext}_R^{j-1}(M, R)$ by putting

$$F_n\mathrm{Ext}_R^{j-1}(M, R) = \mathrm{Ker} f_j^* \cap F_nL_{j-1}^* + \mathrm{Im} f_{j-1}^* / \mathrm{Im} f_{j-1}^*, \ n \in \mathbb{Z}.$$

Then we obtain the associated graded right $G(R)$-module

$$G(\mathrm{Ext}_R^{j-1}(M,R)) = \bigoplus_{n \in \mathbf{Z}} \frac{\mathrm{Ker} f_j^* \cap F_n L_{j-1}^* + \mathrm{Im} f_{j-1}^*}{\mathrm{Ker} f_j^* \cap F_{n-1} L_{j-1}^* + \mathrm{Im} f_{j-1}^*}$$

Note that, since every f_j is a filtered morphism of degree zero, each f_j^* is also a filtered morphism of degree zero.

3. Lemma. Let the notation be as above.
(1). $F\mathrm{Ext}_R^{j-1}(M,R)$ is a good filtration, $j = 1, 2, \cdots$.
(2). There is an isomorphism of complexes in $G(R)$-gr (graded right modules!):

$$
\begin{array}{ccccccccc}
0 & \to & G(L_0^*) & \overset{G(f_1^*)}{\longrightarrow} & G(L_1^*) & \overset{G(f_2^*)}{\longrightarrow} & G(L_2^*) & \to & \cdots \\
 & & \cong \downarrow & & \cong \downarrow & & \cong \downarrow & & \\
0 & \to & G(L_0)^* & \underset{G(f_1)^*}{\longrightarrow} & G(L_1)^* & \underset{G(f_2)^*}{\longrightarrow} & G(L_2)^* & \to & \cdots
\end{array}
$$

Proof. (1). Since R is also a right Zariski ring the filtration on $\mathrm{Ker} f_j^*$ induced by $F L_{j-1}^*$ is good and thus $F\mathrm{Ext}_R^{j-1}(M,R)$ is good too by definition (quotient filtrations of good filtrations are always good).
(2). Clear, because of Lemma 2.2.2..   □

4. Proposition. With notation as before, $G(\mathrm{Ext}_R^{j-1}(M,R))$ is isomorphic to a subquotient of $\mathrm{Ext}_{G(R)}^{j-1}(G(M),G(R))$, $j = 1, 2, \cdots$.

Proof. Reconsider the filt-complex $F(*)$ as described before Lemma 2.2.3., if we look at the spectral sequence determined by $F(*)$ as constructed in §1. then $G(\mathbf{H}_{j-1}) = G(\mathrm{Ext}_R^{j-1}(M,R))$, and $\mathbf{H}_{gr}^{j-1} \cong \mathrm{Ext}_{G(R)}^{j-1}(G(M),G(R))$ by Lemma 2.2.3.. Since R is a left and right Zariski ring the condition necessary in order to apply Proposition 1.1.5. does hold because of Lemma 2.2.2. and Lemma 2.2.1..   □

Now, we are ready to prove our main result concerning the Auslander regularity of Zariski rings.

5. Theorem. Let R be a left and right Zariski ring with filtration FR. Assume that $G(R)$ is an Auslander regular ring then R is an Auslander regular ring.

Proof. That gl.dim $R < \infty$ follows from Ch.II. Theorem 3.1.4.. Consider any finitely generated R-module M and let N be any nonzero R-submodule of $\mathrm{Ext}_R^k(M,R)$ where $k \geq 0$.

We have to prove that $j_R(N) \geq k$. Choose a good filtration FM on M and construct a good filtration $F\text{Ext}_R^k(M, R)$ on $\text{Ext}_R^k(M, R)$ as before, then in view of Proposition 2.2.4. $G(\text{Ext}_R^k(M, R))$ is a subquotient of $\text{Ext}_{G(R)}^k(G(M), G(R))$. Hence Proposition 2.1.5. yields that $j_{G(R)}(G(\text{Ext}_R^k(M, R))) \geq k$. The good filtration on $\text{Ext}_R^k(M, R)$ induces a good filtration FN on N and thus $G(N) \subseteq G(\text{Ext}_R^k(M, R))$ (Ch.I. Theorem 4.2.4.). By Proposition 2.1.5. again we have $j_{G(R)}(G(N)) \geq j_{G(R)}(G(\text{Ext}_R^k(M, R))) \geq k$. Finally, the Zariskian property of R and Proposition 2.2.4. entail that $j_R(N) \geq j_{G(R)}(G(N)) \geq k$ and the proof is completed.
□

2.3. Auslander regularity of polynomial rings

Let R be a left Noetherian ring and $R[t]$ the polynomial extension of R in the commuting variable t. Consider the natural gradation on $R[t]$: $R[t]_n = Rt^n$, $n \in I\!\!N$, if M is a finitely generated graded left $R[t]$-module then we may write $M = \sum_{i=1}^n R[t]\xi_i$ with ξ_i being homogeneous of degree k_i in M. Thus we have $M_n = \sum_{i=1}^n R[t]_{n-k_i}\xi_i = \sum_{i=1}^n Rt^{n-k_i}\xi_i$ for all $n \in Z\!\!\!Z$. Put $l_1 = \min\{k_i, i = 1, \cdots, n\}$, $l_2 = \max\{k_i, i = 1, \cdots, n\}$, then we obtain the following properties:

(P_1). $M_n = 0$ for all $n < l_1$.

(P_2). Each M_n is a finitely generated R-module and $M = (M_{l_1} \oplus \cdots \oplus M_{l_2-1}) \oplus R[t]M_{l_2}$.

(P_3). The map $\chi_t: M \to M$ with $m \mapsto tm$, is a graded $R[t]$-morphism of degree 1 and $\text{Ker}\chi_t = \oplus_{n \geq l_1}(M_n \cap \text{Ker}\chi_t)$ is a graded submodule of M.

With notations as above, we may now derive the following

1. Lemma. There is an exact sequence

$$0 \longrightarrow M' \longrightarrow M \longrightarrow M'' \longrightarrow 0$$

where $M' = R[t] \otimes_R M_0$ for some finitely generated R-module M_0 and M'' is an $R[t]$-submodule of M such that $t^\omega M'' = 0$ for some positive integer ω. Moreover M'' is a finitely generated R-module.

Proof. By the assumptions $\text{Ker}\chi_t$ is a finitely generated graded $R[t]$-module and hence there is an $l_2' \in Z\!\!\!Z$ such that for all $q > 0$ we have: $\text{Ker}\chi_t \cap M_{q+l_2'} = R[t]_q(\text{Ker}\chi_t \cap M_{l_2'}) = 0$. Choose $l > \max\{l_2, l_2'\}$, then for $v \geq 0$, $M_{v+l} = t^v M_l \cong t^v R \otimes_R M_l$. Put $M' = R[t]M_l$ and $M'' = M_{l_1} \oplus \cdots \oplus M_{l-1}$ then obviously $M' \cong R[t] \otimes_R M_l$ and M'' is a finitely generated R-module such that $t^\omega M'' = 0$ for some ω.   □

2. Lemma. Let R be a Auslander regular ring and let M be an $R[t]$-module. If M is finitely generated as an R-module then M satisfies the Auslander condition as an $R[t]$-module.

Proof. By the well known Rees theorem (cf. [Rot], P.248, Theorem 9.37.) we have $\mathrm{Ext}_R^k(M, R) \cong \mathrm{Ext}_{R[t]}^{k+1}(M, R[t])$ for every $k \geq 0$. It follows from the Auslander regularity of R that M, as an $R[t]$-module, satisfies the Auslander condition.   □

3. Lemma. Let R be an Auslander regular ring, and let M be an $R[t]$-module such that $M = R[t] \otimes_R M_0$ for some finitely generated R-module M_0, then M satisfies the Auslander condition as an $R[t]$-module.

Proof. Consider a free R-resolution of M_0:

$$\cdots \longrightarrow L_1 \longrightarrow L_0 \longrightarrow M_0 \longrightarrow 0$$

where each L_i has finite rank. Then we obtain a gr-free $R[t]$-resolution of M:

$$\cdots \longrightarrow R[t] \otimes_R L_1 \longrightarrow R[t] \otimes_R L_0 \longrightarrow R[t] \otimes_R M_0 \longrightarrow 0$$

and one may easily check (use the flatness of $R[t]$ over R and a light modification of [Rot] Lemma 3.83.)

$$\mathrm{Ext}_{R[t]}^k(M, R[t]) \cong \mathrm{Ext}_R^k(M_0, R) \otimes_R R[t]$$
$$= \oplus_{p \geq 0}(\mathrm{Ext}_R^k(M_0, R) \otimes_R R[t]_p)$$

Since R is left and right Noetherian , the (natural) grading filtration (see Ch.I. §2) on $R[t]$: $F_n R[t] = \sum_{k \leq n} R t^k$, $n \in \mathbb{Z}$, makes $R[t]$ into a left and right Zariski ring (see Ch.II. §2). Let N be a nonzero submodule of $\mathrm{Ext}_{R[t]}^k(M, R[t])$, where $k \geq 0$. Since the latter is a graded $R[t]$-module we may take the natural filtration on it: $F_n \mathrm{Ext}_{R[t]}^k(M, R[t]) = \oplus_{p \leq n}(\mathrm{Ext}_R^k(M_0, R) \otimes_R R[t]_p)$, $n \in \mathbb{Z}$. This filtration is a good filtration and since $FR[t]$ is Zariskian it induces a good filtration on N, denoted FN. Moreover, $G(N) \subseteq G(\mathrm{Ext}_{R[t]}^k(M, R[t])) = \mathrm{Ext}_{R[t]}^k(M, R[t])$ and $G(N)_p \subseteq \mathrm{Ext}_{R[t]}^k(M, R[t])_p = \mathrm{Ext}_R^k(M_0, R) \otimes_R t^p R \cong \mathrm{Ext}_R^k(M_0, R)$. Of course, for the grading filtration $FR[t]$ on $R[t]$ we have $G(R[t]) \cong R[t]$ and so $G(N)_p$ is an R-module and it follows that $j_R(G(N)_p) \geq k$ because R is assumed to be Auslander regular. Put $G(N) = T$. By Lemma 2.3.1. there is an exact sequence of $R[t]$-modules

$$0 \longrightarrow T' \longrightarrow T \longrightarrow T'' \longrightarrow 0$$

where T'' is a finitely generated R-module and $T' = R[t] \otimes_R T_0'$ for some finitely generated R-module T_0'. The well known Rees theorem now yields: $\mathrm{Ext}_R^k(T'', R) \cong \mathrm{Ext}_{R[t]}^{k+1}(T'', R[t])$ and consequently $j_{R[t]}(T'') = j_R(T'') + 1$. The proof of Lemma 2.3.1. combined with the obvious induction leads to $j_R(T'') \geq k$ (note that $T'' = T_{l_1} \oplus \cdots \oplus T_{l-1}$), where M is replaced by T in the proof of Lemma 2.3.1.), hence $j_{R[t]}(T'') \geq k+1$. However, from the isomorphism $\text{Ext}^q_{R[t]}(R[t] \otimes_R T'_0, R[t]) \cong R[t] \otimes_R \text{Ext}^q_R(T'_0, R)$ we may deduce that $j_{R[t]}(T') = j_R(T'_0) \geq k$ since $T'_0 = T_v$ for some $v \in \mathbb{Z}$ (see the proof of Lemma 2.3.1.). It follows from Lemma 2.1.2. that $j_{R[t]}(T) \geq k$. Finally, since $FR[t]$ is Zariskian we may use Proposition 2.2.4. to derive $j_{R[t]}(N) \geq j_{R[t]}(G(N)) = j_{R[t]}(T) \geq k$. This completes the proof. □

4. Corollary. If R is an Auslander regular ring then any finitely generated graded $R[t]$-module M satisfies the Auslander condition.

Proof. Easy from the foregoing lemma. □

5. Theorem. Let R be an arbitrary ring, then the polynomial ring extension $R[t]$ of R is an Auslander regular ring if and only if R is an Auslander regular ring.

Proof. Suppose that R is Auslander regular. It is well known that $\text{gl.dim} R[t] = \text{gl.dim} R + 1 < \infty$. Hence we only need to check that every finitely generated $R[t]$-module satisfies the Auslander condition. To this end, let $R[t]$ be endowed with the (natural) grading filtration $FR[t]$ as in the proof of Lemma 2.3.3. and let M be a finitely generated $R[t]$-module. Choose any good filtration FM on M, then Proposition 2.2.4. entails that $j_{R[t]}(M) \geq j_{R[t]}(G(M))$ because $R[t]$ is now a Zariski ring.

Take any $R[t]$-submodule $N \subseteq \text{Ext}^k_{R[t]}(M, R[t])$ where $k \geq 0$. Choose a good filtration on $\text{Ext}^k_{R[t]}(M, R[t])$ as in Lemma 2.2.3. and consider the induced good filtration FN on N, then $G(N) \subseteq G(\text{Ext}^k_{R[t]}(M, R[t]))$. From $G(R[t]) = R[t]$ and Proposition 2.2.4. again, it follows that $G(\text{Ext}^k_{R[t]}(M, R[t]))$ is a subquotient of $\text{Ext}^k_{R[t]}(G(M), R[t])$. Observe that $\text{Ext}^k_{R[t]}(G(M), R[t])$ as well as $\text{Ext}^q_{R[t]}(\text{Ext}^k_{R[t]}(G(M), R[t]), R[t])$ are graded $R[t]$-modules for all $q \geq 0$. So by the above corollary and Proposition 2.1.5. we arrive at: $j_{R[t]}(G(N)) \geq j_{R[t]}(\text{Ext}^k_{R[t]}(G(M), R[t])) \geq k$. Consequently $j_{R[t]}(N) \geq j_{R[t]}(G(N)) \geq k$ as desired.

Conversely, let $R[t]$ be Auslander regular. If M_0 is a finitely generated R-module and N is a submodule of $\text{Ext}^k_R(M_0, R)$, then from the proof of Lemma 2.3.3. we see that $\text{Ext}^k_R(M_0, R) \otimes_R R[t] \cong \text{Ext}^k_{R[t]}(R[t] \otimes_R M_0, R[t])$. It follows that $j_R(N) \geq k$. Hence R is Auslander regular. □

6. Corollary. A ring R is Auslander regular if and only if the polynomial ring extension $R[t_1, \cdots, t_n]$ of R in the commuting variables t_i, where $n \geq 1$, is Auslander regular.

2.4. Examples of Auslander regular rings

1. Any field is Auslander regular.

2. Any commutative Noetherian regular local ring R is Auslander regular.

Proof. Let Ω be the unique maximal ideal of R and $k = R/\Omega$ the residue field of R. Consider the Ω-adic filtration on R, then R becomes a (classical) Zariski ring. Since $G(R) \cong k[t_1, \cdots, t_n]$ where $n = \mathrm{K.dim}\,R = \mathrm{gl.dim}\,R$, it follows from Theorem 2.3.5. that $G(R)$ is Auslander regular. Hence R is Auslander regular by Theorem 2.2.5.   □

3. Any commutative Noetherian ring R with finite global dimension is Auslander regular.

Proof. By the assumption and example 2. the regular local ring R_p is Auslander regular, where R_p is the localization of R at any prime ideal p of R. Hence the Auslander regularity of R follows from the well known formula (cf. [Rot], Theorem 9.50.): $\mathrm{Ext}^k_{R_p}(M_p, R_p) \cong \mathrm{Ext}^k_R(M, R)_p$ and the localization principle with $k \geq 0$, where M is any finitely generated R-module.   □

4. Recalling the examples given in Ch.II. §2, then R is Auslander regular if R is one of the following rings:

- (a). The universal enveloping algebra $U(\mathbf{g})$ of a finite n-dimensional k-Lie algebra $\mathbf{g}$ (where k is a field) with the standard filtration.
- (b). The n-th Weyl algebra $A_n(k)$ over a field k with the Bernstein filtration (or Σ-filtration).
- (c). The ring $\mathcal{D}(V)$ of $\mathbb{C}$-linear differential operators on V, where V is an irreducible smooth subvariety of the affine n-space $\mathbb{C}^n$.
- (d). The ring $\mathcal{D}_n$ of $\mathbb{C}$-linear differential operators on $\mathcal{O}_n$, where $\mathcal{O}_n$ is the regular local ring of convergent power series in n variables with coefficients in $\mathbb{C}$.
- (e). The stalks $\mathcal{E}_p$ of the sheaf of microlocal differential operators.

Proof. From Ch.II. §2. we have seen that all rings mentioned above are left and right Zariski rings. Moreover we have

Case (a). $G(U(\mathbf{g})) \cong k[t_1, \cdots, t_n]$;

Case (b). $G(A_n(k)) \cong k[t_1, \cdots, t_{2n}]$ for both filtrations;

Case (c). $G(\mathcal{D}(V)) \cong A(T^*(V))$, where the latter one is the coordinate ring of the non-singular $2n$-dimensional cotangent bundle $T^*(V)$ over V;

Case (d). $G(\mathcal{D}_n) \cong \mathcal{O}_n[t_1, \cdots, t_n]$;

Case (e). $G(\mathcal{E}_p) \cong \mathcal{O}_{2n-1}[t, t^{-1}]$.

Hence by Theorem 2.3.5. and Theorem 2.2.5. all rings in consideration are Auslander regular. □

5. Let $A_{n+1}(k)$ be the $n+1$-st Weyl algebra over a field k. Consider the Fuchsian filtration on $A_{n+1}(k)$, then as we have seen in Ch.II. §2. this filtration is not Zariskian. However, from Ch.I. §4. we have known that the Rees ring $\widetilde{A_{n+1}(k)}$ of $A_{n+1}(k)$ associated to the Fuchsian filtration is isomorphic to the polynomial ring $A_{n+1}(k)[t]$, hence $\widetilde{A_{n+1}(k)}$ is Auslander regular.

Remark. (1). A detailed discussion of the Auslander regularity for Rees rings and skew polynomial rings will be given in next section.

(2). There exists a noncommutative Noetherian ring with finite global dimension not satisfying the Auslander condition (cf [FGR]).

2.5. $j_R(M) = j_{G(R)}(G(M)) = j_{\widetilde{R}}(\widetilde{M}) = j_{\widehat{R}}(\widehat{M})$

In view of Theorem 2.2.4. if R is a Zariskian filtered ring then for any $M \in R$-filt with a good filtration FM we have $j_R(M) \geq j_{G(R)}(G(M))$. We now intend to show that the equality $j_R(M) = j_{G(R)}(G(M)) = j_{\widetilde{R}}(\widetilde{M}) = j_{\widehat{R}}(\widehat{M})$ holds incase $G(R)$ is an Auslander regular ring.

We start the argumentation with a useful lemma.

1. Lemma. Let R be an Auslander regular ring. Let M be a finitely generated R-module.
(1). If $\text{Ext}_R^k(\text{Ext}_R^k(M, R), R) \neq 0$, then $j_R(\text{Ext}_R^k(M, R)) = k$.
(2). $j_R(\text{Ext}_R^{j_R(M)}(M, R)) = j_R(M)$.

Proof. (1). Put $N = \text{Ext}_R^k(M, R)$. Since R is Auslander regular we have $j_R(N) \geq k$. So if $\text{Ext}_R^k(\text{Ext}_R^k(M, R), R) \neq 0$ then $\text{Ext}_R^k(N, R) \neq 0$ and hence $j_R(N) = k$.

(2). From (1) it suffices to prove that $\text{Ext}_R^{j_R(M)}(\text{Ext}_R^{j_R(M)}(M, R), R) \neq 0$. But this follows from Theorem 1.4.2., Proposition 2.1.5. and the minimality of $\delta(M)$ (see Proposition 2.1.5.), where $\delta(M)$ is given by taking a projective resolution of M. □

In what follows let R be a left and right Zariski ring with filtration FR. We also assume that $G(R)$ is an Auslander regular ring (hence R is an Auslander regular ring by Theorem 2.2.5.).

Consider a filt-complex in R-filt:

$(\mathbf{M}, f, 0) \qquad 0 \longrightarrow M_0 \xrightarrow{f_1} M_1 \longrightarrow \cdots \longrightarrow M_{j-1} \xrightarrow{f_j} M_j \longrightarrow \cdots$

with all FM_j being good filtrations, then we have the corresponding complex in $G(R)$-gr:

$(\mathbf{GM}, G(f), 0)$ $0 \longrightarrow G(M_0) \overset{G(f_1)}{\longrightarrow} G(M_1) \longrightarrow \cdots \longrightarrow G(M_{j-1}) \overset{G(f_j)}{\longrightarrow} G(M_j) \longrightarrow \cdots$

Let the notation be given as in §1., then we recall from Observation 1.1.(7). that if $E_{j-1}^1 = 0$ then for each $k \geq 1$ there is an exact sequence

(Δ)
$$0 \longrightarrow E_j^{k+1} \longrightarrow E_j^k \longrightarrow S_{j+1}^k \longrightarrow 0$$

where $S_{j+1}^k = \mathrm{Im}\chi_{j+1}(-k) \subseteq E_{j+1}^k$ and $\mathrm{Ker}\chi_{j+1}(-k) \cong E_j^{k+1}$, i.e., in this case we may view E_j^{k+1} as a submodule of E_j^k. Note that $E_j^1 = \mathbf{H}_{gr}^{j-1}(\mathbf{GM})$ is finitely generated in $G(R)$-gr then Corollary 2.1.6. gives rise to

$$j_{G(R)}(E_j^k) = \inf\{j_{G(R)}(E_j^{k+1}), \; j_{G(R)}(S_{j+1}^k)\}, \; k \geq 1.$$

Now, for a filtered R-module M with good filtration FM we may replace $(\mathbf{M}, f, 0)$ by the filt-complex $F(*)$ as given in subsection 2.2. (i.e., $F(*)$ is the derived sequence from a given filt-free resolution of M). It turns out that $\mathbf{H}_{j-1} = \mathrm{Ext}_R^{j-1}(M, R)$, $E_j^1 = \mathbf{H}_{gr}^{j-1} = \mathrm{Ext}_{G(R)}^{j-1}(G(M), G(R))$, $j \geq 1$. Hence if we put $j_{G(R)}(G(M)) = l = j - 1$, then

$$\begin{aligned} E_{j-1}^1 = \mathbf{H}_{gr}^{j-2} &= \mathrm{Ext}_{G(R)}^{j-2}(G(M), G(R)) \\ &= \mathrm{Ext}_{G(R)}^{l-1}(G(M), G(R)) \\ &= 0 \end{aligned}$$

Next consider the short exact sequence (Δ) given above. Since $S_{j+1}^k \subseteq E_{j+1}^k$ and E_{j+1}^k is a subquotient of $E_{j+1}^1 = \mathrm{Ext}_{G(R)}^j(G(M), G(R)) = \mathrm{Ext}_{G(R)}^{l+1}(G(M), G(R))$, we have by Proposition 2.1.5. that for all $k \geq 1$

$(*)$
$$j_{G(R)}(S_{j+1}^k) \geq l + 1.$$

Thus, on the one hand, Proposition 2.2.4. entails that $j_R(M) \geq l$; On the other hand, if $j_R(M) > l$ then $\mathrm{Ext}_R^l(M, R) = 0$ and hence $G(\mathrm{Ext}_R^l(M, R)) = G(\mathbf{H}_{j-1}) = 0$ (note that $l = j - 1$). But then $E_j^k = 0$ by Lemma 2.2.1. and Corollary 1.1.7. where $k \geq \omega \geq 0$ for some $\omega \in \mathbb{Z}$. Repeating use of (Δ) and $(*)$ yields $j_{G(R)}(E_j^1) \geq l + 1$. But from Lemma 2.5.1. we have $j_{G(R)}(E_j^1) = l$, a contradiction. Hence we have proved the following

2. Theorem. Let R be a left and right Zariski ring with filtration FR. Assume that $G(R)$ is an Auslander regular ring. Then for any $M \in R$-filt with good filtration FM we have $j_R(M) = j_{G(R)}(G(M))$.

In order to prove the equality $j_R(M) = j_{\widetilde{R}}(\widetilde{M})$ we need some homological properties of a strongly $\mathbb{Z}$-graded ring.

Let $A = \oplus_{n \in \mathbb{Z}} A_n$ be a strongly $\mathbb{Z}$-graded ring. Then for any finitely generated graded left (right) R-module T we have

(1). $T \cong A \otimes_{A_0} T_0$ $(T \cong T_0 \otimes_{A_0} A)$;

(2). $\text{Hom}_A(T, A)$ is a $\mathbb{Z}$-graded A-module (Ch.I. Lemma 4.1.1.).

Noticing that $(A \otimes_{A_0} T_0)_0 = A_0 \otimes_{A_0} T_0 = T_0$, from (1) and (2) it is not hard to check

(3). the map $\vartheta \colon \text{Hom}_{A_0}(T_0, A_0) \to (\text{Hom}_A(A \otimes_{A_0} T_0, A))_0$ defined by $\vartheta(\alpha) = \tilde{\alpha}$ such that $\tilde{\alpha}(a \otimes t_0) = a \cdot \alpha(t_0)$ is an isomorphism of A_0-modules.

Moreover, since A is a left and right flat A_0-module, if we start with a free resolution of T_0 then we obtain the following

3. Lemma. Let A be a strongly $\mathbb{Z}$-graded Noetherian ring and T a finitely generated graded left A-module. Then $\text{Ext}_{A_0}^k(T_0, A_0) \otimes_{A_0} A \cong \text{Ext}_A^k(T, A)$ as graded A-modules, for all $k \geq 0$ and hence $j_A(T) = j_{A_0}(T_0)$.

Now we return to the situation where R is a Zariski ring R with $G(R)$ being Auslander regular. Let M be in R-filt with filtration FM. Let $\widetilde{R}_{(X)}$ resp. $\widetilde{M}_{(X)}$ be the localization of $\widetilde{R}$ resp $\widetilde{M}$ at the (central) Ore set $\{1, X, X^2, \cdots\}$ where X is the canonical central regular homogeneous element of degree 1 in $\widetilde{R}$, then from Ch.I. we know that $\widetilde{R}_{(X)} \cong R[t, t^{-1}]$, $\widetilde{M}_{(X)} \cong M[t, t^{-1}]$. If we view $R[t, t^{-1}]$ as a strongly $\mathbb{Z}$-graded ring with the natural gradation then $M[t, t^{-1}]$ is a graded $R[t, t^{-1}]$-module with the obvious gradation, in particular, $(R[t, t^{-1}])_0 \cong R$ and $(M[t, t^{-1}])_0 \cong M$.

4. Lemma. Let the assumptions and notation be given as before.

(1). $j_{\widetilde{R}}(\widetilde{M}) \leq j_R(M)$.

(2). For all $k < j_R(M)$ the cohomology groups $\text{Ext}_{\widetilde{R}}^k(\widetilde{M}, \widetilde{R})$ are X-torsion modules.

Proof. By Lemma 2.5.3. we have

$$
\begin{aligned}
\text{Ext}_R^k(M, R)[t, t^{-1}] &= \text{Ext}_R^k(M, R) \otimes_R R[t, t^{-1}] \\
&\cong \text{Ext}_{R[t,t^{-1}]}^k(M[t, t^{-1}], R[t, t^{-1}]) \\
&\cong \text{Ext}_{\widetilde{R}_{(X)}}^k(\widetilde{M}_{(X)}, \widetilde{R}_{(X)}) \\
&\cong \text{Ext}_{\widetilde{R}}^k(\widetilde{M}, \widetilde{R}) \otimes_{\widetilde{R}} \widetilde{R}_{(X)}.
\end{aligned}
$$

Hence the statements (1) and (2) follow again from Lemma 2.5.3.. $\qquad \square$

5. Theorem. Let R be a left and right Zariski ring with filtration FR. Assume that $G(R)$ is an Auslander regular ring. Then the equality $j_R(M) = j_{\widetilde{R}}(\widetilde{M})$ holds for every $M \in R$-filt with a good filtration FM.

Proof. By Lemma 2.5.4. it is clear that if $j_R(M) = 0$ then $j_{\widetilde{R}}(\widetilde{M}) = 0$. It remains to prove the inequality $j_{\widetilde{R}}(\widetilde{M}) \geq j_R(M)$ whenever $j_R(M) \geq 1$. To this end, consider the exact sequence in $\widetilde{R}$-gr:

$$0 \longrightarrow \widetilde{M} \xrightarrow{\mu_X} \widetilde{M} \longrightarrow \widetilde{M}/X\widetilde{M} \longrightarrow 0$$

where μ_X is left multiplication by X, if we look at the long exact Ext-sequence:

$$\longrightarrow \operatorname{Ext}^k_{\widetilde{R}}(\widetilde{M}/X\widetilde{M}, \widetilde{R}) \longrightarrow \operatorname{Ext}^k_{\widetilde{R}}(\widetilde{M}, \widetilde{R}) \xrightarrow{\mu_X^*} \operatorname{Ext}^k_{\widetilde{R}}(\widetilde{M}, \widetilde{R}) \longrightarrow$$

then for $k = 0$ Lemma 2.5.4. (2) entails that $\operatorname{Hom}_{\widetilde{R}}(\widetilde{M}, \widetilde{R})$ is an X-torsion module and hence equal to zero because $j_R(M) \geq 1$. Note that $\widetilde{R}/X\widetilde{R} \cong G(R)$, $\widetilde{M}/X\widetilde{M} \cong G(M)$ and $j_R(M) = j_{G(R)}(G(M))$ we obtain for $1 \leq k \leq j_R(M)$

$$\operatorname{Ext}^k_{\widetilde{R}}(\widetilde{M}/X\widetilde{M}, \widetilde{R}) \cong \operatorname{Ext}^{k-1}_{G(R)}(G(M), G(R)) = 0$$

by Rees theorem. Consequently if $k < j_R(M)$ then $\operatorname{Ext}^k_{\widetilde{R}}(\widetilde{M}, \widetilde{R}) = 0$ since it is X-torsion by Lemma 2.5.4.(2). and μ_X^* is right multiplication by X (see [Rot] Theorem 7.16.). This proves $j_{\widetilde{R}}(\widetilde{M}) \geq j_R(M)$ as desired.   □

Finally let us proceed to establish the equality $j_R(M) = j_{\widehat{R}}(\widehat{M})$.

6. Theorem. Let R be a left and right Zariski ring with filtration FR. Assume that $G(R)$ is an Auslander regular ring. Then the equality $j_{\widehat{R}}(\widehat{M}) = j_R(M)$ holds for every $M \in R$-filt with good filtration FM.

Proof. Since R is a left and right Zariski ring it follows from Ch.II. §2. that $\widehat{R}$ is a faithful left and right R-module and $\widehat{N} \cong \widehat{R} \otimes_R N$ (or $\widehat{N} \cong N \otimes_R \widehat{R}$) for every left (right) R-module N with a good filtration FN. Moreover, the Noetherian condition on R and the flatness of $\widehat{R}$ entail that the morphisms $\vartheta\colon \operatorname{Hom}_R(M, \widehat{R}) \to \operatorname{Hom}_{\widehat{R}}(\widehat{R} \otimes M, \widehat{R})$ and $\varphi\colon \operatorname{Hom}_R(M, R) \otimes_R \widehat{R} \to \operatorname{Hom}_R(M, \widehat{R})$ are isomorphisms, where $\vartheta(f) = \widetilde{f}$ with $\widetilde{f}(b \otimes m) = bf(m)$ and $\varphi(g \otimes b) = g_b$ with $g_b(m) = g(m)b$ (cf. [Rot] Lemma 3.83). Hence we obtain

$$\operatorname{Ext}^k_{\widehat{R}}(\widehat{M}, \widehat{R}) \cong \operatorname{Ext}^k_R(M, R) \otimes_R \widehat{R}, \ k \geq 0$$

It follows from the faithful flatness of $\widehat{R}$ that $j_R(M) = j_{\widehat{R}}(\widehat{M})$.   □

§3. Auslander Regularity of Graded Rings

In this section we give some applications of the foregoing results from §§1, 2 to Rees rings and general $\mathbb{Z}$-graded rings. In particular, by using the Zariskian properties and introducing the " good graded filtration " on graded modules we prove that the gr-Auslander regularity defined in the category of graded modules is the same as in the ungraded case.

3.1. Auslander regularity of Rees rings

We start with an arbitrary ring R. For any R-module M, put

$$J_s(M) = \inf\{j_R(N), \ N \subseteq M \text{ an } R\text{-submodule}\}$$

and then put

$$\mathcal{C}_n^R = \{M \in R\text{-mod}, \ J_s(M) \geq n\}.$$

Consider an exact sequence in R-mod, say

(*) $$0 \longrightarrow M_1 \longrightarrow M \xrightarrow{\ \varepsilon\ } M_2 \longrightarrow 0$$

if N is any R-submodule of M then we have the exact sequence

(Δ) $$0 \longrightarrow \operatorname{Ker}\varepsilon \cap N \longrightarrow N \longrightarrow N/\operatorname{Ker}\varepsilon \cap N \longrightarrow 0$$

By taking the long (Ext-) exact sequences for (*) and (Δ), and recalling the definition of grade number and the definition of $J_s(M)$, one easily checks the following

1. Lemma. Let the notation be given as above.

(1). $\mathcal{C}_n^R$ is closed under the following operations: taking submodules, quotients, extension and direct limits;

(2). Suppose that $M_1 \in \mathcal{C}_{n_1}^R$, $M_2 \in \mathcal{C}_{n_2}^R$, $M \in \mathcal{C}_n^R$. If we put $t_1 = \min\{n_1, n_2\}$, $t_2 = \min\{n_1, n\}$ then $M \in \mathcal{C}_{t_1}^R$, $M_2 \in \mathcal{C}_{t_2}^R$;

(3). If for each $i \geq 0$ we have that $J_s(\operatorname{Ext}_R^i(M_1, R)) \geq i$ and $J_s(\operatorname{Ext}_R^i(M_2, R)) \geq i$ then $J_s(\operatorname{Ext}_R^i(M, R)) \geq i$.

Obviously, if $J_s(\operatorname{Ext}_R^i(M, R)) \geq i$ for every $i \geq 0$ then the R-module M satisfies the Auslander condition.

In what follows we assume that R is a **left and right Noetherian** ring and let X be a **noninvertible central regular** element in R.

2. Lemma. Assume that we are given an exact sequence of finitely generated R-modules

(**) $$0 \longrightarrow M_1 \longrightarrow M \xrightarrow{\mu_X} M \longrightarrow M_2 \longrightarrow 0$$

where μ_X is multiplication by X, then the following statements hold:

(1). If $N \subseteq M$ is a submodule of M such that $X^t N = 0$ for some $t \in I\!N$ then N has a finite filtration such that the corresponding subquotients are submodules of M_1.

(2). If Q is a quotient module of M such that $X^t Q = 0$ for some $t \in I\!N$ then Q has a finite filtration such that the corresponding subquotients are quotients of M_2.

Proof. For $i = 0, \cdots, t$ we may define $N_i = \{n \in N, X^i a = 0\}$ and obtain a finite filtration such that multiplication by X induces an injective morphism

$$0 \longrightarrow N_{i+1}/N_i \longrightarrow N_i/N_{i-1}.$$

Since $N_1 \subseteq M_1$ this completes the proof of (1).

For (2) we can give the dual of the above argumentation by defining $Q_i = X^i Q, i = 0, \cdots, t$, and the morphism from M_2 onto Q_i/Q_{i+1} is given by multiplication of X^i (note that $M_2 \cong M/XM$). $\qquad\square$

3. Theorem. Given an exact sequence $(**)$ as in Lemma 3.2. but such that

(a). $M_1 \in \mathcal{C}_n^R$;

(b). $M_{(X)} \in \mathcal{C}_n^{R_{(X)}}$, where $M_{(X)}$ resp. $R_{(X)}$ is the localization of M resp. R at the (central) Ore set $\{1, X, X^2, \cdots\}$;

(c). $M_2 \in \mathcal{C}_{n+1}^R$,

then $M \in \mathcal{C}_n^R$.

Proof. Let N be any R-submodule of M, we have to prove $j_R(N) \geq n$. By the assumption (a), Lemma 3.1.1. and Lemma 3.1.2. we may assume that M is X-torsionfree. Then the proof will be finished after we consider the following two cases:

(i). M/N is an X-torsionfree R-module.

Now N/XN is isomorphic to an R-submodule of M/XM and hence $j_R(N/XN) \geq n + 1$ (note that $M_2 \cong M/XM$). On the other hand, since $N_{(X)} \subseteq M_{(X)}$ we see that $\mathrm{Ext}_R^j(N, R)$ is an X-torsion R-module for every $j \leq n - 1$. Take the long (Ext-) exact sequence for the exact sequence

$$0 \longrightarrow N \xrightarrow{\mu_X} N \longrightarrow N/XN \longrightarrow 0$$

we obtain that $\mathrm{Ext}_R^j(N, R) = 0$ for every $j \leq n - 1$. It follows that $j_R(N) \geq n$.

(ii). M/N has a nonzero X-torsion R-submodule.

Put $T_X(N) = \{m \in M, X^t m \in N \text{ for some } t \in I\!N\}$. If $T_X(N) \neq M$ then by the proof of (i) above we have $j_R(T_X(N)) \geq n$ since $M/T_X(N)$ is X-torsionfree. Note that $T_X(N)/XT_X(N)$ is isomorphic to a submodule of M/XM, it follows that $j_R(T_X(N)/XT_X(N)) \geq n+1$. Since

R is Noetherian there exists an integer $\omega > 0$ such that $X^\omega(T_X(N)/N) = 0$, hence by Lemma 3.1.1. and Lemma 3.1.2.(2). we have $j_R(T_X(N)/N) \geq n + 1$. Finally, by taking the long (Ext-) exact sequence for the exact sequence

$$0 \longrightarrow N \longrightarrow T_X(N) \longrightarrow T_X(N)/N \longrightarrow 0$$

we arrive at $j_R(N) \geq n$. □

4. Lemma. If an R/XR-module M satisfies the Auslander condition then M satisfies the Auslander condition as an R-module.

Proof. Use the Rees theorem: $\text{Ext}^i_{R/XR}(M, R/XR) \cong \text{Ext}^{i+1}_R(M, R)$. □

5. Lemma. Let M be a finitely generated R-module that is X-torsionfree. Assume that $M_{(X)}$ resp. M/XM satisfies the Auslandr condition as an $R_{(X)}$-module resp. an R/XR-module, then M satisfies the Auslander condition as an R-module.

Proof. From the exact sequence

$$0 \longrightarrow M \xrightarrow{\mu_X} M \longrightarrow M/XM \longrightarrow 0$$

we obtain an exact sequence

$$\text{Ext}^n_R(M/XM, R) \longrightarrow \text{Ext}^n_R(M, R) \xrightarrow{\mu^*_X} \text{Ext}^n_R(M, R) \longrightarrow \text{Ext}^{n+1}_R(M/XM, R)$$

By the assumption we have $\text{Ext}^n_R(M/XM, R) \in C^R_n$, $\text{Ext}^{n+1}_R(M/XM, R) \in C^R_{n+1}$, and $(\text{Ext}^n_R(M, R))_{(X)} \cong \text{Ext}^n_{R_{(X)}}(M_{(X)}, R_{(X)}) \in C^{R_{(X)}}_n$. Hence by Theorem 3.1.3. $\text{Ext}^n_R(M, R) \in C^R_n$. This shows that M satisfies the Auslander condition. □

6. Theorem. With assumptions and notation as before, if $R_{(X)}$ and R/XR are Auslander regular, then R is Auslander regular.

Proof. From Ch.I. Theorem 7.2.4. it is clear that R has finite global dimension. Let M be any finitely generated R-module. If we consider the exact sequence of R-modules

$$0 \longrightarrow T_X(M) \longrightarrow M \longrightarrow M/T_X(M) \longrightarrow 0$$

where $T_X(M)$ is the X-torsion R-submodule of M, then it follows from Lemma 3.1.4., Lemma 3.1.5. and Lemma 2.1.4. that M satisfies the Auslander condition. Hence R is an Auslander regular ring. □

Now, Let R be a filtered ring with filtration FR and $\tilde{R}$ the Rees ring of R. Let X be the canonical central regular homogeneous element of degree 1 in $\tilde{R}$, then $\tilde{R}_{(X)} \cong R[t, t^{-1}]$ and $G(R) \cong \tilde{R}/X\tilde{R}$ and hence we may use the above theorem to $\tilde{R}$.

7. Theorem. Let R be a filtered ring with filtration FR. Suppose that the Rees ring $\tilde{R}$ of R is left and right Noetherian. If R and $G(R)$ are Auslander regular rings then $\tilde{R}$ is an Auslander regular ring.

Proof. If R is an Auslander regular ring then the polynomial ring $R[t]$ is an Auslander regular ring by Theorem 2.3.5.. Since $R[t, t^{-1}]$ is the localization of $R[t]$ at the (central) Ore set $\{1, t, t^2, \cdots\}$ one easily sees that $R[t, t^{-1}]$ is Auslander regular. Consequently $\tilde{R}$ is Auslander regular by Theorem 3.1.6.. $\square$

8. Theorem. Let R be a left and right Zariski ring with filtration FR. Suppose that $G(R)$ is an Auslander regular ring then $\tilde{R}$ is an Auslander regular ring.

Proof. Use Theorem 2.2.5. and Theorem 3.1.6.. $\square$

3.2. Good graded filtrations

Let R be a $\mathbb{Z}$-graded ring with double gradations, i.e., there are two gradations on R:

$$g(\mathrm{I}): \quad R = \oplus_{n \in \mathbf{Z}} R_n^{(I)} \qquad\qquad g(\mathrm{II}): \quad R = \oplus_{n \in \mathbf{Z}} R_n^{(II)}$$

Then there are two corresponding categories of graded R-modules, denoted R-gr(I), R-gr(II) respectively.

1. Definition. With notation as above, we say that $g(\mathrm{I})$ ($g(\mathrm{II})$) is **decomposible** by $g(\mathrm{II})$ ($g(\mathrm{I})$) if for all $n \in \mathbb{Z}$ we have $R_n^{(I)} = \sum_{j \in \mathbf{Z}} R_n^{(I)} \cap R_j^{(II)}$ ($R_n^{(II)} = \sum_{j \in \mathbf{Z}} R_n^{(II)} \cap R_j^{(I)}$).

From now on we assume that $g(\mathrm{I})$ is decomposible by $g(\mathrm{II})$.

Take the (natural) grading filtration FR on R as defined in Ch.I. §2., i.e., $F_n R = \oplus_{p \le n} R_p^{(I)}$, $n \in \mathbb{Z}$, then obviously $G(R) \cong R$ with respect to $g(\mathrm{I})$.

Let $M = \oplus_{v \in \mathbf{Z}} M_v^{(II)}$ be any finitely generated graded R-module in R-gr(II), say $M = \sum_{i=1}^s R\xi_i$ with $\xi_i \in M$ and $\deg \xi_i = k_i$, $i = 1, \cdots, s$, then we may define the good filtration FM on M as follows:

$$F_n M = \sum_{i=1}^s F_{n-k_i} R\xi_i, \quad n \in \mathbb{Z}.$$

2. Lemma. With notations and assumptions as above, we have $G(M) \in R\text{-gr(I)}$ and $G(M) \in R\text{-gr(II)}$.

Proof. Since $G(R) \cong R$ with respect to g(I) we immediately have $G(M) \in R\text{-gr(I)}$. On the other hand, since g(I) is decomposible by g(II) we have for any $n \in \mathbb{Z}$,

$$
\begin{aligned}
F_n M &= \sum_{i=1}^{s} F_{n-k_i} R\xi_i \\
&= \sum_{i=1}^{s} (\oplus_{p \le n-k_i} R_p^{(I)})\xi_i \\
&= \sum_{i=1}^{t} \Big(\oplus_{p \le n-k_i} (\sum_{j \in \mathbb{Z}} R_p^{(I)} \cap R_j^{(II)})\Big)\xi_i.
\end{aligned}
$$

Note that $R_p^{(I)} \cap R_j^{(II)} \subseteq R_j^{(II)}$ and every ξ_i is a homogeneous element in M, it follows that $F_n M$ is a graded subgroup of M. Hence we may put $F_n M = \oplus_{v \in \mathbb{Z}} M_{n,v}$ where $M_{n,v}$ is a subgroup of $M_v^{(II)}$, $n \in \mathbb{Z}$. It follows that $F_{n-1} M = \oplus_{v \in \mathbb{Z}} M_{n-1,v}$ such that $M_{n-1,v} \subseteq M_{n,v}$ and $F_n M / F_{n-1} M = \oplus_{v \in \mathbb{Z}} (M_{n,v} + F_{n-1} M)/F_{n-1} M$. Consequently we obtain

$$
\begin{aligned}
G(M) = \oplus_{n \in \mathbb{Z}} F_n M / F_{n-1} M &= \oplus_{n \in \mathbb{Z}} (\oplus_{v \in \mathbb{Z}} (M_{n,v} + F_{n-1} M)/F_{n-1} M) \\
&= \oplus_{v \in \mathbb{Z}} (\oplus_{n \in \mathbb{Z}} (M_{n,v} + F_{n-1} M)/F_{n-1} M) \\
&= \oplus_{v \in \mathbb{Z}} S_v.
\end{aligned}
$$

Note that $a_t \in R_t^{(II)}$, then $a_t = \sum_{k=1}^{m} a_{k,t}$ where $a_{k,t} \in R_k^{(I)} \cap R_t^{(II)}$. Furthermore, since the module operation of $G(R)$ on $G(M)$ is induced by the module operation of R on M and $G(R) \cong R$, it follows from the structure of S_v that $a_t S_v \subseteq S_{v+t}$. Therefore $G(M) \in R\text{-gr(I)}$ with gradation $G(M) = \oplus_{n \in \mathbb{Z}} F_n M / F_{n-1} M$ and $G(M) \in R\text{-gr(II)}$ with gradation $G(M) = \oplus_{v \in \mathbb{Z}} S_v$. $\qquad\square$

3. Definition. Let M be any finitely generated graded R-module in $R\text{-gr(II)}$ with good filtration FM. FM is said to be a **good graded filtration** if it is of the form $F_n M = \sum_{i=1}^{s} F_{n-k_i} R\xi_i$, $n \in \mathbb{Z}$, such that all ξ_i are homogeneous elements of M.

4. Proposiiton. Let R be a $\mathbb{Z}$-graded ring with double gradations g(I) and g(II). Suppose that g(I) is decomposible by g(II) and the grading filtration FR on R is Zariskian, then good graded filtrations on objects of $R\text{-gr(II)}$ induce good graded filtrations on R-subobjects of $R\text{-gr(II)}$.

Proof. Let $M = \oplus_{v \in \mathbb{Z}} M_v^{(II)}$ be a finitely generated graded module in $R\text{-gr(II)}$ with good graded filtration FM. Let N be a graded submodule of M in $R\text{-gr(II)}$, say $N = \oplus_{v \in \mathbb{Z}} N_v$, where $N_v = N \cap M_v^{(II)}$. Consider the good filtration FN on N induced by FM (since FR is Zariskian), then $G(N) \subseteq G(M)$ and $G(N)$ is finitely generated, say $G(N) = \sum_{j=1}^s G(R)\sigma(u_j)$ with $u_j \in F_{k_j}N = F_{k_j}M \cap N$ such that $u_j \notin F_{k_j-1}N$. Since $F_{k_j}M \cap N = (\oplus_{v \in \mathbb{Z}} M_{k_j,v}) \cap N$ we may put $u_j = \sum m_{k_j,v}$ where $m_{k_j,v} \in M_{k_j,v}$. It follows that $m_{k_j,v} \in F_{k_j}M \cap N = F_{k_j}N$ because N is a graded submodule of M in R-gr(II). More precisely $m_{k_j,v} \in N_v$. Without loosing generality we may assume that all $m_{k_j,v} \notin F_{k_j-1}N$ (otherwise $u_j \in F_{k_j-1}N$). Hence $\sigma(m_{k_j,v}) \in G(N)_{k_j}$ and $\sigma(u_j) = \sum \sigma(m_{k_j,v})$. Now replacing $\sigma(u_j)$ by $\sigma(m_{k_j,v})$, we may assume that all u_j, $j = 1, \cdots, s$, are homogeneous elements of N. Since FN is good it follows from Ch.II. §2. that $F_nN = \sum_{j=1}^s F_{n-k_j}Ru_j$, $n \in \mathbb{Z}$. This shows that the induced filtration FN on N is good graded.   □

5. Corollary. Let R be a $\mathbb{Z}$-graded ring as in the above proposition. For any finitely generated object M in R-gr(II), there exists a gr-free resolution of M in R-gr(II), say

$$\cdots \longrightarrow L_j \xrightarrow{f_j} L_{j-1} \longrightarrow \cdots \longrightarrow L_1 \longrightarrow L_0 \longrightarrow M \longrightarrow 0$$

such that the sequence is also a filt-free resolution of M in R-filt with respect to the grading filtration FR and the good graded filtration FM, and moreover all f_j are tstrict filtered morphisms.

Proof. Let $M = \sum_{i=1}^s R\xi_i$ with $\xi_i \in M$ and $\deg \xi_i = k_i$, $i = 1, \cdots, s$, then $M_n^{(II)} = \sum_{i=1}^s R_{n-k_i}^{(II)}\xi_i$, $n \in \mathbb{Z}$. Construct a gr-free R-module $L = \oplus_{i=1}^s Re_i$ with gradation $L_n = \sum_{i=1}^s R_{n-k_i}^{(II)}e_i$, $n \in \mathbb{Z}$, such that there is an exact sequence in R-gr(II): $L \xrightarrow{\varepsilon_0} M \to 0$ where ε_0 is given by $\varepsilon_0(e_i) = \xi_i$. Let FM be the good graded filtration on M defined as before; On L we also construct FL by putting

$$F_nL = \sum_{i=1}^s F_{n-k_i}Re_i = \sum_{i=1}^s (\oplus_{p \leq n-k_i} R_p^{(I)})e_i, \quad n \in \mathbb{Z}.$$

It is clear that FL is a good graded filtration making ε_0 into a filtered epimorphism of degree zero in R-filt. Moreover, $K = \mathrm{Ker}\varepsilon_0$ is a graded submodule of L in R-gr(II) and Proposition 3.2.4. entails that the filtration FK induced on K by FL is a good graded filtration. Arguing as before we then arrive at a gr-free R-module L_1 in R-gr(II) having a good graded filtration FL_1 such tha the sequence $L_1 \xrightarrow{f_1} K \to 0$ is exact in R-gr(II) as well as R-filt. This leads to the strict filtered and graded resolution as required.   □

6. Remark. Let R be a $\mathbb{Z}$-graded ring as before. Suppose that $g(I)$ is decomposible by $g(II)$ and FR (the grading filtration on R determined by $g(I)$) is left and right Zariskian.

(1). If $L = \oplus_{i=1}^{s} Re_i$ is a gr-free R-module in R-gr(II) with gradation $L_n = \sum_{i=1}^{s} R_{n-k_i}^{(II)} e_i$, $n \in \mathbb{Z}$, for some $k_i \in \mathbb{Z}$, and if we construct the good graded filtration FL on L in R-filt: $F_n L = \sum_{i=1}^{s} F_{n-k_i} Re_i$, $n \in \mathbb{Z}$, then by Ch.I. Lemma 4.1.1. and Proposition 6.6. $L^* = \mathrm{Hom}_R(L, R) = \mathrm{HOM}_R(L, R)$ is both filtered in R-filt and graded in R-gr(II), and $F_p L^* = \sum_{i=1}^{s} e_i^* F_{p+k_i} R$, $p \in \mathbb{Z}$, is a good graded filtration on L^*, where $\{e_i^*\}$ is the dual basis of L.

(2). Let M be as in Corollary 3.2.5. and

$$\cdots \longrightarrow L_j \xrightarrow{f_j} L_{j-1} \longrightarrow \cdots \longrightarrow L_1 \longrightarrow L_0 \longrightarrow M \longrightarrow 0$$

the strict filtered and graded free resolution of M in R-gr(II) and R-filt. Then by the proof of Lemma 3.2.2. and Ch.I. Theorem 4.2.4. we have the associated exact sequence

$$\cdots \longrightarrow G(L_j) \xrightarrow{G(f_j)} G(L_{j-1}) \longrightarrow \cdots \longrightarrow G(L_1) \longrightarrow G(L_0) \longrightarrow G(M) \longrightarrow 0$$

in R-gr(I) and R-gr(II). If we consider the dual sequences

$$F(*) \qquad 0 \longrightarrow L_0^* \longrightarrow L_1^* \longrightarrow \cdots \longrightarrow L_{j-1}^* \xrightarrow{f_j^*} L_j^* \longrightarrow \cdots$$

$$0 \longrightarrow G(L_0)^* \longrightarrow G(L_1)^* \longrightarrow \cdots \longrightarrow G(L_{j-1})^* \xrightarrow{G(f_j)^*} G(L_j)^* \longrightarrow \cdots$$

then $F(*)$ is in R-filt and R-gr(II), and the second one is in R-gr(I) and R-gr(II).

(3). Let us further consider the spectral sequence determined by $F(*)$ given in (2). From Proposition 3.2.4. it is clear that each cohomology group $\mathrm{Ext}_R^{j-1}(M, R) = \mathbf{H}_{j-1}$ has a good graded filtration $F_n \mathbf{H}_{j-1} = (\mathrm{Ker} f_j^* \cap F_n L_{j-1}^*) + \mathrm{Im} f_{j-1}^* / \mathrm{Im} f_{j-1}^*$, $n \in \mathbb{Z}$, $j = 1, 2, \cdots$. Moreover, write $L_{j-1}^* = \oplus_{v \in \mathbb{Z}} L_v^*$ in R-gr(II), then $F_n L_{j-1}^* = \sum_{i=1}^{s} e_i^* F_{n+k_i} R = \oplus_{v \in \mathbb{Z}} L_{j-1,n}^{*v}$ with $L_{j-1,n}^{*v} \subseteq L_v^*$. Hence we have

$$Z_n^k(j) = \oplus_{v \in \mathbb{Z}} Z_{n,v}^k(j) \text{ with } Z_{n,v}^k(j) \subseteq L_{j-1,n}^{*v}$$
$$B_n^k(j) = \oplus_{v \in \mathbb{Z}} B_{n,v}^k(j) \text{ with } B_{n,v}^k(j) \subseteq L_{j-1,n}^{*v}$$

and consequently each E_j^k in R-gr(II) is of the form

$$E_j^k = \bigoplus_{n \in \mathbb{Z}} \frac{Z_n^k(j) + F_{n-1} L_{j-1}^*}{B_n^k(j) + F_{n-1} L_{j-1}^*}$$

$$= \bigoplus_{v \in \mathbb{Z}} \left(\bigoplus_{n \in \mathbb{Z}} \frac{Z_{n,v}^k(j) + L_{j-1,n-1}^{*v}}{B_{n,v}^k(j) + L_{j-1,n-1}^{*v}} \right)$$

Therefore by the definition of $\chi_j(-k)$ it is easy to see that $\chi_j(-k) \colon E_{j-1}^k \to E_j^k$ is of degree $-k$ in R-gr(I) and of degree zero in R-gr(II).

Finally, since FR is left and right Zariskian we may use a similar argumentation as given in the proof of Proposition 1.1.5. and Proposition 2.2.4. in order to obtain the following result: $G(\mathrm{Ext}_R^{j-1}(M,R))$ is isomorphic to a subquotient of $\mathrm{Ext}_{G(R)}^{j-1}(G(M),G(R))$ in both R-gr(I) and R-gr(II).

3.3. gr-Auslander regular rings are Auslander regular

Let R be a left and right Noetherian $\mathbb{Z}$-graded ring and $M \in R$-gr a finitely generated R-module. We say that M satisfies the **graded Auslander condition** in R-gr if for any **graded submodule** N of $\mathrm{Ext}_R^k(M,R)$ with $k \geq 0$ we have $j_R(N) \geq k$; If R has finite global dimension and every finitely generated graded R-module satisfies the graded Auslander condition in R-gr, then we say that R is a **gr-Auslander regular ring**.

Note. In the above definition, the fact that $\mathrm{Ext}_R^k(M,R)$ is a graded R-module follows from Ch.I. Lemma 4.1.1..

Our aim in this subsection is to prove that a gr-Auslander regular ring is really an Auslander regular ring. So from now on we are dealing with a left and right Noetherian $\mathbb{Z}$-graded ring R.
We first consider a special case.

1. Theorem. If R is positively graded or left limited then R is Auslander regular if and only if R is gr-Auslander regular.

Proof. Consider the polynomial ring $R[t]$ over R in the commuting variable t. If we take the mixed gradation on $R[t]$: $R[t]_n = \{\sum_{i+j=n} r_i t^j, \ r_i \in R_i\}$, $n \in \mathbb{Z}$, then $t \in R[t]_1$ and this gradation on $R[t]$ is still positive or left limited. It follows that the filtered ring (or the dehomogenization of $R[t]$ with respect to t) $A = R[t]/(1-t)R[t]$ with filtration $F_n A = R[t]_n + (1-t)R[t]/(1-t)R[t]$, $n \in \mathbb{Z}$, is Zariskian because $G(A) \cong R[t]/tR[t] \cong R$ as graded rings (see Ch.I. Example 4.3.9.(b)). Hence by Theorem 2.2.5. (indeed by its proof) A is Auslander regular. But Since A is also isomorphic to R as an ungraded ring, R is Auslander regular, too. $\square$

To investigate the general situation, once again we consider the polynomial ring $R[t]$ over R in the commuting variable t, then $R[t]$ has two different gradations, i.e., the **natural gradation** $R[t]^{(I)}$ defined by putting

$$R[t]_p^{(I)} = Rt^p, \quad p \in \mathbb{N}$$

and the **mixed gradation** $R[t]^{(II)}$ defined by putting

$$R[t]_p^{(II)} = \Big\{ \sum_{i+j=p} r_i t^j,\ r_i \in R_i \Big\}, \quad p \in \mathbb{Z}.$$

It is easy to see that $R[t]^{(I)}$ is decomposible by $R[t]^{(II)}$ in the sense of last subsection, and the (natural) grading filtration $FR[t]$ on $R[t]$: $F_n R[t] = \oplus_{p \leq n} R[t]_p^{(I)} = \oplus_{p \leq n} Rt^p$, $n \geq 0$, is left and right Zariskian. Moreover, let $M \in R[t]$-gr(II) be any finitely generated graded $R[t]$-module, say $M = \sum_{i=1}^s R[t]\xi_i$ with $\xi_i \in M$ and $\deg\xi_i = k_i$ for some $k_i \in \mathbb{Z}$, $i = 1, \cdots, s$. Take the good graded filtration FM on M with respect to ξ_i as in the last subsection, then from Lemma 3.2.2. and the fact that R is a graded subring of $R[t]$ with respect to the gradation $R[t]^{(II)}$ the following lemma is clear

2. Lemma. With notation as above, $G(M) \in R[t]$-gr(I), $G(M) \in R[t]$-gr(II). Moreover each $G(M)_n = (\sum_{i=1}^s Rt^{n-k_i}\xi_i + F_{n-1}M)/F_{n-1}M$ is a finitely generated graded R-module.

3. Lemma. Let M be a finitely generated graded $R[t]$-module in $R[t]$-gr(II) with good graded filtration FM as before. Assume that R is a gr-Auslander regular ring, then $G(M)$ satisfies the graded Auslander condition.

Proof. Since $G(M) \in R[t]$-gr(I) it follows from the proof of Lemma 2.3.1. and Lemma 3.3.2. that there is an exact sequence in $R[t]$-gr(I) and $R[t]$-gr(II)

$$0 \longrightarrow G(M)' \longrightarrow G(M) \longrightarrow G(M)'' \longrightarrow 0$$

with the following properties:

(a). $G(M)' = R[t]G(M)_l \cong R[t] \otimes_R G(M)_l$ in $R[t]$-gr(I) and $R[t]$-gr(II) for some part $G(M)_l$ of degree l of $G(M)$, where the gradation on $R[t] \otimes_R G(M)_l$ is the tensor-gradation (see the proof of Lemma 3.2.2. for the graded structure of $G(M)_l$ with respect to the gradation $G(M)^{(II)}$).

(b). $G(M)''$ is a finitely generated $R[t]$-module such that $t^\omega G(M)'' = 0$ for some integer $\omega > 0$; and in particular, $G(M)''$ is also a finitely generated graded R-module.

First of all, we prove that $G(M)''$ satisfies the graded Auslander condition in $R[t]$-gr(II). If $\omega = 1$, then since $R[t]/tR[t] \cong R$ as graded rings (where R has the original gradation! see Ch.I. §4. Example 9.(b).) it follows from the theorem of Rees (indeed a graded version of that) we have for every $k \geq 0$

$$\mathrm{Ext}_R^k(G(M)'', R) \cong \mathrm{Ext}_{R[t]}^{k+1}(G(M)'', R[t]).$$

Hence the graded Auslander regularity of R entails that $G(M)''$ satisfies the graded Auslander condition in $R[t]$-gr(II). If $\omega > 1$, then by a graded version of Lemma 2.1.4. and an easy induction on ω we may conclude that $G(M)''$ satisfies the graded Auslander condition in $R[t]$-gr(II).

Next, consider a graded R-free resolution of $G(M)_l$ in R-gr:

$$\cdots \longrightarrow L_1 \longrightarrow L_0 \longrightarrow G(M)_l \longrightarrow 0$$

where each L_j has finite rank in R-gr since $G(M)_l$ is a finitely genrated graded R-module (Lemma 3.3.2.), then we obtain a graded $R[t]$-free resolution of $G(M)'$ in $R[t]$-gr(II):

$$\cdots \longrightarrow R[t] \otimes_R L_1 \longrightarrow R[t] \otimes_R L_0 \longrightarrow R[t] \otimes_R G(M)_l \longrightarrow 0$$

where $R[t] \otimes_R G(M)_l$ resp. each $R[t] \otimes_R L_j$ has the tensor-gradation. By the Noetherian condition on R and the flatness of $R[t]$ over R we may derive from a light modification of ([Rot] Lemma 3.83) the following $R[t]$-isomorphism in $R[t]$-gr(I) and $R[t]$-gr(II):

$$\begin{aligned}
\mathrm{Ext}^k_{R[t]}(G(M)', R[t]) &\cong \mathrm{Ext}^k_R(G(M)_l, R) \otimes_R R[t] \\
&= \oplus_{p \geq 0}(\mathrm{Ext}^k_R(G(M)_l, R) \otimes_R Rt^p).
\end{aligned}$$

Working on the second gradation of $\mathrm{Ext}^k_{R[t]}(G(M)', R[t])$, i.e., the tensor-gradation on $\mathrm{Ext}^k_R(G(M)_l, R) \otimes_R R[t]$, if we consider the (natural) grading filtration

$$F_n(\mathrm{Ext}^k_R(G(M)_l, R) \otimes_R R[t]) = \oplus_{p \leq n}(\mathrm{Ext}^k_R(G(M)_l, R) \otimes_R Rt^p), \quad n \in \mathbb{Z},$$

and choose a homogeneous generating system of the graded R-module $\mathrm{Ext}^k_R(G(M)_l, R)$, then it is easy to see that this filtration is a good graded filtration with respect to the grading filtration $FR[t]$. If N is any graded $R[t]$-submodule of $\mathrm{Ext}^k_{R[t]}(G(M)', R[t])$ in $R[t]$-gr(II), then the filtration FN on N induced by $F\mathrm{Ext}^k_{R[t]}(G(M)', R[t])$ is a good graded filtration by Proposition 3.2.4.. It follows from Lemma 3.3.2. that $G(N)$ is a graded submodule of $G(\mathrm{Ext}^k R[t](G(M)', R[t])) = \mathrm{Ext}^k_{R[t]}(G(M)', R[t])$ in $R[t]$-gr(II) and moreover $G(N)_p \subseteq G(\mathrm{Ext}^k_{R[t]}(G(M)', R[t]))_p = \mathrm{Ext}^k_R(G(M)_l, R) \otimes_R Rt^p$. Note that $\mathrm{Ext}^k_R(G(M)_l, R) \otimes_R Rt^p$ is a finitely generated graded R-module and $G(N)_p$ is a graded R-submodule of it, by the graded Auslander regularity of R we have $j_R(G(N)_p) \geq j_R(\mathrm{Ext}^k_R(G(M)_l, R) \otimes_R Rt^p)$ (a graded version of Proposition 2.1.5.). But $\mathrm{Ext}^k_R(G(M)_l, R) \otimes_R Rt^p \cong \mathrm{Ext}^k_R(G(M)_l, R)$ as ungraded R-modules and $j_R(\mathrm{Ext}^k_R(G(M)_l, R)) \geq k$, it follows that $j_R(G(N)_p) \geq k$ (since the grade number is independent of the choice of gradations and projective resolutions!).

Put $G(N) = H$. Again by Lemma 2.3.1. and Lemma 3.3.2. there is an exact sequence of $R[t]$-modules:

$$0 \longrightarrow H' \longrightarrow H \longrightarrow H'' \longrightarrow 0$$

in $R[t]$-gr(I) and $R[t]$-gr(II), where H'' is a finitely generated graded R-module and $H' = H_q \otimes_R R[t]$ for some part of degree q of H. Finally, note that $H_q = G(N)_q$, a similar argumentation as given in the proof of Lemma 2.3.3. works and hence we arrive at $j_{R[t]}(G(N)) = j_{R[t]}(H) \geq k$. Since $R[t]$ is a Zariski ring with respect to the grading filtration $FR[t]$ it follows from Proposition 2.2.4. that $j_{R[t]}(N) \geq j_{R[t]}(G(N)) \geq k$. This shows that $G(M)'$ satisfies the graded Auslander condition in $R[t]$-gr(II). Therefore by a graded version of Lemma 2.1.4. we can conclude that $G(M)$ satisfies the graded Auslander condition in $R[t]$-gr(II). □

4. Proposition. Let the notations be as before. If R is a gr-Auslander regular ring then $R[t]$ is a gr-Auslander regular ring with respect to the mixed gradation $R[t]^{(II)}$.

Proof. $R[t]$ has finite global dimension is clear. Let M be any finitely generated object in $R[t]$-gr(II). Consider the grading filtration $FR[t]$ on $R[t]$ and take the good graded filtration FM on M as before, then $G(M) \in R[t]$-gr(I) and $G(M) \in R[t]$-gr(II) and $G(M)$ satisfies the graded Auslander condition in $R[t]$-gr(II) by Lemma 3.3.3.. Moreover, from Remark 3.2.6. we know that there is a good graded filtration $F\mathrm{Ext}^k_{R[t]}(M, R[t])$ on $\mathrm{Ext}^k_{R[t]}(M, R[t])$ such that $G(\mathrm{Ext}^k_{R[t]}(M, R[t]))$ is a subquotient of $\mathrm{Ext}^k_{R[t]}(G(M), R[t])$ in both $R[t]$-gr(I) and $R[t]$-gr(II) since $FR[t]$ is Zariskian and $G(R[t]) = R[t]$. If N is any graded $R[t]$-submodule of $\mathrm{Ext}^k_{R[t]}(M, R[t])$ in $R[t]$-gr(II), then after working on the good graded filtration FN induced by $F\mathrm{Ext}^k_{R[t]}(M, R[t])$ we have $G(N) \subseteq G(\mathrm{Ext}^k_{R[t]}(M, R[t]))$ is a (graded in both gradations) subquotient of $\mathrm{Ext}^k_{R[t]}(G(M), R[t])$. It follows from Remark 3.2.6. and a graded version of Proposition 2.1.(2). that $j_{R[t]}(G(N)) \geq k$, and hence $j_{R[t]}(N) \geq j_{R[t]}(G(N)) \geq k$ by Proposition 2.2.4. since $R[t]$ is a Zariski ring with respect to the grading filtration $FR[t]$. This shows that every finitely generated graded $R[t]$-module in $R[t]$-gr(II) satisfies the graded Auslander condition and consequently $R[t]$ is a gr-Auslander regular ring. □

5. Theorem. Let R be a $\mathbb{Z}$-graded ring. If R is a gr-Auslander regular ring then it is an Auslander regular ring.

Proof. By Proposition 3.3.4. we know from the assumption that $R[t]$ is gr-Auslander regular where $R[t]$ has the mixed gradation as before. Consider the localization $R[t]_{(t)}$ of $R[t]$ at the (central) Ore set $\{1, t, t^2 \cdots\}$, then $R[t]_{(t)} \cong R[t, t^{-1}]$ where the latter also has the mixed gradation: $R[t, t^{-1}]_n = \{\sum_{i+j=n} r_i t^j, \ r_i \in R_i\}$, $n \in \mathbb{Z}$. From the definition it is clear that $R[t]$ is a graded subring of $R[t, t^{-1}]$ and $R[t, t^{-1}]_0 = \sum_{i+j=0} R_i t^j \cong R$ as ungraded rings. Hence if we can prove that $R[t, t^{-1}]$ is gr-Auslander regular then the equivalence of categories: $R[t, t^{-1}]_0$-mod $\leftrightarrow R[t, t^{-1}]$-gr, or Lemma 2.5.3. will yield the Auslander regularity of $R[t, t^{-1}]_0$ and hence the Auslander regularity of R (since $R[t, t^{-1}]$ is strongly $\mathbb{Z}$-graded with respect to the mixed gradation).

First, $R[t, t^{-1}]$ is Noetherian and it has finite global dimension by the assumption and a classical theory. If M is any finitely generated graded $R[t, t^{-1}]$-module, say $M = \sum_{i=1}^s R[t, t^{-1}]\xi_i$ where all ξ_i are homogeneous elements in M, then $M_0 = \sum_{i=1}^s R[t]\xi_i$ is a finitely generated graded $R[t]$-submodule of M such that the localization $M_{0(t)}$ of M_0 is equal to M. Hence we have

$$\mathrm{Ext}^k_{R[t,t^{-1}]}(M, R[t, t^{-1}]) \cong \mathrm{Ext}^k_{R[t]}(M_0, R[t])_{(t)}.$$

If N is any graded $R[t, t^{-1}]$-submodule of $\mathrm{Ext}^k_{R[t,t^{-1}]}(M, R[t, t^{-1}])$, then N is finitely generated. Let $N = \sum_{j=1}^m R[t, t^{-1}]h_j$, where all h_j are homogeneous elements of N, then there is an integer $\omega > 0$ such that $t^\omega h_j$, $j = 1, \cdots, m$, are homogeneous elements contained in $\mathrm{Ext}^k_{R[t]}(M_0, R[t])$. Hence $\sum_{j=1}^m R[t]t^\omega h_j$ is a graded submodule of $\mathrm{Ext}^k_{R[t]}(M_0, R[t])$ such that $(\sum_{j=1}^m R[t]t^\omega h_i)_{(t)} = N$. Again by the graded isomorphism

$$\mathrm{Ext}^k_{R[t,t^{-1}]}(N, R[t, t^{-1}]) \cong \mathrm{Ext}^k_{R[t]}(N_0, R[t])_{(t)}$$

where $N_0 = \sum_{j=1}^m R[t]t^\omega h_j$, we have that $j_{R[t,t^{-1}]}(N) \geq j_{R[t]}(N_0) \geq k$ since N_0 is a graded submodule of $\mathrm{Ext}^k_{R[t]}(M_0, R[t])$ and $R[t]$ is gr-Auslander regular by Proposition 3.3.3.. This shows that $R[t, t^{-1}]$ is gr-Auslander regular with respect to the mixed gradation, as desired. $\square$

3.4. Applications to invertible ideals

We first apply the foregoing results to a strongly $\mathbb{Z}$-graded ring.

1. Theorem. Let $R = \oplus_{n \in \mathbb{Z}} R_n$ be a strongly $\mathbb{Z}$-graded ring. The following statements are equivalent:

(1). R_0 is an Auslander regular ring;

(2). R is an Auslander regular ring;

(3). $R^+ = \oplus_{n \geq 0} R_n$ is an Auslander regular ring;

(4). $R^- = \oplus_{n \leq 0} R_n$ is an Auslander regular ring.

Proof. First, it is well known that R_0 is Noetherian if and only if R is Noetherian if and only if R^+ is Noetherian if and only if R^- is Noetherian.

Secondly, by Ch.II. Proposition 5.1.12. we know that R_0 has finite global dimension if and only if R has finite global dimension if and only if R^+ has finite global dimension if and only if R^- has finite global dimension.

Finally, if we consider the $+$-gradation (or $-$-gradation) of the polynomial ring $R[t]$ in the sense of Ch.I. Example 4.3.9.(d). then $R[t]$ becomes a strongly $\mathbb{Z}$-graded ring with $R[t]_0^+ = R^+$ ($R[t]_0^- = R^-$). Hence all equivalences mentioned in the theorem now follow from Theorem 2.3.5., Lemma 2.5.3. and Theorem 3.3.5.. $\square$

Next, let us consider general invertible ideals.

2. Lemma. Let R be an arbitrary ring and I an invertible ideal of R with respect to some overring S of R. Consider the I-adic filtration FR on R. If R/I is an Auslander regular ring then $G(R)$ is an Auslander regular ring.

Proof. Note that the graded ring $A = \oplus_{n \in \mathbb{Z}} I^n/I^{n+1}$ with $A_n = I^{-n}/I^{-n+1}$ is strongly $\mathbb{Z}$-graded (indeed it is the associated graded ring of the filtered ring $S' = \cup_{n \in \mathbb{Z}} I^n \subseteq S$ with filtration $F_n S' = I^{-n}$), and in paricular $A^- = G(R)$, $A_0 = R/I = G(R)_0$. It follows from Theorem 3.4.1. that $G(R)$ is Auslander regular. $\square$

3. Theorem. Let R be a left and right Noetherian ring and I an invertible ideal of R with respect to some overring S of R. Suppose that I is contained in the Jacobson radical $J(R)$ of R. With notation as in Ch.I. Example 4.3.9.(c)., the following statements hold:

(1). If gl.dim$(R/I) < \infty$ then

    a. gr.gl.dim$\check{R}(I) = 1+$ gr.gl.dim$G(A) = 1+$ gl.dim(R/I), where $A = \check{R}(I)/(1-y)\check{R}(I)$;

    b. gl.dim$R = 1+$ gl.dim(R/I);

    c. gl.dim$\check{R}(I) = 1+$ gl.dim$G(A) \leq 2+$ gl.dim(R/I), gl.dim$\tilde{R} = 1+$ gl.dim$G(R) \leq 2+$ gl.dim(R/I).

(2). If R/I is an Auslander regular ring then R, $G(R)$, $\tilde{R}$ and $\check{R}(I)$ are Auslander regular rings.

Proof. This follows from Theorem 2.2.13. and Theorem 5.2.5. of Ch.II., and the foregoing results concerning the Auslander regularity of Zariski ring, Rees ring and (strongly) graded ring. $\square$.

Now, we turn to the case of $\mathbb{Z}$-graded ring where the graded Jacobson radical is in consideration.

4. Theorem. Let $R = \oplus_{n \in \mathbb{Z}} R_n$ be a left and right Noetherian $\mathbb{Z}$-graded ring and I a graded ideal of R. Suppose that I is invertible with respect to some overring of R and that

I is contained in the graded Jacobson radical $J^g(R)$ of R. If R has finite global dimension and R/I is an Auslander regular ring then R and $\tilde{R}$ are Auslander regular rings, where $\tilde{R}$ is the Rees ring of R associated to the I-adic filtration on R.

Proof. Considering the I-adic filtration FR on R, the associated graded ring $G(R)$ of R is Auslander regular by Lemma 3.4.2.. Let M be any finitely generated graded R-module. Note that the cohomology group $\text{Ext}_R^k(M, R)$ with $k \geq 0$ is independent of the choice of projective resolution for M. If N is a graded submodule of $\text{Ext}_R^k(M, R)$, then by Ch.II. Lemma 3.2.3. and the assumption a similar argumentation as given in the proof of Proposition 2.2.4. yields that $j_R(N) \geq j_{G(R)}(G(N))$ and $G(N)$ is isomorphic to a subquotient of $\text{Ext}_{G(R)}^k(G(M), G(R))$, where N has the good filtration FN induced by $F\text{Ext}_R^k(M, R)$. Consequently $j_R(N) \geq j_{G(R)}(G(N)) \geq k$ because $G(R)$ is Auslander regular. This shows that R is a gr-Auslander regular ring and hence an Auslander regular ring by Theorem 3.3.5.. Finally, the Auslander regularity of $\tilde{R}$ follows from Theorem 3.1.7. and Ch.I. Corollary 7.2.6..
□

5. Corollary. Let R be a left and right Noetherian $\mathbb{Z}$-graded ring and X a regular noninvertible homogeneous normalizing element of R. Suppose that $X \in J^g(R)$ and that R/XR is an Auslander regular ring, then R and $\tilde{R}$ are Auslander regular rings, where $\tilde{R}$ is the Rees ring of R associated to the XR-adic filtration on R.

Proof. By the assumption and Ch.II. Theorem 5.2.3. we have gl.dim $R < \infty$. Therefore the conditions of Theorem 3.4.4. are fulfilled.   □

6. Theorem. Let A be an Auslander regular ring, then each one of the following rings is Auslander regular:

(1). $A[t, \sigma]$, the skew polynomial ring over A in variable t, where σ is an automorphism of A;

(2). $A[t, t^{-1}, \sigma]$, the localization of $A[t, \sigma]$ at the Ore set $\{1, t, t^2, \cdots\}$;

(3). $A[[t, \sigma]]$, the t-completion of $A[t, \sigma]$;

(4). $A[t, \sigma, \delta]$, where δ is a σ-derivation of A;

(5). The crossed product $A * G$ of A by G, where G is a poly-infinite cyclic group;

(6). The crossed product $A * U(\mathbf{g})$ of A by $U(\mathbf{g})$, where A is now a k-algebra over a field k and $\mathbf{g}$ a k-Lie algebra of finite dimension.

Proof. Since $A[t, \sigma] = A[t, t^{-1}, \sigma]^+$ and $A[t, t^{-1}, \sigma]$ is a strongly $\mathbb{Z}$-graded ring with $A[t, t^{-1}, \sigma]_0 = A$, (1) and (2) follow from Theorem 3.4.1.. Note that the standard filtrations on $A[t, \sigma, \delta]$ and $A[[t, \sigma]]$ are Zariskian with the associated graded rings which are isomorphic to $A[t, \sigma]$, (3) and (4) follow from Theorem 2.2.5.. From ([M-R] 1.5.11.) we know that $A * G$ can be obtained from A by iterated extensions as in (2). So induction on (2) yields (5). Again from ([M-R] 1.7.14.) we know that with respect to the standard filtration on $A * U(\mathbf{g})$ the associated graded ring $G(A * U(\mathbf{g}))$ is isomorphic to a polynomial ring over A in finitely many commuting variables. Hence (6) follows from Corollary 2.3.6. and Theorem 2.2.5.. $\qquad\square$

§4. Dimension Theory and Pure Module Theory over Zariskian Filtered Rings with Auslander Regular Associated Graded Rings

4.1. Generalized Roos theorem and an application to Zariski rings

We first recall some basic properties of a commutative Noetherian local ring. For the classical theory concerning multiplicity and (Krull-)dimension theory for modules over commutative Noetherian rings we refer to [Ser], [Z-S] and [Nor1], etc, $\cdots$.

Let A be a commutative Noetherian local ring with the unique maximal ideal Ω. Consider the Ω-adic filtration FA on A, then the associated graded ring $G(A)$ of A is of a well described type, i.e., $G(A) \cong k[t_1, \cdots, t_n]/I$ where $k = A/\Omega$ is the residue field of A, $k[t_1, \cdots, t_n]$ is the polynomial ring over k in n variables and I is some graded ideal of $k[t_1, \cdots, t_n]$ with respect to the natural gradation on $k[t_1, \cdots, t_n]$. For any A-module M, if we define FM to be the Ω-adic filtration on M then $G(M)$ may be viewed as graded $k[t_1, \cdots, t_n]$-module which is annihilated by the graded ideal I. If M is a finitely generated A-module then after taking a generating system of M the Ω-adic filtration FM is obviously a good filtration on M and hence $G(M)$ is a finitely generated graded $k[t_1, \cdots, t_n]$-module. Moreover, for a finitely generated A-module the length of $M/\Omega^r M$ is finite, and the length $l_A(M/\Omega^{r+1}M) = \sum_{j=0}^{r} \dim_k(\Omega^i M/\Omega^{i+1}M)$ may be calculated, for large values r, by the Hilbert polynomial of M, say $a_d x^d + \cdots + a_0$ with rational coefficients. The degree d of this Hilbert polynomial associated to M is equal to the **Krull dimension** K.dim$M$ of M while the **multiplicity** of M is defined to be $e(M) = d!a_d$. A Noetherian local ring A is called **regular** if gl.dim$A < \infty$.

1. Theorem. (Characterizations of Noetherian regular local ring) Let A be a Noetherian local ring of (Krull-)dimension n. Let Ω be the unique maximal ideal of A and $k = A/\Omega$ the residue field of A. The following statements are equivalent:

(1). A is a regular local ring;

(2). Ω may be generated by n elements;

(3). The k-dimension of the vector space Ω/Ω^2 over k equals n;

(4). $G(A)$ is isomorphic to the polynomial ring $k[t_1, \cdots, t_n]$ where $G(A)$ is the associated graded ring of A with respect to the Ω-adic filtration FA on A.

2. Proposition. Let A be an n-dimensional Noetherian regular local ring with the unique maximal ideal Ω.

(1). A is an integral closed domain.

(2). (Auslander-Buchsbaum) A is factorial.

(3). A is Cohen-Macaulay.

(4). $\mathrm{gl.dim}A = n = \mathrm{K.dim}A$.

(5). $\mathrm{Ext}_A^i(A/\Omega, A) = 0$ for all $i < n$.

3. Lemma. Let A be an n-dimensional Noetherian regular local ring with the unique maximal ideal Ω. If M is any zero-dimensional (i.e., $\mathrm{K.dim}_A M = 0$) finitely generated A-module then $j_A(M) = n$, where $j_A(M)$ is the grade number of M as defined in §2.

Proof. Since $\mathrm{K.dim}_A M = 0$ it is well known that M has finite length, or in other words there is a finite chain of A-submodules:

$$0 \subseteq M_0 \subseteq M_1 \subseteq \cdots \subseteq M_t = M$$

such that $M_v/M_{v-1} \cong A/\Omega$, $v = 1, \cdots, t$. Hence by taking the long exact Ext-sequence for each short exact sequence

$$0 \longrightarrow M_{v-1} \longrightarrow M_v \longrightarrow M_v/M_{v-1} \longrightarrow 0$$

Proposition 4.1.2. entails that $\mathrm{Ext}_A^i(M, A) = 0$ for all $i < n$. Therefore we must have $j_A(M) = n$. □

4. Lemma. Let A be an n-dimensional Noetherian regular local ring with the unique maximal ideal Ω. If M is any finitely generated A-module such that there is an M-regular element $x \in \Omega$ (i.e., $m \in M$ such that $xm = 0$ implies $m = 0$), then $j_A(M) + \mathrm{K.dim}_A M = n$.

Proof. On the one hand, choose p to be a minimal prime divisor of $\sqrt{\mathrm{Ann}_A M}$ in $\mathrm{spec}(A)$ (the prime spectrum of A) such that $\mathrm{K.dim}_A M = \mathrm{K.dim}(A/p)$, then since A is Cohen-Macaulay and A_p (the localization of A at p) is a regular local ring the following formula is well known (cf. [Mats] Theorem 29.):

$$(*) \qquad\qquad n = \mathrm{K.dim}(A/p) + \mathrm{gl.dim}A_p.$$

Note that M_p is zero-dimensional as an A_p-module it follows from Lemma 4.1.3. that $j_{A_p}(M_p) = \mathrm{gl.dim}A_p$. Put $\mathrm{gl.dim}A_p = s$. Then we have

$$\mathrm{Ext}_A^s(M, A)_p \cong \mathrm{Ext}_{A_p}^s(M_p, A_p) \neq 0$$

This means that $\mathrm{Ext}_A^s(M, A) \neq 0$, i.e., $j_A(M) \leq s$. Therefore $(*)$ yields the equality $j_A(M) + \mathrm{K.dim}_A M \leq n$.

On the other hand, it is well known that (cf [Mats] P.105, Lemma 4.) $\mathrm{K.dim}_A(M/xM) = \mathrm{K.dim}_A M - 1$. Consider the long exat Ext-sequence from the short exact sequence

$$0 \longrightarrow M \xrightarrow{\mu_x} M \longrightarrow M/xM \longrightarrow 0$$

where μ_x is multiplication by x, we obtain the exact sequence

$$\mathrm{Ext}_A^{j_A(M)}(M/xM, A) \longrightarrow \mathrm{Ext}_A^{j_A(M)}(M, A) \xrightarrow{\mu_x^*}$$
$$\xrightarrow{\mu_x^*} \mathrm{Ext}_A^{j_A(M)}(M, A) \longrightarrow \mathrm{Ext}_A^{j_A(M)+1}(M/xM, A)$$

where μ_x^* is again multiplication by x. If $j_A(M) < j_A(M/xM) - 1$, then μ_x^* becomes an isomorphism, contradicting $x \in \Omega$. Hence $j_A(M) \geq j_A(M/xM) - 1$. So if K.dim$_A(M/xM) = 0$ then by Lemma 4.1.3. we have $n = j_A(M/xM) +$ K.dim$_A(M/xM)$, and consequently, $j_A(M) \geq j_A(M/xM) - 1 = n-$ K.dim$_A(M/xM) - 1 = n - ($K.dim$_A(M/xM) + 1) = n-$ K.dim$_A M$, i.e., $j_A(M) +$ K.dim$_A M \geq n$. Suppose that K.dim$_A(M/xM) > 0$. Since we know that (cf. [Mats] P.97) dep$(M/xM) = 0$ if and only if $\Omega \in \mathrm{Ass}(M/xM)$ where dep(M/xM) denotes the depth of M/xM and Ass(M/xM) is the set of associated primes of M/xM in spec(A), we may find a zero-dimensional submodule N/xM of M/xM such that there is an $M/N(\cong (M/xM)/(N/xM))$-regular element $y \in \Omega$ (this can be done because K.dim$_A(M/xM) > 0$ and A is Noetherian). It follows that there is an exact sequence

$$0 \longrightarrow N/xM \longrightarrow M/xM \longrightarrow M/N \longrightarrow 0$$

and by Lemma 4.1.3. we have $j_A(N/xM) = n$. Hence Corollary 2.1.6. yields $j_A(M/xM) = j_A(M/N) \leq n$ since A is Auslander regular (Example 2.4.2.). It is also obvious that K.dim$_A(M/N) \leq $ K.dim$_A(M/xM) = $ K.dim$_A M - 1$. Hence by Lemma 4.1.3. we can make the induction hypothesis as follows: $j_A(M/N) +$ K.dim$_A(M/N) = n$. Then the foregoing formulas lead to the following

$$\begin{aligned} j_A(M) \geq j_A(M/xM) - 1 &= j_A(M/N) - 1 \\ &= n - \mathrm{K.dim}_A(M/N) - 1 \\ &= n - (\mathrm{K.dim}_A(M/N) + 1) \\ &\geq n - (\mathrm{K.dim}_A(M/xM) + 1) \\ &= n - \mathrm{K.dim}_A M. \end{aligned}$$

This shows that $j_A(M) +$ K.dim$_A M \geq n$, finishing the proof. $\square$

5. Theorem. (Roos theorem) Let A be an n-dimensional Noetherian regular local ring with the unique maximal ideal Ω. If M is any nonzero finitely generated A-module then $j_A(M) +$ K.dim$_A M = n$.

Proof. If K.dim$_A M = 0$, then Roos theorem follows from Lemma 4.1.3.. Let K.dim$_A M > 0$ Then by a similar argumentation as given in the above proof we may find a zero-dimensional submodule N of M such that there is an M/N-regular element $x \in \Omega$. It follows that there is an exact sequence

$$0 \longrightarrow N \longrightarrow M \longrightarrow M/N \longrightarrow 0$$

and by Lemma 4.1.3. we have $j_A(N) = n$. Hence Corollary 2.1.6. entails $j_A(M) = j_A(M/N) \leq n$. Moreover we also have $\mathrm{K.dim}_A M = \mathrm{K.dim}_A(M/N)$ because $\mathrm{K.dim}_A N = 0$. Finally, the proof is completed by Lemma 4.1.4.. $\qquad\square$

In order to generalize Roos theorem to a commutative Noetherian ring with finite global dimension we first need the following

6. Lemma. Let A be a Noetherian ring with finite global dimension, say $\mathrm{gl.dim} A = n$. If M is a finitely generated A-module then

(1). $\mathrm{K.dim}_A(\mathrm{Ext}_A^k(M, A)) \leq n - k$ for all $k \geq 0$;

(2). $j_A(M) = s = \inf\{j_{A_p}(M_p), \ p \in \mathrm{spec}(A)\} = \inf\{j_{A_P}(M_P), \ P \in \mathrm{spec}(A), \ P \text{ maximal}\}$.

Proof. (1). Since A is a commutative Auslander regular ring by Example 2.4.3. it follows that $j_{A_p}(\mathrm{Ext}_A^k(M, A)_p) \geq j_A(\mathrm{Ext}_A^k(M, A)) \geq k$ for any $p \in \mathrm{spec}(A)$. But on the other hand, the theorem of Roos says that $\mathrm{K.dim}_{A_p}\mathrm{Ext}^k(A(M, A)_p + j_{A_p}(\mathrm{Ext}_A^k(M, A)_p) = \mathrm{gl.dim} A_p \leq n$, $p \in \mathrm{spec}(A)$. Therefore $\mathrm{K.dim}_A \mathrm{Ext}_A^k(M, A) \leq n - k$.

(2). Since for each $p \in \mathrm{spec}(A)$ and each $k < s$ we have $\mathrm{Ext}_A^k(M, A)_p \cong \mathrm{Ext}_{A_p}^k(M_p, A_p) = 0$, it follows from the localization principle that $\mathrm{Ext}_A^k(M, A) = 0$. Hence $j_{A_p}(M_p) \geq j_A(M) \geq s$, $p \in \mathrm{spec}(A)$. By the definition of s we must have $j_A(M) = s$. $\qquad\square$

7. Theorem. (Generalized Roos theorem) Let A be a commutative Noetherian ring with $\mathrm{gl.dim} A = n < \infty$. Suppose that all maximal ideals P in $\mathrm{spec}(A)$ have the same height, then for every finitely generated A-module M we have $n = j_A(M) + \mathrm{K.dim}_A M$.

Proof. Since $\mathrm{K.dim}_A M = \sup\{\mathrm{K.dim}_{A_P} M_P, \ P \text{ maximal in } \mathrm{spec}(A)\}$, if we pass to localization then the theorem follows from Roos theorem and Lemma 4.1.6.. $\qquad\square$

Let $A = \oplus_{n \in \mathbb{Z}} A_n$ be a commutative Noetherian $\mathbb{Z}$-graded ring with $\mathrm{gl.dim} A = n < \infty$. To obtain a graded version of Theorem 4.1.7., for a finitely generated graded A-module M we define

$$d^g(M) = \sup\{\mathrm{K.dim}_{A_P} M_P, \ P \text{ maximal in } \mathrm{spec}^g(A)\}$$

where $\mathrm{spec}^g(A) = \{p \in \mathrm{spec}(A), \ p \text{ graded}\}$. Comparing with $\mathrm{K.dim}_A M$, it is obvious that $d^g(M) \leq \mathrm{K.dim}_A M$.

Note that $j_A(M)$ is independent of the choice of the projective resolution of M, $\mathrm{Ext}_A^k(M, A)$

is a graded A-module, and A is Auslander regular, if we use the graded localization principle in A-gr then a similar argumentation as before yields the following

8. Theorem. (Graded version of generalized Roos theorem) Let A be a commutative Noetherian $\mathbb{Z}$-graded ring with gl.dim $A = n < \infty$. Let M be any nonzero finitely generated graded A-module. Suppose that every maximal element in $\operatorname{spec}^g(A)$ has the same height in $\operatorname{spec}(A)$, say ω, then

(1). $j_A(M) = \inf\{j_{A_P}(M_P),\ P \text{ maximal in } \operatorname{spec}^g(A)\}$;

(2). $\omega = j_A(M) + d^g(M)$;

(3). if $\omega = n = $ gl.dim A, then $d^g(M) = $ K.dim$_A M$ and $n = j_A(M) + $ K.dim$_A M$.

Now, Let R be a filtered ring with filtration FR such that $G(R)$ is commutative.

9. Lemma. Let $M \in R$-filt have two good filtrations FM and $F'M$. If $G_F(M)$ and $G_{F'}(M)$ denote the associated graded modules of M with respect to FM and $F'M$ respectively, then

$$I = \sqrt{\operatorname{Ann}_{G(R)} G_F(M)} = \sqrt{\operatorname{Ann}_{G(R)} G_{F'}(M)} = I'$$

and consequently K.dim$_{G(R)} G_F(M) = $ K.dim$_{G(R)} G_{F'}(M)$.

Proof. Since good filtrations on M are equivalent there is an integer $c > 0$ such that

$$F_{k-c}M \subseteq F'_k M \subseteq F_{k+c}M, \quad k \in \mathbb{Z}.$$

Note that I and I' are graded ideals of $G(R)$. If $\sigma(r) \in G(R)_n$ and $\sigma(r) \neq 0$ is such that $\sigma(r)^m G_F(M) = 0$, where $r \in F_n R$, $m > 0$, then $r^m F_k M \subseteq F_{mn+k-1}M$, $k \in \mathbb{Z}$. This means that $r^{3cm} F_k M \subseteq F_{3cmn+k-3c}M$ and consequently $r^{3cm} F'_k M \subseteq r^{3cm} F_{k+c}M \subseteq F_{3cmn+k+c-3c}M \subseteq F'_{3cmn+k-c}M$. Since $c > 0$ it follows that $\sigma(r)^{3cm} G_{F'}(M) = 0$, i.e., $\sigma(r) \in I'$. Hence $I \subseteq I'$. Similarly one can prove $I' \subseteq I$, and then $I = I'$ follows.   □

From Lemma 4.1.9. above it is clear that for any $M \in R$-filt with a good filtration FM, if we put

$$d(M) = \text{K.dim}_{G(R)} G_F(M), \quad d^g(M) = d^g(G_F(M))$$

then $d(M)$ resp. $d^g(M)$ is independent of the choice of good filtrations on M. Moreover, since every finitely generated R-module has a good filtration the $d(M)$ and $d^g(M)$ make sense for all finitely generated R-modules.

10. Theorem. Let R be a left and right Zariski ring with filtration FR such that $G(R)$ is commutative with finite global dimension n.

(1). If every maximal ideal of $G(R)$ has the same height then for every finitely generated R-module M we have $j_R(M) + d(M) = n$.

(2). If every P which is maximal in $\text{spec}^g(G(R))$ has the same height ω in $\text{spec}(G(R))$, then for every finitely generated R-module M we have $j_R(M) + d^g(M) = \omega$.

(3). If the number ω in (2) is equal to $n = \text{gl.dim} G(R)$, then $\text{K.dim}_{G(R)} G(M) = d^g(G(M))$ and $j_R(M) + \text{K.dim}_{G(R)} G(M) = n$, where FM is taken to be any good filtration on M.

Proof. Since R is left and right Zariskian and $G(R)$ is a commutative Auslander regular ring it follows from Theorem 2.5.2. that $j_{G(R)}(G(M)) = j_R(M)$ for any finitely generated R-module M with good filtration FM. Hence Theorem 4.1.7. and Theorem 4.1.8. yield the asserted equalities. □

To give some applications of the foregoing results we also need the following well known

11. Lemma. Let $A = \oplus_{n \in \mathbb{Z}} A_n$ be a left limited commutative graded ring. Then any maximal element P in $\text{spec}^g(A)$ is of the type

$$(\oplus_{n<0} A_n) \oplus M_0 \oplus (\oplus_{n>0} A_n),$$

where M_0 is a maximal ideal of A_0. Moreover every maximal element P in $\text{spec}^g(A)$ is also maximal in $\text{spec}(A)$.

12. Lemma. Let $A = \oplus_{n \in \mathbb{Z}} A_n$ be a commutative strongly $\mathbb{Z}$-graded ring. Then there is an one to one correspondence between $\text{spec}(A_0)$ and $\text{spec}^g(A)$, and under this correspondence maximal elements are one-one corresponding.

13. Proposition. (cf. [Mats] P.79, Theorem 19.) Let $\varphi : A \to B$ be a homomorphism of commutative Noetherian rings. Let $P \in \text{spec}(B)$ and $p = P \cap A$ (the contraction of P). If φ is flat then $\text{ht}(P) = \text{ht}(p) + \text{ht}(P/pB)$.

14. Theorem. Let A be a commutative Noetherian ring with finite global dimension n. Let $R = A[t_1, \cdots, t_m]$ be the polynomial ring over A in m variables, $m \geq 1$. Suppose that every maximal ideal of A has the same height, then every maximal element P in $\text{spec}^g(R)$ has the same height $n + m$ in $\text{spec}(R)$, where R has the natural gradation with $R_0 = A$. Hence for any nonzero finitely generated graded R-module M all equalities in Theorem 4.1.8. hold (here $\omega = n + m$).

Proof. Since A has finite global dimension n it follows from the assumption that every maximal ideal of A has the same height that is equal to n. Let R be equiped with the natural gradation then $R_0 = A$. If P is any maximal element in $\text{spec}^g(R)$, then by Lemma

4.1.11. $P = M_0 \oplus (\oplus_{n>0} R_n)$, where M_0 is some maximal ideal of A. Hence by Proposition 4.1.13. we have $\mathrm{ht}(P) = \mathrm{ht}(M_0) + \mathrm{ht}(P/M_0 R) = n + m$.   □

15. Theorem. Let A be a commutative Noetherian ring with finite global dimension n. Let $R = A[t, t^{-1}]$ be the ring of finite Laurent series over A in variable t. Suppose that every maximal ideal of A has the same height, then every maximal element P in $\mathrm{spec}^g(R)$ has the same height $n + 1$ in $\mathrm{spec}(R)$, where R has the natural gradation with $R_0 = A$. Hence for any nonzero finitely generated graded R-module M all equalities in Theorem 4.1.8. hold (here $\omega = n + 1$).

Proof. With respect to the natural gradation, R is a strongly $\mathbb{Z}$-graded ring with $R_0 = A$. If M_0 is any maximal ideal of A then by the assumption we see that $\mathrm{ht}(M_0) = n$. Put $P = M_0 R$, then this is a maximal element in $\mathrm{spec}^g(R)$. It follows from Proposition 4.1.13. that $\mathrm{ht}(P) = \mathrm{ht}(M_0) + \mathrm{ht}(P/M_0 R) = n + 1$. Since every maximal element in $\mathrm{spec}^g(R)$ is of the type $M_0 R$ where M_0 is a maximal ideal of A (Lemma 4.1.12.) we have proved the theorem.   □

16. Corollary. If R is one of the following filtered rings :
(a) The universal enveloping algebra $U(\mathbf{g})$ of a finite n-dimensional k-Lie algebra $\mathbf{g}$ (where k is a field) with the standard filtration,
(b) The n-th Weyl algebra $A_n(k)$ over a field k with the Bernstein filtration (or Σ-filtration),
(c) The ring $\mathcal{D}(V)$ of $\mathbb{C}$-linear differential operators on V, where V is an irreducible smooth subvariety of the affine n-space $\mathbb{C}^n$,
(d) The ring $\mathcal{D}_n$ of $\mathbb{C}$-linear differential operators on $\mathcal{O}_n$, where $\mathcal{O}_n$ is the regular local ring of convergent power series in n variables with coefficients in $\mathbb{C}$,
(e) The stalks $\mathcal{E}_p$ of the sheaf of microlocal differential operators.

Then for any nonzero finitely generated R-module M with good filtration FM we have
(i). $d^g(M) = \mathrm{K.dim}_{G(R)} G(M)$;
(ii). $j_R(M) + \mathrm{K.dim}_{G(R)} G(M) = \mathrm{gl.dim} G(R)$.

Proof. See the proof of Example 2.4.3. and use Theorem 4.1.14. and Theorem 4.1.15..   □

17. Remark. (1). By using Proposition 4.1.13. one can easily prove the following result:

Theorem. If A is a commutative Noetherian local domain of (Krull-)dimension n, then every maximal ideal of the polynomial ring $A[t_1, \cdots, t_n]$ (or $A[t, t^{-1}]$) has the same height if and only if A is a field.

Hence maximal ideals in the rings $G(\mathcal{D}_n)$ and $G(\mathcal{E}_p)$ as given in Corollary 4.1.16. do not have the same height, contradicting what Björk asserted in ([Bj1] P.138, P.197, and P.100). This has been noticed by Wu Quanshui in [Wu].

(2). The original proof for $\mathrm{gl.dim}\mathcal{E}_p = n$ given in ([Bj1] PP.179–180) has gaps:

  (a). The results given in ([Bj1] Ch.2. section 4. and section 7.) depend on the assumption that the ring in consideration has finite global dimension, but $\mathrm{gl.dim}\mathcal{E}_p < \infty$ was never proved in [Bj1].

  (b). The formula $j_R(M) + d(M) = 2n$ was obtained in ([Bj1] Ch.2.) in the case where all rings are assumed to be positively filtered and the proof depends on the assumption that every maximal ideal of the ring considered has the same height, but $\mathcal{E}_p$ has non-discrete filtration and does not satisfy the latter condition as we have seen above.

Nevertheless, we can save the result $\mathrm{gl.dim}\mathcal{E}_p = n$ now by using Ch.II. Theorem 3.1.4. and the foregoing results.

4.2. Pure (holonomic) modules over Zariski rings

We start this subsection with an (not necessarily commutative) Auslander regular ring R. Let $\mathrm{gl.dim}R = \mu$.

1. Definition. A finitely generated left (right) R-module M is called **holonomic** if $j_R(M) = \mu$.

In view of Roos theorem it is clear that any zero-dimensional finitely generated module over a commutative Noetherian regular local ring is holonomic. More holonomic modules over noncommutative Auslander regular rings (in particular over rings of differential operators) may be found in [Ka], [Bj1], $\cdots$.

If we put

$$\beta_l = \{M, M \text{ holonomic left } R\text{-module}\}$$
$$\beta_r = \{M, M \text{ holonomic right } R\text{-module}\}$$

then it is easy to see that β_l resp. β_r is a full subcategory of R-mod resp. mod-R.

2. Proposition. Let the notation be given as above.

(1). The class β_l resp. β_r is closed under taking submodules, factor modules and extensions.

(2). $M^* = \mathrm{Ext}_R^\mu(M, R)$ (the dual of M) is holonomic for any holonomic left (right) R-module M.

Proof. (1). This follows from Proposition 2.1.5. since R is Auslander regular.

(2). This follows from Lemma 2.5.1..   □

3. Theorem. Let the notation be given as in Proposition 4.2.2..

(1). ( )*: $M \mapsto M^*$ is an exact functor from β_l to β_r and moreover it defines an equivalence between β_l and β_r.

(2). If $M \in \beta_l$ (β_r) then M has finite length as an R-module.

Proof. (1). If

$$0 \longrightarrow M_1 \longrightarrow M \longrightarrow M_2 \longrightarrow 0$$

is an exact sequence in β_l, then

$$0 \longrightarrow \operatorname{Ext}_R^\mu(M_2, R) \longrightarrow \operatorname{Ext}_R^\mu(M, R) \longrightarrow \operatorname{Ext}_R^\mu(M_1, R) \longrightarrow 0$$

is the only nontrivial part in the long exact Ext-sequence. Hence ( )* is exact.

Since gl.dim$R = \mu$ and R is Noetherian we may choose a projective resolution of M in R-mod:

$$0 \longrightarrow P_\mu \longrightarrow \cdots \longrightarrow P_0 \longrightarrow M \longrightarrow 0$$

where all P_j are finitely generated projective R-modules. It follows that the derived sequence

$$0 \longrightarrow \operatorname{Hom}_R(P_0, R) \longrightarrow \cdots \longrightarrow \operatorname{Hom}_R(P_\mu, R) \longrightarrow M^* \longrightarrow 0$$

is exact. It is well known that if P is a finitely generated projective left (right) R-module then $\operatorname{Hom}_R(P, R)$ is a finitely generated projective right (left) module. Repeating the first step with M^* instead of M leads to the exact sequence

$$0 \longrightarrow \operatorname{Hom}_R(\operatorname{Hom}_R(P_\mu, R), R) \longrightarrow \cdots \longrightarrow \operatorname{Hom}_R(\operatorname{Hom}_R(P_0, R), R) \longrightarrow M^{**} \longrightarrow 0$$

because pojective modules are reflexive. Thus

$$0 \longrightarrow P_\mu \longrightarrow \cdots \longrightarrow P_0 \longrightarrow M^{**} \longrightarrow 0$$

is exact. This sequence of the P_j is the same as the first one, hence $M^{**} \cong M$.

(2). Obviously M satisfies the ascending chain condition since M is finitely generated over the Noetherian ring R. It remains to prove the descending chain condition for M. Let

$$M \supset M_1 \supset M_2 \supset \cdots$$

be a descending sequence of submodules of M. Then from the exact sequence

$$0 \longrightarrow M_v \longrightarrow M \longrightarrow M/M_v \longrightarrow 0$$

we obtain the exact sequence

$$0 \longrightarrow (M/M_v)^* \longrightarrow M^* \longrightarrow (M_v)^* \longrightarrow 0$$

and similarly the exact sequence

$$0 \longrightarrow M_v/M_{v+1} \longrightarrow M/M_{v+1} \longrightarrow M/M_v \longrightarrow 0$$

yields the exact sequnce

$$0 \longrightarrow (M/M_v)^* \longrightarrow (M/M_{v+1})^* \longrightarrow (M_v/M_{v+1})^* \longrightarrow 0$$

for all v. This shows that

$$(M/M_1)^* \subset (M/M_2)^* \subset \cdots$$

is an increasing sequence of submodules of the right (left) R-module M^*. Since R is left and right Noetherian, there exists an integer s such that $(M/M_v)^* = (M/M_s)^*$ for all $v \geq s$. These equalities entail $(M_s/M_v)^* = 0$ for all $v \geq s$ and thus $M_s/M_v = 0$ by (1).. It follows that $M_s = M_v$ for all $v \geq s$, or in other words M satisfies the descending chain condition.□

4. Remark. (1). If $R = A_n(k)$ is the n-th Weyl algebra over a field k, then β_l (β_r) is just the Bernstein class as defined in [Bj1] or [Bor].

(2). Let R be a holonomic R-module then the projective dimension of M is equal to gl.dim$R = j_R(M)$. In other words, if a finitely generated R-module M has projective dimension strictly less than gl.dim$R$ then it cannot be a holonomic module.

(3). Let $R = A_n(k)$ be the n-th Weyl algebra over a field k. If $n \geq 2$, then from [Sta] we know that there are simple $A_n(k)$-modules that are not holonomic. For $n = 1$ it is known that every cyclic $A_1(k)$-, $\mathcal{D}_1$- or $\mathcal{E}_p$-module that is not isomorphic to the ring is itself holonomic.

If M is a finitely generated holonomic R-module then we have seen that

$$\text{Ext}_R^{j_R(M)}(\text{Ext}_R^{j_R(M)}(M,R), R) = M$$
$$\text{Ext}_R^k(\text{Ext}_R^k(M,R), R) = 0 \text{ for all } k \neq j_R(M) = \mu.$$

Consequently, if we choose a projective resolution of length μ for M then we obtain the canonical filtration FM on M as given in Theorem 1.4.2., say

$$FM : \qquad F_0M \subseteq F_1M \subseteq \cdots \subseteq F_\mu M = M$$

Let $\delta(M)$ be the smallest positive integer such that $F_{\delta(M)}M = M$, then $F_{\delta(M)-1}M = 0$.

Conversely, let M be a nonzero finitely generated R-module such that $F_{\delta(M)-1}M = .0$ with respect to a choosen projective resolution of length μ, then we have

5. Lemma. With notation and assumptions as above, then for all $1 \leq s \leq \delta(M)$

$$j_R(\text{Ext}_R^{j_R(M)+s}(\text{Ext}_R^{j_R(M)+s}(M,R), R)) \geq j_R(M) + s + 2.$$

Proof. Choose a projective resolution of M:

$$0 \longrightarrow P_\mu \longrightarrow \cdots \longrightarrow P_0 \longrightarrow M \longrightarrow 0$$

and consider the associated double complex $(\mathbf{Q}^*)$ as in §1.. If we look at the (first) bounded filt-complex $(\mathbf{M}^I, \Delta, 1)$ with respect to $(\mathbf{Q}^*)$ then we obtain the following table from the spectral sequence determined by $(\mathbf{M}^I, \Delta, 1)$:

$$
\begin{array}{ccccc}
0 & \to & E^2_{\mu+1} & \to & E^2_{\mu+2} & \to \\
  &   & \cdots & & \cdots & \\
0 & \to & E^k_{\mu+1} & \overset{\chi(1-k)}{\longrightarrow} & E^k_{\mu+2} & \to \\
0 & \to & E^{k+1}_{\mu+1} & \to & E^{k+1}_{\mu+2} & \to \\
  &   & \cdots & & \cdots &
\end{array}
$$

since $E^2_j = \oplus_{p=0}^{j-1} \mathrm{Ext}_R^{j-1-p}(\mathrm{Ext}_R^{\mu-p}(M,R), R)$ and R is Auslander regular. It follows that there is an exact sequence

(*) $$0 \longrightarrow (E^{k+1}_{\mu+1})_p \longrightarrow (E^k_{\mu+1})_p \longrightarrow (S^k_{\mu+2})_{p+1-k} \longrightarrow 0$$

for every $0 \le p \le \mu$, $k \ge 2$ (note that $\chi(1-k)$ is of degree $1-k$), where $(S^k_{\mu+2})_{p+1-k}$ is a subquotient of $(E^2_{\mu+2})_{p+1-k} = \mathrm{Ext}_R^{\mu-p+k}(\mathrm{Ext}_R^{\mu-p-1+k}(M,R), R)$. It follows from Proposition 2.1.5. that

(**) $$j_R((S^k_{\mu+2})_{p+1-k}) \ge \mu + k - p, \quad k \ge 2$$

Moreover, we have by Lemma 1.3.2. that for each $j \ge 1$, $G(\mathbf{H}_{j-1}) \cong E^q_j$ for all $q \ge \omega$ and some integer $\omega \ge 0$, in particular we have $G(M) = G(\mathbf{H}_\mu) \cong E^q_{\mu+1}$ for large q (also see Proposition 1.4.1.). Hence the assumption $F_{\delta(M)-1}M = 0$ entails that $(E^q_{\mu+1})_{\delta(M)-s} = 0$ for every $1 \le s \le \delta(M)$. But then the above (*), (**), Lemma 2.1.2. and Proposition 2.1.5. lead to the following

$$
\begin{aligned}
& j_R(\mathrm{Ext}_R^{j_R(M)+s}(\mathrm{Ext}_R^{j_R(M)+s}(M,R), R)) \\
= \ & j_R(\mathrm{Ext}_R^{\mu-(\delta(M)-s)}(\mathrm{Ext}_R^{\mu-(\delta(M)-s)}(M,R), R)) \\
= \ & j_R((E^2_{\mu+1})_{\delta(M)-s}) \\
\ge \ & j_R(M) + s + 2.
\end{aligned}
$$

This completes the proof. $\qquad\qquad\qquad\qquad\qquad\qquad\qquad\qquad\qquad\qquad\qquad\qquad$ $\square$

6. Theorem. With notation as before, the following statements are equivalent for a finitely generated R-module:

(1). $\mathrm{Ext}_R^k(\mathrm{Ext}_R^k(M,R),R) = 0$ if $k \neq j_R(M)$;

(2). $F_{\delta(M)-1}M = 0$.

Proof. The implication $(1) \Rightarrow (2)$ is clear.

$(2) \Rightarrow (1)$. Suppose that $\mathrm{Ext}_R^{j_R(M)+s}(\mathrm{Ext}_R^{j_R(M)+s}(M,R),R) \neq 0$ for some $s \geq 1$. Note that by Theorem 1.4.2. for any nonzero R-module N one has: $\mathrm{Ext}_R^{j_R(N)}(\mathrm{Ext}_R^{j_R(N)}(N,R),R) \neq 0$. Using Lemma 2.5.1. twice we can deduce that

$$j_R(\mathrm{Ext}_R^{j_R(M)+s}(\mathrm{Ext}_R^{j_R(M)+s}(M,R),R)) = j_R(M) + s$$

contradicting Lemma 4.2.5.. □

7. Definition. Let M be a nonzero and finitely generated left (right) R-module. M is said to be a **pure** R-module if it satisfies the equivalent conditions in Theorem 4.2.6..

By the definition it is obvious that every holonomic R-module is pure, and every projective R-module is pure.

Moreover, Let R be a commutative Noetherian regular local ring of dimension n. Then by Roos theorem the formula $j_R(M)+ \mathrm{K.dim}_R M = n$ holds for every nonzero finitely generated R-module M. It is also well known that the formula $\mathrm{p.dim}_R M + \mathrm{depth}\, M = n$ (Auslander-Buchsbaum) holds for every finitely generated R-module M. It follows that $\mathrm{p.dim}_R M - j_R(M) = \mathrm{K.dim}_R M - \mathrm{depth}\, M$. From this we may derive the following

(1). If R is a commutative Noetherian regular local ring of dimension n, then a finitely generated R-module M is Cohen-Macaulay if and only if $j_R(M) = \mathrm{p.dim}_R M$ (hence M is pure).

(2). Let R be a commutative Noetherian ring with finite global dimension. If $j_R(M) = \mathrm{p.dim}_R M$ then M is (locally) Cohen-Macaulay; If M is a (locally) Cohen-Macaulay R-module, then $\mathrm{p.dim}_R M = \max\{j_{R_P}(M_P),\ P \text{ maximal ideal of } R\}$ and $j_R(M) = \min\{j_{R_P}(M_P),\ P \text{ maximal ideal of } R\}$.

8. Proposition. Let R be an Auslander regular ring and let M be a nonzero and finitely generated R-module. Then

(1). $\mathrm{Ext}_R^{j_R(M)}(M,R)$ is a pure R-module;

(2). If $\mathrm{Ext}_R^q(\mathrm{Ext}_R^q(M,R),R) \neq 0$ then it is pure.

Proof. (1). By Lemma 2.5.1. we have $j_R(\mathrm{Ext}_R^{j_R(M)}(M,R)) = j_R(M)$. In view of Theorem 4.2.6. it remains to prove

$$j_R(\mathrm{Ext}_R^{j_R(M)+s}(\mathrm{Ext}_R^{j_R(M)}(M,R),R)) > j_R(M) + s, \quad \text{all } s \geq 1$$

Again look at the (first) bounded filt-complex $(\mathbf{M}^I, \Delta, 1)$ with respect to the double complex $(\mathbf{Q}^*)$ as before (see §1.), then in the spectral sequence determined by $(\mathbf{M}^I, \Delta, 1)$ we have for $r \geq 2$ the complexes

$$\cdots \to (E^r_{\mu+s})_{\delta(M)-1+r} \xrightarrow{x(1-r)} (E^r_{\mu+s+1})_{\delta(M)} \xrightarrow{x(1-r)} (E^r_{\mu+s+2})_{\delta(M)+1-r} \to \cdots$$

Noticing that $E^2_j = \oplus_{p=0}^{j-1} \mathrm{Ext}_R^{j-1-p}(\mathrm{Ext}_R^{\mu-p}(M, R), R)$ and $(E^r_{\mu+s})_{\delta(M)-1+r}$ is a subquotient of $(E^2_{\mu+s})_{\delta(M)-1+r}$, it follows from the equality $\mu = j_R(M) + \delta(M)$ (Proposition 2.1.5.) that $(E^r_{\mu+s})_{\delta(M)-1+r} = 0$. Hence there is an exact sequence

$$(*) \qquad 0 \longrightarrow (E^{r+1}_{\mu+s+1})_{\delta(M)} \longrightarrow (E^r_{\mu+s+1})_{\delta(M)} \longrightarrow S^r_{\mu+s+2} \longrightarrow 0$$

where $S^r_{\mu+s+2}$ is a subquotient of

$$(E^2_{\mu+s+2})_{\delta(M)+1-r} = \mathrm{Ext}_R^{j_R(M)+s+r}(\mathrm{Ext}_R^{j_R(M)-1+r}(M, R), R)$$

and consequently

$$(**) \qquad j_R(S^r_{\mu+s+2}) \geq j_R(M) + s + r, \quad r \geq 2$$

because R is Auslander regular.

Since $(\mathbf{M}^I, \Delta, 1)$ is bounded it follows from Corollary 1.3.3. that $G(\mathbf{H}_{\mu+s}) \cong E^q_{\mu+s+1}$ for all $q \geq \omega$ and some $\omega \geq 0$. But note that $\mathbf{H}_{\mu+s}$ is also the $\mu + s$-th cohomology group of $(\mathbf{M}^{II}, \Delta, 1)$ and $s \geq 1$, so if we look at the table determined by the spectral sequence with respect to $(\mathbf{M}^{II}, \Delta, 1)$ (see Proposition 1.4.1.) we find that the associated graded module $G^{II}(\mathbf{H}_{\mu+s})$ (with respect to $F^{II}M$) equals zero, i.e., $\mathbf{H}_{\mu+s} = 0$ because the filtration $F^{II}\mathbf{H}_{\mu+s}$ on $\mathbf{H}_{\mu+s}$ is discrete. Accordingly $E^q_{\mu+1+s} = 0$. Finally our assertion follows from $(*)$, $(**)$ and a simple induction procedure (use the fact that $(E^2_{\mu+s+1})_{\delta(M)} = \mathrm{Ext}_R^{j_R(M)+s}(\mathrm{Ext}_R^{j_R(M)}(M, R), R))$.

(2). This follows from Lemma 2.5.1. and (1).. $\square$

9. Proposition. Let R be an Auslander regular ring. If M is a pure R-module then every R-submodule N of M is pure and $j_R(M) = j_R(N)$.

Proof. Let N be a nonzero R-submodule of M. Looking at the exact sequence

$$(*) \qquad 0 \longrightarrow N \longrightarrow M \longrightarrow M/N \longrightarrow 0$$

and the associated long exact Ext-sequence we obtain an exact sequence

$$0 \longrightarrow K_q \longrightarrow \mathrm{Ext}_R^q(N, R) \longrightarrow S_{q+1} \longrightarrow 0$$

where K_q is a quotient of $\text{Ext}_R^q(M, R)$ and $S_{q+1} \subseteq \text{Ext}_R^{q+1}(M/N, R)$. If $q > j_R(M)$, then by the purity of M and Proposition 2.1.5. we have $j_R(K_q) > q+1$. We also have $j_R(S_{q+1}) \geq q+1$ since R is Auslander regular. It folllows from Lemma 2.1.2. that

$$(**) \qquad\qquad j_R(\text{Ext}_R^q(N, R)) \geq q + 1$$

Suppose that $j_R(N) > j_R(M)$. Then by $(**)$ we have $j_R(\text{Ext}_R^{j_R(N)}(N, R)) \geq j_R(N) + 1$, contradicting Lemma 2.5.1.. Therefore $j_R(N) = j_R(M)$. To prove the purity of N, let us put $q = j_R(N) + s = j_R(M) + s$ with $s \geq 1$. Then we have $j_R(\text{Ext}_R^q(N, R)) \geq q + 1$. It follows from $(**)$ that $\text{Ext}_R^q(\text{Ext}_R^q(N, R), R) = 0$ as desired. $\qquad\square$

10. Theorem. Let R be an Auslander regular ring. A finitely generated R-module M is pure if and only if $j_R(M) = j_R(N)$ for every R-submodule N of M.

Proof. By considering the canonical filtration FM on M, this is a consequence of the foregoing results. $\qquad\square$

Comparing with holonomic module, it is clear that if R is Auslander regular then every **simple** R-module is pure.

Now let us consider a left and right Zariski ring with filtration FR. Suppose that $G(R)$ is an Auslander regular ring. Then it follows from §§2., 3. that R and $\tilde{R}$ are Auslander regular rings. So we may assume that $\text{gl.dim} R = \mu < \infty$. As usual the canonical homogeneous regular element of degree 1 in $\tilde{R}$ is denoted by X.

11. Lemma. If T is any finitely generated X-torsionfree graded $\tilde{R}$-module, then $\text{Ext}_{\tilde{R}}^p(T, \tilde{R})$ and $\text{Ext}_{\tilde{R}}^p(\text{Ext}_{\tilde{R}}^p(T, \tilde{R}), \tilde{R})$ are X-torsionfree graded $\tilde{R}$-module with $p = j_{\tilde{R}}(T)$.

Proof. Indeed, since T is a finitely generated graded $\tilde{R}$-module, by taking a gr-free resolution of T in $\tilde{R}$-gr, we see that $\text{Ext}_{\tilde{R}}^q(T, \tilde{R})$ is a graded $\tilde{R}$-module for each $q \geq 0$. Moreover, from Ch.I. we know that there exists a filtered R-module H with good filtration FH such that $\overline{H} = T$. Hence it follows from subsection 2.5. that $j_R(H) = j_{G(R)}(G(H)) = j_{\tilde{R}}(\overline{H}) = j_{\tilde{R}}(T) = p$. Consider the exact sequence

$$0 \longrightarrow T \xrightarrow{\mu_X} T \longrightarrow T/XT \longrightarrow 0$$

where μ_X is multiplication by X, if we take the long exact Ext-sequence

$$\longrightarrow \text{Ext}_{\tilde{R}}^p(T/XT, \tilde{R}) \xrightarrow{\mu_X} \text{Ext}_{\tilde{R}}^p(T, \tilde{R}) \xrightarrow{\mu_X^*} \text{Ext}_{\tilde{R}}^p(T, \tilde{R}) \longrightarrow$$

then μ_X^* is again mutiplication by X and moreover the theorem of Rees reads the following

$$\text{Ext}^p_{\tilde{R}}(T/XT, \tilde{R}) \cong \text{Ext}^{p-1}_{\tilde{R}/X\tilde{R}}(T/XT, \tilde{R}/X\tilde{R}) = 0$$

since $\tilde{R}/X\tilde{R} \cong G(R)$, $T/XT \cong G(H)$ and $j_R(H) = j_{G(R)}(G(H)) = j_{\tilde{R}}(T) = p$. Hence μ_X^* is injective and consequently $\text{Ext}^p_{\tilde{R}}(T, \tilde{R})$ is X-torsionfree. $\qquad\square$

12. Lemma. If FM and $F'M$ are two good filtrations on an R-module M, then $\widetilde{M}$ is pure if and only if $\widetilde{M}'$ is pure as $\tilde{R}$-module, where $\widetilde{M}'$ is the Rees module of M with respect to $F'M$.

Proof. Since any two good filtrations on M are equivalent we may view $\widetilde{M}$ resp. $\widetilde{M}'$ as a submodule of $\widetilde{M}'$ resp. $\widetilde{M}$. Hence it follows from Proposition 4.2.9. that if one of them is pure then so is another one. $\qquad\square$

13. Theorem. Let M be a finitely generated R-module.

(1). M is pure if and only if $\widetilde{M}^F$, the Rees module of M with respect to any good filtration FM on M, is a pure $\tilde{R}$-module.

(2). M is pure if and only if $\widehat{M}^F$, the completion of M with respect to any good filtration FM on M, is a pure $\hat{R}$-module.

(3). Let FM be a good filtration on M. If $G(M)$ is a pure $G(R)$-module then M is a pure R-module.

(4). If M is a pure R-module then there exists a good filtration FM such that $G(M)$ is a pure $G(R)$-module.

Proof. (1). Let $\tilde{R}_{(X)}$ resp. $\widetilde{M}_{(X)}$ be the localization of $\tilde{R}$ resp. $\widetilde{M}$ at the (central) Ore set $\{1, X, X^2, \cdots\}$. Then on the one hand, by Lemma 2.5.3. we have for all $k \geq 0$

$$\begin{aligned}
&\text{Ext}^k_R(\text{Ext}^k_R(M, R), R)[t, t^{-1}] \\
=\ &R[t, t^{-1}] \otimes_R \text{Ext}^k_R(\text{Ext}^k_R(M, R), R) \\
\cong\ &\text{Ext}^k_{R[t,t^{-1}]}(\text{Ext}^k_{R[t,t^{-1}]}(M[t, t^{-1}], R[t, t^{-1}]), R[t, t^{-1}]) \\
\cong\ &\text{Ext}^k_{\tilde{R}_{(X)}}(\text{Ext}^k_{\tilde{R}_{(X)}}(\widetilde{M}_{(X)}, \tilde{R}_{(X)}), \tilde{R}_{(X)}) \\
\cong\ &(\text{Ext}^k_{\tilde{R}}(\text{Ext}^k_{\tilde{R}}(\widetilde{M}, \tilde{R}), \tilde{R}))_{(X)} \\
=\ &\tilde{R}_{(X)} \otimes_{\tilde{R}} \text{Ext}^k_{\tilde{R}}(\text{Ext}^k_{\tilde{R}}(\widetilde{M}, \tilde{R}), \tilde{R}),
\end{aligned}$$

hence if $\widetilde{M}$ is a pure $\widetilde{R}$-module then M is a pure R-module. On the other hand, put $N = \text{Ext}_R^{j_R(M)}(\text{Ext}_R^{j_R(M)}(M,R),R)$, if M is pure then it is a submodule of N (see Theorem 4.2.6. and Theorem 1.4.2.). In view of Lemma 4.2.12. we may assume that the good filtration FM given on M is induced by the good filtration on N. Hence we have $\widetilde{M} \subseteq \widetilde{M}_{(X)} \subseteq \widetilde{N}_{(X)} \cong N[t,t^{-1}]$. It follows from Lemma 2.5.1., Proposition 4.2.8., Proposition 4.2.9. and Lemma 4.2.11.that $\widetilde{M}$ is pure because $\widetilde{R}$ is left and right Noetherian and $\widetilde{M}$ is a finitely generated X-torsionfree $\widetilde{R}$-module.

(2). This follows from the proof of Theorem 2.5.6.

(3). Suppose that $G(M)$ is a pure $G(R)$-module. Then by definition

$$\text{Ext}_{G(R)}^{j_R(M)+s}(\text{Ext}_{G(R)}^{j_R(M)+s}(G(M),G(R)),G(R)) = 0$$

for any $s \geq 1$ (note that $j_R(M) = j_{G(R)}(G(M))$). This means that

$$j_{G(R)}(\text{Ext}_{G(R)}^{j_R(M)+s}(G(M),G(R))) \geq j_R(M) + s + 1$$

since $G(R)$ is an Auslander regular ring. On the other hand, from Proposition 2.2.4. we know that $G(\text{Ext}_R^{j_R(M)+s}(M,R))$ is a subquotient of $\text{Ext}_{G(R)}^{j_R(M)+s}(G(M),G(R))$. It follows from Proposition 2.1.5. that

$$j_{G(R)}(G(\text{Ext}_R^{j_R(M)+s}(M,R))) \geq j_R(M) + s + 1$$

and hence

$$\text{Ext}_{G(R)}^{j_R(M)+s}(G(\text{Ext}_R^{j_R(M)+s}(M,R)),G(R)) = 0.$$

But again by Proposition 2.2.4. we see that $G(\text{Ext}_R^{j_R(M)+s}(\text{Ext}_R^{j_R(M)+s}(M,R),R))$ is a subquotient of $\text{Ext}_{G(R)}^{j_R(M)+s}(G(\text{Ext}_R^{j_R(M)+s}(M,R)),G(R))$, therefore

$$\text{Ext}_R^{j_R(M)+s}(\text{Ext}_R^{j_R(M)+s}(M,R),R) = 0$$

since R is left and right Zariskian. This shows that M is pure.

(4). Put $N = \text{Ext}_R^{j_R(M)}(M,R)$, then N is a finitely generated R-module. take the good filtration FN on N as before, then it follows from Theorem 2.5.2. that $j_R(N) = j_{G(R)}(G(N))$ since R is also right Zariskian and $G(R)$ is Auslander regular. Next, consider in R-filt a filt-free resolution of N:

$$\cdots \longrightarrow L_j \xrightarrow{f_j} L_{j-1} \longrightarrow \cdots \longrightarrow L_0 \xrightarrow{\varepsilon} N \longrightarrow 0$$

we have a spectral sequence determined by the dualized sequence $F(*)$ (see subsection 2.2., replace M by N) in which $\mathbf{H}_{j-1} = \mathrm{Ext}_R^{j-1}(N, R)$ and $E_j^1 = \mathbf{H}_{gr}^{j-1} = \mathrm{Ext}_{G(R)}^{j-1}(G(N), G(R))$, $j = 1, 2, \cdots$. It follows that $E_{j_R(N)}^1 = 0$ and consequently we have the following table:

$$
\begin{array}{ccccccc}
\cdots & \to & 0 & \to & E_{j_R(N)+1}^1 & \to & E_{j_R(N)+2}^1 & \to & \cdots \\
\cdots & \to & 0 & \to & E_{j_R(N)+1}^2 & \to & E_{j_R(N)+2}^2 & \to & \cdots \\
 & & & & \cdots & & \cdots & & \\
\cdots & \to & 0 & \to & E_{j_R(N)+1}^q & \to & E_{j_R(N)+2}^q & \to & \cdots \\
 & & & & \cdots & & \cdots & &
\end{array}
$$

Furthermore, by Corollary 1.1.7. and Lemma 2.2.1. there exists an $\omega \geq 0$ such that $G(\mathbf{H}_k) = E_{k+1}^q$ for all $q \geq \omega$. Hence $G(\mathbf{H}_{j_R(N)}) = G(\mathrm{Ext}_R^{j_R(N)}(N, R)) = E_{j_R(N)+1}^q$ is a submodule of $E_{j_R(N)+1}^1 = \mathrm{Ext}_{G(R)}^{j_R(N)}(G(N), G(R))$. Now the fact that $j_R(N) = j_{G(R)}(G(N))$, Proposition 4.2.8. and Proposition 4.2.9. entail that $G(\mathrm{Ext}_R^{j_R(N)}(N, R))$ is a pure $G(R)$-module with grade number $j_R(N)$. Finally, by the definition of N we have $\mathrm{Ext}_R^{j_R(N)}(N, R) = \mathrm{Ext}_R^{j_R(N)}(\mathrm{Ext}_R^{j_R(M)}(M, R), R)$ and $j_R(N) = j_R(M)$ by Lemma 2.5.1., it follows from the purity of M that M is a submodule of $\mathrm{Ext}_R^{j_R(N)}(N, R)$. If we take the good filtration FM on M induced by the good filtration $F\mathrm{Ext}_R^{j_R(N)}(N, R)$ then $G(M)$ is a submodule of $G(\mathrm{Ext}_R^{j_R(N)}(N, R))$. Hence by Proposition 4.2.9. $G(M)$ is a pure $G(R)$-module. $\square$

Let R be a left and right Zariski ring with commutative associated graded ring $G(R)$. In order to give a characterization of holonomic modules over R we have to compare the grade number $j_R(M)$ and $j_R(M/N)$ for a pure R-module M and any submodule N of M, let us first pass to the commutative case.

14. Theorem (Gabber-Kashiwara theorem) Let A be a commutative Noetherian ring with finite global dimension (hence A is Auslander regular). Let Q be a nonzero finitely generated and pure A-module, $J(Q) = \sqrt{\mathrm{Ann}_A Q}$ and $\min(J(Q))$ the set of minimal prime divisors of $J(Q)$ in $\mathrm{spec}(A)$. Then the equality

$$
\mathrm{gl.dim} A_p = j_A(Q)
$$

holds for every $p \in \min(J(Q))$, where A_p is the localization of A at p.

Proof. It is well known that for each $p \in \min(J(Q))$ the A_p-module Q_p is zero-dimensional. Hence it follows from Roos theorem that $j_{A_p}(Q_p) = \mathrm{gl.dim} A_p$ since A_p is now a regular local ring. On the other hand, it is also well known that

$$\text{Ext}_{A_p}^k(\text{Ext}_{A_p}^k(Q_p, A_p), A_p) \cong (\text{Ext}_A^k(\text{Ext}_A^k(Q, A), A))_p$$

for all $k \geq 0$. So Lemma 2.5.1. yields that $\text{Ext}_A^{j_{A_p}(Q_p)}(\text{Ext}_A^{j_{A_p}(Q_p)}(Q, A), A) \neq 0$. But then by the purity of M we arrive at $j_A(Q) = j_{A_p}(Q_p) = \text{gl.dim} A_p$.   □

15. Theorem. Let A be a commutative Noetherian ring with finite global dimension and

$$0 \longrightarrow N \longrightarrow M \longrightarrow K \longrightarrow 0$$

an exact sequence of finitely generated A-modules. Let the notation be given as in Theorem 4.2.14..

(1). If M is a pure A-module and $\min(J(M)) \cap \min(J(K)) \neq \emptyset$ then $j_A(M) = j_A(K)$.

(2). Suppose that $j_A(M) = j_A(K)$. If either K is a pure A-module or K satisfies: for any $p \in \min(J(K))$ it follows that $\text{gl.dim} A_p = j_A(K)$, then $\min(J(K)) \subseteq \min(J(M))$.

Proof. (1). Let $p \in \min(J(M)) \cap \min(J(K))$. Since M is pure it follows from the theorem of Roos and Theorem 4.2.14. that

$$j_A(M) = \text{gl.dim} A_p = j_{A_p}(K_p) \geq j_A(K) \geq j_A(M)$$

Hence $j_A(M) = j_A(K)$.

(2). From the given exact sequence it is not hard to see that $J(M) = J(N) \cap J(K)$. So if $p \in \min(J(K))$ then $p \supseteq J(M)$. Accordingly, on the one hand we have $\text{gl.dim} A_p = j_A(K) = j_A(M)$ by Theorem 4.2.14.; and on the other hand, it follows from Roos theorem that $j_{A_p}(M_p) + \text{K.dim}_{A_p} M_p = \text{gl.dim} A_p$. But then we arrive at

$$\begin{aligned} j_A(M) = j_A(K) &= \text{gl.dim} A_p \\ &= j_{A_p}(M_p) + \text{K.dim}_{A_p} M_p \\ &\geq j_A(M) + \text{K.dim}_{A_p} M_p \end{aligned}$$

This entails that $\text{K.dim}_{A_p} M_p = 0$, i.e., M_p is zero-dimensional as an A_p-module. Therefore $p \in \min(J(M))$ as desired.   □

Let us now turn to the situation where R is again a left and right Zariski ring with commutative Auslander regular $G(R)$. If M is a finitely generated R-module with good filtration FM, then in view of Lemma 4.1.9. we can define $J(G(M))$ and $\min(J(G(M)))$ as above and put

$$J(M) = J(G(M)), \quad \min(J(M)) = \min(J(G(M)))$$

16. Corollary. With notation and assumptions as above, if M is a pure R-module then the equality

$$\text{gl.dim} G(R)_p = j_R(M)$$

holds for every $p \in \min(J(M))$.

Proof. This follows from Theorem 4.2.13., Theorem 4.2.14. and the fact that $j_R(M) = j_{G(R)}(G(M))$. $\square$

17. Corollary. If M is an pure R-module and N is any R-submodule of M such that $\min(J(M)) \cap \min(J(M/N)) \neq \emptyset$, then $j_R(M) = j_R(M/N)$.

Proof. This follows from Theorem 4.2.13., Corollary 4.2.16. and Theorem 4.2.15.. $\square$

Now, by using the foregoing results we can give a characerization of holonomic modules over a left and right Zariski ring with $G(R)$ being commutative and Auslander regular.

18. Theorem. Put $\text{gl.dim} R = \mu$. With assumptions and notation as before, the following statements are equivalent for a finitely generated R-module M:
(1). M is a holonomic R-module;
(2). $\text{ht} p (= \text{height of } p) = \mu$, for every $p \in \min(J(M))$.

Proof. (1) $\Rightarrow$ (2). Note that any holonomic module is pure, it follows from Corollary 4.2.16. that $\text{ht} p = \text{gl.dim} G(R)_p = j_R(M) = \mu$ for every $p \in \min(J(M))$.
(2) $\Rightarrow$ (1). If M is a pure R-module then (1) is just a consequence of Corollary 4.2.16.. Now suppose that M is not pure. Looking at the canonical filtration FM on M as defined in Theorem 1.4.2. we have an exact sequence

$$0 \longrightarrow F_{v-1}M \longrightarrow F_v M \longrightarrow F_v M / F_{v-1} M \longrightarrow 0$$

for every $0 \leq v \leq \delta(M)$, where $F_{-1}M = 0$, $F_{\delta(M)}M = M$ and $F_v M / F_{v-1} M \subseteq \text{Ext}_R^{\mu-v}(\text{Ext}_R^{\mu-v}(M,R),R)$. It follows from Proposition 4.2.8. and Proposition 4.2.9. that $F_v M / F_{v-1} M$ is a pure R-module with grade number $\mu - v$ (unless $F_v M / F_{v-1} M = 0$!). Consequently, by Corollary 4.2.16. we have

$$\text{gl.dim} G(R)_p = j_R(F_v M / F_{v-1} M) = \mu - v, \quad p \in \min(J(F_v M / F_{v-1} M))$$

in particular, for $v = \delta(M)$ we have by Proposition 2.1.5. that $\text{gl.dim} G(R)_p = \mu - \delta(M) = j_R(M)$, $p \in \min(J(F_{\delta(M)}M / F_{\delta(M)-1}M))$. But on the other hand it follows from Theorem 4.2.15. that $\min(J(F_{\delta(M)}M / F_{\delta(M)-1}M)) \subseteq \min(J(M))$. Hence the assumption entails that $\mu = \text{ht} p = \text{gl.dim} G(R)_p = j_R(M)$. This shows that M is a holonomic R-module. $\square$

19. Remark. In [Ve2] Van den Essen has pointed out (without proof) that by using the so called " decomposition theorem " of [Bj5] the result mentioned in above theorem can be obtained. Because [Bj5] is an unpublished lecture note and its result has been used for publication, it is necessary to recall the " decomposition theorem " word by word as follows:

([Bj5] Ch.i. Theorem 7.3.) Let (R, Σ) be a filtered ring where Σ is Zariskian and $gr(R)$ is commutative and regular. If M is any finitely generated R-module, then

$$J(F_v M / F_{v-1} M) = \cap p$$

with $\cap$ taken over those minimal prime divisors of $J(M)$ satisfying $\rho(p) = \mu - v$ for all $0 \le v \le \mu = \text{gl.dim} R$.

In the above statement, Σ is a filtration on R, $gr(R)$ is the associated graded ring of R, the assumption that $gr(R)$ is regular means that $gr(R)$ is Noetherian with finite global dimension, $F_v M$ is a term from the canonical filtration FM on M as defined in Theorem 1.4.2. and $\rho(p)$ is the height of p. Moreover our Zariski rings satisfy the Zariskian condition as defined by Björk in [Bj5].

In [Bj5] the existence of the required p has never been established, in particular when M is not a pure module. Indeed, it is easily seen that the " decomposition theorem " is not correct. For example, let $R = A[t_1, \cdots, t_n]$ be the polynomial ring over a commutative Noetherian domain A where A has finite global dimension and $n \ge 1$. Consider the natural gradation and the grading filtration on R, i.e., $F_n R = \oplus_{k \le n} R_k$, $n \ge 0$, then FR is Zariskian and $G(R) \cong R$ as graded rings. Hence if p is any graded prime ideal which is **not minimal** in $\text{spec}(R)$, then under taking the induced filtration on p we have $G(p) \cong p$ as graded R-modules, and moreover $G(R/p) \cong G(R)/G(p) \cong R/p$ as graded R-modules. Now, let $M = R \oplus R/p$ be the graded R-module with filtration $F_n M = F_n R \oplus F_n(R/p)$, $n \ge 0$, where R/p has the quotient filtration, then $G(M) = G(R \oplus R/p) \cong G(R) \oplus G(R/p) \cong R \oplus R/p$. Obviously M is not a pure R-module because $j_R(R/p) > j_R(M) = 0$. But $J(M) = \{p\}$, $\text{ht}p \ne 0$ since it is not a minimal prime in $\text{spec}(R)$.

Nevertheless, Van den Essen's conclusion has a complete proof now.

4.3. Codimension calculation of characteristic varieties

In the foregoing subsection we have proved that many left and right Zariski rings R with Commutative Auslander regular associated graded rings $G(R)$ fit the generalized Roos theorem, i.e., the formula

$$(*) \qquad\qquad j_R(M) + d(M) = n$$

holds for every finitely generated R-module M, where $n = \text{gl.dim} G(R)$. Since by definition $d(M) = \text{K.dim}_{G(R)} G(M)$ for any good filtration FM on M, the geometrical meaning of the formula $(*)$ will be clear once the associated graded ring $G(R)$ of R is a polynomial ring over a field k of characteristic 0 and we are considering the **characteristic variety** $V(J(M))$ in k^n determined by the **characteristic ideal** $J(M)$ of M, i.e., in this case we have $d(M) = \dim V(J(M))$ and $j_R(M) = \operatorname{codim} V(J(M))$ where $\operatorname{codim} V(J(M))$ denotes the codimension of $V(J(M))$. In the present subsection we calculate $j_R(M)$ (and hence the codimension of $V(J(M))$) by using the finite decomposition of $J(M)$ into minimal prime ideals.

We start with the commutative case.

Let A be a commutative Noetherian ring with finite global dimension μ. Let M be a finitely generated R-module and let

$$J(M) = \sqrt{\operatorname{Ann}_A M} = p_1 \cap \cdots \cap p_s$$

be the decomposition of the radical of M by **minimal prime divisors** in $\operatorname{spec}(A)$. Put

$$\min(J(M)) = \{p_1, \cdots, p_s\}$$

If we take notions as in the classical algebraic geometry or the theory of rings of differential operators, then $J(M)$ is called the **characteristic ideal** of M and $\min(J(M))$ is called the **characteristic variety** of M. Also we know that the dimension of $\min(J(M))$ is defined to be the Krull dimension of the ring $A/J(M)$, i.e.,

$$
\begin{aligned}
\dim(\min(J(M))) &= \operatorname{K.dim}(A/J(M)) \\
&= \sup\{\operatorname{K.dim}(A/p_i), \ p_i \in \min(J(M))\} \\
&= \operatorname{K.dim}_A M.
\end{aligned}
$$

If $\dim(\min(J(M))) = \operatorname{K.dim}(A/p_i)$ for all $p_i \in \min(J(M))$, then the characteristic variety $\min(J(M))$ is said to be **geometrically pure** (or have a pure dimension).

1. Lemma. Let M be a pure A-module in the sense of last subsection. If H is any nonzero A-submodule of M then $\min(J(H)) \subseteq \min(J(M))$.

Proof. Since $H \subseteq M$ it is sufficient to prove $\operatorname{K.dim} M_p = 0$ for any $p \in \min(J(H))$, where M_p is the localization of M at p. By Theorem 4.2.14. we have $j_A(H) = \operatorname{gl.dim} A_p$ for all $p \in \min(J(H))$ (note that H is pure by Proposition 4.2.9.). But on the other hand it follows from the theorem of Roos we have $j_{A_p}(M_p) + \operatorname{K.dim}_{A_p} M_p = \operatorname{gl.dim} A_p$. It follows from $j_A(H) = j_A(M)$ that $j_A(H) + \operatorname{K.dim}_{A_p} M_p \leq j_{A_p}(M_p) + \operatorname{K.dim}_{A_p} M_p = j_A(H)$. This shows that $\operatorname{K.dim}_{A_p} M_p = 0$ as desired.   □

2. Theorem. Let M be an arbitrary finitely generated A-module. Then there exists a $p \in \min(J(M))$ such that $j_A(M) = \text{gl.dim} A_p$, or more precisely

$$j_A(M) = \inf\{\text{gl.dim} A_p, \ p \in \min(J(M))\}.$$

Proof. If M is a pure A-module then we are done already by Theorem 4.2.14.. Now suppose that M is not pure. Consider the canonical filtration FM on M as defined in Theorem 1.4.2., then since M is not pure we have $F_{\delta(M)-1}M \neq 0$ and $F_{\delta(M)}M/F_{\delta(M)-1}M = M/F_{\delta(M)-1}M \neq 0$. By Theorem 1.4.2. we know that $M/F_{\delta(M)-1}M$ is a submodule of $N = \text{Ext}_A^{\mu-\delta(M)}(\text{Ext}_A^{\mu-\delta(M)}(M,A),A)$ because A is Auslander regular. It follows from Proposition 2.1.5., Lemma 2.5.1. and Proposition 4.2.9. that $M/F_{\delta(M)-1}M$ is a pure A-module with $j_A(M/F_{\delta(M)-1}M) = j_A(M)$. Hence by Theorem 4.2.15. we obtain

$$\min(J(M)) \cap \min(J(M/F_{\delta(M)-1}M)) \neq \emptyset.$$

Let p be in above intersection, then by Lemma 4.3.1. we have $p \in \min(J(N))$. Thus we may derive from Theorem 4.2.14. that $j_A(M) = j_A(N) = \text{gl.dim} A_p = j_{A_p}(N_p)$ because by Roos theorem $\text{gl.dim} A_p = j_{A_p}(N_p) + \text{K.dim}_{A_p} N_p = j_{A_p}(N_p)$. Note that Roos theorem also reads the equalities $j_{A_p}(M_p) + \text{K.dim}_{A_p} M_p = \text{gl.dim} A_p$ for all $p \in \min(J(M))$, this shows that $\text{gl.dim} A_p \geq j_{A_p}(M_p) \geq j_A(M)$. Hence $j_A(M) = \inf\{\text{gl.dim} A_p, \ p \in \min(J(M))\}$. $\square$

Since $\min(J(M))$ has only a finite number of prime ideals it follows from the above theorem that the grade number $j_A(M)$ of M may be calculated easily. In particular we have

3. Theorem. Let R be a left and right Zariski ring such that $G(R)$ is a Commutative Auslander regular ring. Then for every finitely generated R-module M,

$$j_R(M) = \inf\{\text{ht} p, \ p \in \min(J(M))\}.$$

Proof. Since by Theorem 2.5.2. we have $j_R(M) = j_{G(R)}(G(M))$ for every good filtration FM on M, the theorem follows immediately from Theorem 4.3.2.. $\square$

Finally, we describe the relation between the purity of a module and the geometrical purity of its characteristic variety.

4. Theorem. Let A be a commutative Noetherian ring with finite global dimension. Then a finitely generated A-module M is pure if and only if it satisfies the following conditions:
(1). $j_A(A/p) = j_A(M)$ for all $p \in \min(J(M))$,

(2). $\text{Ass}(M) = \min(J(M))$, where $\text{Ass}(M)$ denotes the set of associated primes of M in $\text{spec}(A)$.

Proof. Suppose that M satisfies the condition (1) and (2). Let N be any nonzero A-submodule of M, then $\text{Ass}(N) \neq \emptyset$. If $p \in \text{Ass}(N)$ then A/p is isomorphic to a submodule of N and hence of M, say N'. Since A is Auslander regular we have by Proposition 2.1.5. that $j_A(N) \leq j_A(N') = j_A(A/p) = j_A(M) \leq j_A(N)$, i.e., $j_A(N) = j_A(M)$. It follows from Theorem 4.2.10. that M is pure.

Conversely, let M be a pure A-module. Then it follows from Theorem 4.2.10. that $j_A(A/p) = j_A(M)$ for all $p \in \min(J(M)) \subseteq \text{Ass}(M)$. It remains to prove the inclusion $\text{Ass}(M) \subseteq \min(J(M))$. But this is clear by Lemma 4.3.1. since for any $p \in \text{Ass}(M)$ there is a submodule $N \subseteq M$ such that $A/p \cong N$. $\square$

5. Proposition. Let A be a commutative Noetherian ring with finite global dimension n. Suppose that every maximal ideal of A has the same height in $\text{spec}(A)$. For any finitely generated A-module M the following statements are equivalent:

(1). M is a pure A-module;

(2). The characteristic variety $\min(J(M))$ of M is geometrically pure and $\text{Ass}(M) = \min(J(M))$.

Proof. (1) $\Rightarrow$ (2). If M is pure then by theorem 4.3.4. we have $j_A(A/p) = j_A(M)$ for all $p \in \min(J(M))$. It follows from the generalized Roos theorem that $n = j_A(A/p) + \text{K.dim}(A/P) = j_A(M) + \text{K.dim}(A/p)$. Hence $\text{K.dim}(A/p) = n - j_A(M)$ for all $p \in \min(J(M))$, i.e., $\min(J(M))$ is geometrically pure.

(2) $\Rightarrow$ (1). By Theorem 4.3.4. it remains to prove $j_A(A/p) = j_A(M)$ for all $p \in \min(J(M))$. But since $\min(J(M))$ is geometrically pure, i.e., $\text{K.dim}_A M = \text{K.dim}(A/p)$ for all $p \in \min(J(M))$, it follows from the generalized Roos theorem that $\text{K.dim}(A/p) + j_A(A/p) = n = \text{K.dim}(A/p) + j_A(M)$. Hence $j_A(A/p) = j_A(M)$ as desired. $\square$

6. Corollary. Let R be a left and right Zariski ring such that $G(R)$ is a comutative Auslander regular ring. Suppose that every maximal element in $\text{spec}^g(R)$ has the same height $(= \text{gl.dim}G(R))$ in $\text{spec}(G(R))$. Let M be a finitely generated R-module, then the following statements are equivalent:

(1). M is a pure R-module;

(2). The characteristic variety $\min(J(M))$ is geometrically pure and $\text{Ass}(G(R)) = \min(J(M))$.

Proof. By Theorem 4.2.13. we know that M is a pure R-module if and only if $G(M)$ is a pure $G(R)$-module for some good filtration FM on M. Hence the assertion follows from Theorem 4.1.10. and the proof of Proposition 4.3.5..   □

In view of the foregoing results we see that although there are so many " bad " simple modules over Weyl algebras and enveloping algebras of finite dimensional Lie algebras which are not holonomic, they are still " good " enough because their characteristic varieties are geometrically pure.

CHAPTER IV
Microlocalization of Filtered Rings and Modules, Quantum Sections and Gauge Algebras

§1. Algebraic Microlocalization

The results used in this section stem mainly from §§5, 6 of Ch.I. and §2 of Ch.II..

Let R be a filtered ring with filtration FR, and $G(R)$ resp. $\tilde{R}$ the associated graded ring of R resp. the Rees ring of R with respect to FR. In what follows X always denotes the classical homogeneous element of degree 1 in $\tilde{R}$ as before.

If an R-module M has a good filtration FM then all good filtrations on M are (algebraically) equivalent in the sense of Ch.I. §3. If $\tilde{R}$ is left Noetherian we know that good filtrations induce good filtrations on R-submodules, moreover if FM is a good filtration on an R-module M and $F'M$ is (algebraically) equivalent to FM then $F'M$ is good too. We point out that Proposition 1.2.8., Theorem 1.2.10. and Theorem 2.1.2. of Ch.II. will be used implicitly in the sequel of this chapter. Recall in particular that under that condition that $\tilde{R}$ is (left) Noetherian FM is good exactly when $\widetilde{M}$ is finitely generated. Filtered properties do translate well into (graded) properties of the Rees objects, e.g., Lemma 6.4. of Ch.I., Lemma 1.2.9. of Ch.II.. In particular for any filtered R-module M with filtration FM we have $(\widehat{M})^\sim \cong (\widetilde{M})^{\wedge_g}$ where $(\widetilde{M})^{\wedge_g} = \varprojlim^g \widetilde{M}/X^n \widetilde{M}$ is the graded $X\tilde{R}$-adic completion of $\widetilde{M}$ in $\tilde{R}$-gr.

Let R be a separated filtered ring. Let S be a multiplicatively closed subset containing 1 but not 0. For $x \in F_n R - F_{n-1} R$ we let $\sigma(x)$ be the image of x in $G(R)_n$ as before. On S we put the condition that $\sigma(S) = \{\sigma(s), s \in S\}$ is multiplicatively too and $0 \notin \sigma(S)$. From the latter it follows that $\sigma(st) = \sigma(s)\sigma(t)$ holds for all s, t in S. If $s \in S$ is such that $\sigma(s) \in G(R)_n$ then we define $\tilde{s} \in \tilde{R}_n$ by putting $\tilde{s} = sX^n$.

1.1. Lemma. The set $\tilde{S}$ is multiplicatively closed in $\tilde{R}$ and if $\sigma(S)$ satisfies the left Ore conditions in $G(R)$ then for each $n \in I\!N$, $\tilde{S}$ maps to a homogeneous left Ore set $\overline{S}$ in $\tilde{R}/X^n \tilde{R}$.

Proof. The first statement follows from $\sigma(st) = \sigma(s)\sigma(t)$ for all s, $t \in S$. In order to check the first Ore condition for $\overline{S}$ in $\tilde{R}/X^n\tilde{R}$ look at $\overline{r}\overline{s} = 0$ in $\tilde{R}/X^n\tilde{R}$ where $\overline{r}$ and $\overline{s}$ are the images of $\tilde{s} \in \tilde{S}$ and $\tilde{r} \in \tilde{R}$. By the left Ore conditions for $\sigma(S)$ it follows that there is an $\tilde{s}_1 \in \tilde{S}$ such that $\overline{s}\overline{r} = \overline{X}\overline{u}_1$ for some $\overline{u}_1 \in \tilde{R}/X^n\tilde{R}$. From $\tilde{s}_1\overline{r}\overline{s} = 0$ we obtain that $Xu_1s = X^nb$ in $\tilde{R}$, hence $\overline{u}_1\overline{s} = 0$ in $\tilde{R}/X^n\tilde{R}$ and so, by using the left Ore conditions again, we obtain an $\overline{s}_2$ with $\tilde{s}_2 \in TS$ and such that $\overline{s}_2\overline{u}_1 = \overline{X}\overline{u}_2$ with $\overline{u}_2 \in \tilde{R}/X^n\tilde{R}$. Consequently $\overline{s}_2\overline{s}_1\overline{r} = \overline{X}^2\overline{u}_2$ and after repeating this argument a finite number of times, taking into account that $\overline{X}^n = 0$ in $\tilde{R}/X^n\tilde{R}$, we do arrive at $\overline{s}_n \cdots \overline{s}_1\overline{r} = 0$ with $\tilde{s}_n, \cdots, \tilde{s}_1 \in \tilde{S}$ and so the first left Ore condition for $\overline{S}$ in $\tilde{R}/X^n\tilde{R}$ has been verified.

For the second left Ore condition, consider $\overline{s} \in \tilde{R}/X^n\tilde{R}$, being an image of $\tilde{s} \in \tilde{S}$, and let $\overline{r} \in \tilde{R}/X^n\tilde{R}$ be arbitrary. We have to establish the existence of $\overline{s}_1, \overline{r}_1$ such that $\overline{s}_1\overline{r} = \overline{r}_1\overline{s}$ where $\overline{s}_1$ is an image of some $\tilde{s}_1 \in \tilde{S}$. From the second left Ore condition in $\tilde{R}/X^n\tilde{R}$ it follows that there exist $\overline{s'}, \overline{t}$ in $\tilde{R}/X^n\tilde{R}$ with $\overline{s'} \in \overline{S}$ such that $\overline{s'}\overline{r} = \overline{t}\overline{s} + \overline{X}\overline{u}_1$ and $\overline{u}_1 \in \tilde{R}/X^n\tilde{R}$. Repeating the argument leads to: $\overline{s''}\overline{u}_1 = \overline{a}_1\overline{s} + \overline{X}\overline{u}_2$ with $\overline{a}_1, \overline{u}_2 \in \tilde{R}/X^n\tilde{R}$. Hence we obtain:

$$\overline{s''}\overline{s'}\overline{r} = \overline{s''}\overline{t}\overline{s} + \overline{X}\overline{a}_1\overline{s} + \overline{X}^2\overline{u}_2 = (\overline{s''}\overline{t} + \overline{X}\overline{a}_1)\overline{s} + \overline{X}^2\overline{u}_2.$$

After at most $n - 1$ steps we do arrive at an equality $\overline{s}_1\overline{r} = \overline{r}_1\overline{s}$ where $\overline{s}_1$ is equal to the product of $\overline{s'}, \overline{s''}, \cdots$, in the correct order and $\overline{u}_1$ being the element in $\tilde{R}/X^n\tilde{R}$ obtained at the final step. $\qquad\square$

Instead of $\overline{S}$ in $\tilde{R}/X^n\tilde{R}$ we shall write $S(n)$ in order to indicate the dependency on n. Now, for every $n \in I\!N$ we may define $Q^g_{S(n)}(\tilde{R}/X^n\tilde{R})$ by inverting the homogeneous left Ore set $S(n)$ of $\tilde{R}/X^n\tilde{R}$ in the classical way. We obtain an inverse system of graded morphisms in $\tilde{R}$-gr:

$$\{Q^g_{S(n)}(\tilde{R}/X^n\tilde{R}) \longrightarrow Q^g_{S(n)}(\tilde{R}/X^{n-1}\tilde{R}), \; n \in I\!N\}$$

and we may take the inverse limit in the graded sense:

$$\tilde{Q}^\mu_{\overline{S}}(\tilde{R}) = \underset{n}{\varprojlim}^g \, Q^g_{S(n)}(\tilde{R}/X^n\tilde{R})$$

The latter is clearly a graded ring (and $\tilde{R}$-module). For a filtered R-module M we define

$$\tilde{Q}^\mu_{\overline{S}}(\widetilde{M}) = \underset{n}{\varprojlim}^g \, Q^g_{S(n)}(\widetilde{M}/X^n\widetilde{M})$$

and one easily verifies that this defines a graded $\tilde{Q}^\mu_{\overline{S}}(\tilde{R})$-module.

1.2. Lemma. With notation as above, $\tilde{Q}^{\mu}_{\overline{S}}(\widetilde{M})$ is X-torsionfree as an $\tilde{R}$-module, consequently we may correspond to it a filtered R-module

$$Q^{\mu}_{S}(M) = \tilde{Q}^{\mu}_{\overline{S}}(\widetilde{M})/(1 - X)\tilde{Q}^{\mu}_{\overline{S}}(\widetilde{M}).$$

Proof. Put $I = X\tilde{R}$. If $a \in \tilde{Q}^{\mu}_{\overline{S}}(\widetilde{M})$ is annihilated by X then $X_{a_{(n)}} = 0$ for $0 \neq a_{(n)} \in Q^{g}_{\overline{S}}(\widetilde{M}/I^{n}\widetilde{M})$ representing a. There is an $\overline{s}$ in the image of $\tilde{S}$ in $\tilde{R}/I^{n}$ such that $X\overline{s}a_{(n)} = 0$ with $\overline{s}a_{(n)} \neq 0$ in $\mathrm{Im}\varphi$, $\varphi\colon \widetilde{M}/I^{n}\widetilde{M} \to Q^{g}_{\overline{S}}(\widetilde{M}/I^{n}\widetilde{M})$ the natural localization homomorphism. If $b_{(n)} \in \widetilde{M}/I^{n}\widetilde{M}$ represents $\overline{s}a_{(n)}$ then $b_{(n)}$ is not $\overline{S}$-torsion since $\overline{s}a_{(n)}$ is not $\overline{S}$-torsion. From $X\overline{s}a_{(n)} = 0$ it then follows that $\overline{s}_{1}Xb_{(n)} = 0$ hence $\overline{s}_{1}b_{(n)} \in I^{n-1}\widetilde{M}/I^{n}\widetilde{M}$ and $b_{(n)}$ as well as $a_{(n)}$ must then be in $Q^{g}_{\overline{S}}(I^{n-1}\widetilde{M}/I^{n}\widetilde{M})$, i.e., $a_{(n-1)} = 0$. Since this argument works for any n it follows that $a = 0$ and therefore $\tilde{Q}^{\mu}_{\overline{S}}(\widetilde{M})$ is X-torsionfree. □

The ring $Q^{\mu}_{S}(R)$, thus obtained, is called the **microlocalization of R at S**. Similarly, $Q^{\mu}_{S}(M)$ is the **microlocalization of M** and it is a filtered R-module. It follows later that $Q^{\mu}_{S}(M)$ is in fact a filtered $Q^{\mu}_{S}(R)$-module in case FM is a good filtration. We have to justify the terminology introduced above by verifying that $Q^{\mu}_{S}(R)$ satisfies the universal property characterizing the microlocalization in the sense of [Vel] a.o..

1.3. Proposition. Let R be a filtered ring with separated filtration FR, let j_{S} be the canonical ring morphism $R \to Q^{\mu}_{S}(R)$. Given a complete filtered ring B with filtration FB and a filtered ring homomorphism $h\colon R \to B$ such that for all $s \in S$, $h(s)$ is invertible in B with $h(s) \in F_{-n}B$ if $\sigma(s) \in G(R)_{n}$. There exists a unique filtered ring homomorphism $g\colon Q^{\mu}_{S}(R) \to B$ such that h factorizes as

$$R \xrightarrow{\ j_S\ } Q^{\mu}_{S}(R) \xrightarrow{\ g\ } B$$

Proof. Since h is a filtered morphism it yields a map of graded $\tilde{R}$-modules $\tilde{h}\colon \tilde{R} \to \tilde{B}$, which is of degree zero. From $h(1_{R}) = 1_{B}$ it follows that $\tilde{h}((X) = X_{B}$, where X_{B} is 1_{B} viewed as element of $\tilde{B}_{1}$. It is trivial to verify that $\tilde{B}$ is in fact a ring and that $\tilde{h}$ is a graded ring morphism. From $\tilde{h}(I^{n}) = X^{n}_{B}\tilde{h}(\tilde{R})$ it follows that $\tilde{h}$ induces graded morphisms $\tilde{R}/I^{n} \to \tilde{h}(\tilde{R})/X^{n}_{B}\tilde{h}(\tilde{R}) \to \tilde{B}/I^{n}_{B}$, where $I_{B} = X_{B}\tilde{B}$.

Now note that an $s \in S$ with $\sigma(s) \in G(R)_{n}$ has $h(s)^{-1}$ in $F_{-n}B - F_{-n-1}B$ because $h(s)^{-1} \in F_{-n-1}B$ leads to $1 = h(s)h(s)^{-1} \in F_{n}BF_{-n-1}B \subseteq F_{-1}B$ a contradiction while $h(s)^{-1} \in F_{-1}B$

holds by assumption. The assumptions on S entail that $\tilde{S}$ maps to left Ore sets $\overline{S} \subset \tilde{R}/I^n$, $\overline{S}_h = \overline{\tilde{h}(\tilde{S})} \subset \tilde{h}(\tilde{R})/X_B^n \tilde{h}(\tilde{R})$ and the image of $h(\tilde{S})$ in $\tilde{B}/I^n B$ is invertible in $\tilde{B}/I^n B$ because $\widetilde{h(S)}$ is invertible in $\tilde{B}$. Note that the latter fact is really depending on the observation that $h(s)^{-1} \notin F_{-n-1} B$ when $\sigma(s) \in G(R)_n$! After localizing at the Ore sets involved we arrive at graded ring homomorphisms:

$$Q_{\overline{S}}^g(\tilde{R}/I^n) \longrightarrow Q_{\overline{S}_h}^g(\tilde{h}(\tilde{R})/X_B^n \tilde{h}(\tilde{R})) \longrightarrow \tilde{B}/I_B^n$$

Taking inverse limits in the graded sense (i.e., taking limits over n for each graded homogeneous part separately and taking the direct sum of these homogeneous parts afterwards) leads to graded morphisms:

$$\tilde{g}: \quad \tilde{Q}_{\overline{S}}^\mu(\tilde{R}) \longrightarrow \tilde{Q}_{\overline{S}_h}^\mu(\tilde{h}(\tilde{R})) \longrightarrow \tilde{B} = \varprojlim_n^g \tilde{B}/I_B^n \tilde{B}$$

The filtered morphism $g: Q_S^\mu(R) \to B$ corresponding to $\tilde{g}$ is also a ring homomorphism (not only an R-module morphism) and from $\tilde{R} \to \tilde{Q}_{\overline{S}}^\mu(\tilde{R}) \to \tilde{B}$ we obtain

$$R \xrightarrow{js} Q_S^\mu(R) \xrightarrow{g} B$$

factorizing h as desired (this is trivial to check). $\qquad\qquad\qquad\qquad\square$

1.4. Corollary. Let the notation be given as before.

(1). $Q_S^\mu(R)$ is the microlocalization of R at S in the sense of [Ve1].

(2). A similar proof establishes the same result for $Q_S^\mu(M)$.

(3). The fact that we keep trace of the intermediate ring $\overline{Q}_{\overline{S}_h}^\mu(\tilde{h}(\tilde{R}))$ in the proof shows that $Q_{S_h}^\mu(h(R))$ is a ring and that $Q_S^\mu(R) \to Q_{S_h}^\mu(h(R))$ is a filtered ring homomorphism.

(4). The equality of norms as in Lemma 5.16. and Corollary 5.20. in [Ve1] is a consequence of our proof above, in fact here it follows from the very simple fact that $h(s)^{-1} \in F_{-n} B$ entails $h(s)^{-1} \notin F_{-n-1} B$. This marks the advantage of the method used here over the cumbersome norm-calculations used in [Ve1].

In what follows we keep on using the foregoing notations.

1.5. Lemma. For any given filtered R-module M with filtration FM, $FQ_S^\mu(M)$ is separated.

Proof. We only have to check whether: $\cap_{n\geq 1} \tilde{j}_{\tilde{S}}(X)^n \tilde{Q}_{\overline{S}}^\mu(\tilde{M}) = \cap_{n \geq 1} \tilde{I}^n = 0$.

It is clear that $\tilde{I} = \varprojlim_n Q_{\overline{S}}^g(I/I^n)$, $\tilde{I}^m = \varprojlim_n^g Q_{\overline{S}}^g(I^m/I^{m+n})$ and it is then also clear that $\cap_{n\geq 1} \tilde{I}^n = 0$, $\cap_{n\geq 1} \tilde{I}^n \tilde{M} = 0$ hence $FQ_S^\mu(R)$ and $FQ_S^\mu(M)$ are separated. $\qquad\square$

At this point we have to introduce a condition that enables us to calculate $\varprojlim$ in an easier way, i.e., making it right exact in some easy recognizable cases. If one is given a sequence of homomorphisms of inverse systems

$$(*) \qquad\qquad 0 \longrightarrow (A_n) \longrightarrow (B_n) \longrightarrow (C_n) \longrightarrow 0$$

then we say that the system is exact if we have exact sequences

$$0 \longrightarrow A_n \longrightarrow B_n \longrightarrow C_n \longrightarrow 0$$

for each n. For an exact sequence $(*)$ we have that

$$0 \longrightarrow \varprojlim_n A_n \longrightarrow \varprojlim_n B_n \longrightarrow \varprojlim_n C_n \longrightarrow 0$$

is also exact.

Recall from ([Har] P. 191) the Mittag-Lefler condition (abbreviated to M.L.): An inverse system (A_n) satisfies M.L. if for each n, the decreasing family of subgroups of A_n $\{\varphi_{n'n'}(A_{n'}) \subseteq A_n,\ n' \geq n\}$ is stationary; in other words for each n there exists an $n_0 \geq n$ such that for all $n', n'' \geq n_0$, $\varphi_{n'n}(A_{n'}) = \varphi_{n''n}(A_{n''})$, where the $\varphi_{n'n}$ are morphisms in the inverse system.

1.6. Lemma. (cf. [Har] Proposition 9.1. and Example 9.1.1.) Let there be given a short exact sequence of inverse systems of abelian groups:

$$0 \longrightarrow (A_n) \longrightarrow (B_n) \longrightarrow (C_n) \longrightarrow 0$$

(1). If (B_n) satisfies M.L. so does (C_n).

(2). If (A_n) satisfies M.L. then

$$0 \longrightarrow \varprojlim_n A_n \longrightarrow \varprojlim_n B_n \longrightarrow \varprojlim_n C_n \longrightarrow 0$$

is an exact sequence.

(3). If all the maps $\varphi_{n'n}\colon A_{n'} \to A_n$ are surjective then (A_n) satisfies M.L..

When considering inverse systems of graded objects one may consider a weaker version of the M.L. condition, i.e., the " graded M.L. condition " by restricting attention to the system $\varphi_{n'n}^t|(A_{n'})_t$ for a fixed degree $t \in \mathbb{Z}$. However in the sequel we will accidently use only the full M.L. condition.

1.7. Proposition. $FQ_S^\mu(M)$ is complete, for $M \in R$-filt.

Proof. We establish that $\tilde{Q}^{\mu}_{\overline{S}}(\widetilde{M})$ is $\tilde{I}$-adically complete in $\tilde{R}$-gr. Look at the exact sequence of inverse systems (for fixed n):

$$0 \longrightarrow I^n/I^{n+m} \longrightarrow \tilde{R}/I^{n+m} \longrightarrow \tilde{R}/I^n \longrightarrow 0$$

Inverting the Ore set $\overline{S}$ yields an exact (!) sequence of inverse systems

$$0 \longrightarrow Q^g_{\overline{S}}(I^n/I^{n+m}) \longrightarrow Q^g_{\overline{S}}(\tilde{R}/I^{n+m}) \longrightarrow Q^g_{\overline{S}}(\tilde{R}/I^n) \longrightarrow 0$$

(we have denoted $\overline{S}$ for the image of $\tilde{S}$ in $\tilde{R}/I^{n+m}$ as well as the image in $\tilde{R}/I^n$ but this creates no confusion and it does not harm the exactness). The maps $\varphi_{m'm}$ in the system $\{Q^g_{\overline{S}}(I^n/I^{n+m}), m\}$ are all surjective (and graded) so we apply Lemma 3.8. to obtain an exact sequence:

$$0 \longrightarrow \tilde{I}^n = \varprojlim_m Q^g_{\overline{S}}(I^n/I^{n+m}) \longrightarrow \tilde{Q}^{\mu}_{\overline{S}}(\tilde{R}) \longrightarrow Q^g_{\overline{S}}(\tilde{R}/I^n) \longrightarrow 0$$

hence we obtain: $\tilde{Q}^{\mu}_{\overline{S}}(\tilde{R})/\tilde{I}^n = Q^g_{\overline{S}}(\tilde{R}/I^n)$. Taking $\varprojlim_n$ in $\tilde{R}$-gr learns that the graded $\tilde{I}$-adic completion of $\tilde{Q}^{\mu}_{\overline{S}}(\tilde{R})$ equals $\varprojlim_n Q^g_{\overline{S}}(\tilde{R}/I^n) = \tilde{Q}^{\mu}_{\overline{S}}(\tilde{R})$, i.e., the claim is proved for $M = R$. For $M \in R$-filt a formally similar proof is valid too. $\qquad \square$

1.8. Proposition. The associated graded module of $Q^{\mu}_S(M)$ is $G(Q^{\mu}_S(M)) = \sigma(S)^{-1}G(M)$.

Proof. To find $G(Q^{\mu}_S(M))$ we have to calculate the module

$$\tilde{Q}^{\mu}_{\overline{S}}(\widetilde{M})/\tilde{I}\tilde{Q}^{\mu}_{\overline{S}}(\widetilde{M}) = \varprojlim_n{}^g\, Q^g_{\overline{S}}(\widetilde{M}/I^n\widetilde{M})/\varprojlim_n{}^g\, Q^g_{\overline{S}}(I\widetilde{M}/I^n\widetilde{M})$$

Since the maps in the inverse system $\{Q^g_{\overline{S}}(I\widetilde{M}/I^n\widetilde{M}), n\}$ are all surjective it follows that we may calculate the quotient term per term, and by using exactness of $Q^g_{\overline{S}}$ we thus arrive at $G(Q^{\mu}_S(M)) = \varprojlim_n{}^g Q^g_{\overline{S}}(\widetilde{M}/I\widetilde{M}) = Q^g_{\overline{S}}(\widetilde{M}/I\widetilde{M}) = \sigma(S)^{-1}G(M)$ because $G(M) = \widetilde{M}/I\widetilde{M}$ and $\overline{S}$ in $\tilde{R}/I = G(R)$ is just $\sigma(S)$ which is assumed to be an Ore set. $\qquad \square$

1.9. Remarks. (1). From the results established above it is also clear that Q^{μ}_S is a functor from R-filt to $Q^{\mu}_S(R)$-filt. Indeed, it is clear that $Q^g_{\overline{S}}(\widetilde{M}/I^n\widetilde{M})$ is a graded $Q^g_{\overline{S}}(\tilde{R}/I^n)$-module for each n, so it is also clear that $Q^{\mu}_S(M)$ is a $Q^{\mu}_S(R)$-module. Similarly one verifies the action of Q^{μ}_S on filtered morphisms.

(2). If S is an Ore set of R and FR is trivial then

$$\tilde{R} = R[t], \quad G(R) = R, \quad \tilde{Q}^{\mu}_{\tilde{S}}(\tilde{R}) = \varprojlim_n{}^g S^{-1}(R[t]/t^n R) = (S^{-1}R)[t]$$

and

$$Q^{\mu}_S(R) = (\dot{S}^{-1}R)[t]/(1-t) = S^{-1}R.$$

(3). If S is an Ore set of R and FR is any separated filtration then $\tilde{Q}^{\mu}_{\tilde{S}}(\tilde{R}) = \varprojlim_n{}^g \overline{S}^{-1}(\tilde{R}/I^n) = \varprojlim_n \overline{S}^{-1}\tilde{R}/\overline{S}^{-1}I^n = (S^{-1}\tilde{R})^{\wedge}$ where $\wedge$ is the garded $S^{-1}I$-adic completion. We obtain that $Q^{\mu}_S(R) = (S^{-1}R)^{\wedge F_S}$ where $\wedge(F_S)$ is a completion corresponding to the S-closure of the filtration F. This F_S corresponds to the pseudo-norm associated to S in [Vel].

(4). If $S = \{1\}$ then $Q^{\mu}_S(R) = \hat{R}^F$ follows from (3).

So far we have obtained short elegant proofs of most important properties of Q^{μ}_S as well as some new results. We now continue the study of (right) exactness of the functor Q^{μ}_S. Here certain finiteness conditions have to be imposed and it will be the condition $\tilde{R}$ is left Noetherian that turns out to be fundamental. However we will not assume that R is a Zariski ring. Indeed, an easy example is obtained by taking an invertible ideal I of a ring R such that I is not contained in the Jacobson radical $J(R)$ of R, or take I such that the Artin-Rees property does not hold, then the strong filtration defined by I on $\cup_{n \in \mathbb{Z}} I^n = Q$ can easily satisfy the condition: $\tilde{R}$ is Noetherian (for example by starting from a Noetherian ring R and R/I being Noetherian too) but it will always fail to be Zariskian.

For a graded $\tilde{R}$-module T, $T(n)$ with $n \in \mathbb{Z}$ stands for the n-th shifted graded $\tilde{R}$-module as defined in Ch.I.. We introduce the notation

$$
\begin{aligned}
t_{X^n}(T) &= \{t \in T, \; X^n t = 0\} \\
t_{X^\infty}(T) &= t \in T, \; X^m t = 0 \text{ for some } m \in \mathbb{Z}\}.
\end{aligned}
$$

We write $(t_{X^n}(T)(-n))_n$ for the inverse system defined by the graded morphisms $\varphi_{n'n}$: $t_{X^n}(T)(-n) \to t_{X^{n'}}(T)(-n')$, being multiplication by $X^{n'-n}$.

We view $t_{X^n}(T)$ as a $\tilde{R}/I^n$-module in the obvious way, then it makes sense to consider the inverse system (again graded maps) $\varphi_{n'n}$: $Q^g_{\tilde{S}}(t_{X^n}(T)(-n)) \to Q^g_{\tilde{S}}(t_{X^{n'}}(T)(-n'))$.

An $\tilde{R}$-module $\widetilde{M}$ is said to be **prefinite** if there is an $m \in \mathbb{N}$ such that $X^m t_{X^\infty}(\widetilde{M}) = 0$.

1.10. Lemma. If $\widetilde{M}$ is prefinite then the inverse system $(Q^g_{\tilde{S}}(t_{X^n}(\widetilde{M})(-n)))_n$ satisfies M.L. and we have that its projective limit is zero.

Proof. Obvious. □

1.11. Lemma. (1). Any finitely generated graded X-torsion module T in $\tilde{R}$-gr is prefinite.
(2). If $f: M \to N$ is a filtered morphism in R-filt such that there exists an $m \in I\!N$ such that $f(F_n M) \supset F_{n-m} N$ then $\text{Coker}(\tilde{f})$ is prefinite.

Proof. (1). Trivial.
(2). $F_p(N/f(M)) = F_p N / F_p f(M) = F_p N / f(F_p M)$, hence we have a surjective map $F_p N / F_{p-m} N \to F_p(N/f(M))$ hence $\tilde{N}/X^m \tilde{N} \to \text{Coker}\tilde{f}$ is surjective but then $\text{Coker}(\tilde{f})$ is prefinite. □

1.12. Corollary. If FM and $F'M$ are filtrations on M such that the identity of M satisfies the condition in (2). of Lemma 1.11., i.e., $F_n M \supset F'_{n-m} M$ for all $n \in \mathbb{Z}$ then the cokernel of $\tilde{1}_M$ is prefinite. Our first result concerning exactness properties of Q^μ_S can now be stated as follows.

1.13. Theorem. Let

$$0 \longrightarrow \tilde{A} \longrightarrow \tilde{B} \longrightarrow \tilde{C} \longrightarrow 0$$

be an exact sequence of graded $\tilde{R}$-modules such that either
1. $\tilde{C}$ is X-torsionfree (hence certainly prefinite)
2. $\tilde{B}$ and $\tilde{C}$ are prefinite
then, with conventions on S as before, the following sequence is exact in $\tilde{R}$-gr:

$$0 \longrightarrow \tilde{Q}^\mu_{\overline{S}}(\tilde{A}) \longrightarrow \tilde{Q}^\mu_{\overline{S}}(\tilde{B}) \longrightarrow \tilde{Q}^\mu_{\overline{S}}(\tilde{C}) \longrightarrow 0$$

Proof. There is an exact sequence (in $\tilde{R}$-gr) of inverse systems:

$$0 \; \to \; (Q^g_{\overline{S}}(t_{X^n}(\tilde{A})(-n)))_n \to (Q^g_{\overline{S}}(t_{X^n}(\tilde{B})(-n)))_n \to$$
$$\to (Q^g_{\overline{S}}(t_{X^n}(\tilde{C})(-n)))_n \to (Q^g_{\overline{S}}(\tilde{A}/I^n\tilde{A}))_n \to$$
$$\to (Q^g_{\overline{S}}(\tilde{B}/I^n\tilde{B}))_n \to (Q^g_{\overline{S}}(\tilde{C}/I^n\tilde{C}))_n \to 0$$

It now suffices to apply Lemma 1.10. and Lemma 1.6., taking into account that in case 2. A is also prefinite, and by taking $\varprojlim$ we obtain the desired exact sequence. □

1.14. Corollary. (1). If

$$0 \longrightarrow A \longrightarrow B \longrightarrow C \longrightarrow 0$$

is a strict exact sequence of filtered R-modules then also the sequence

$$0 \longrightarrow Q_S^\mu(A) \longrightarrow Q_S^\mu(B) \longrightarrow Q_S^\mu(C) \longrightarrow 0$$

is exact.

(2). If FM and $F'M$ are equivalent filtrations on M then $FQ_S^\mu(M)$ and $F'Q_S^\mu(M)$ are equivalent, i.e., the R-module $Q_S^\mu(M)$ does not depend on the chosen equivalent filtration on M.

(3). If $\tilde{R}$ is left Noetherian and

$$0 \longrightarrow A \longrightarrow B \longrightarrow C \longrightarrow 0$$

is an exact sequence of filtered R-modules such that $\tilde{A}, \tilde{B}, \tilde{C}$ are finitely generated $\tilde{R}$-modules then the sequence

$$0 \longrightarrow Q_S^\mu(A) \longrightarrow Q_S^\mu(B) \longrightarrow Q_S^\mu(C) \longrightarrow 0$$

is an exact sequence of filtered $Q_S^\mu(R)$-modules.

Proof. (1). Follows from Theorem 1.13. because strictness of the sequence entails that $\tilde{C}$ is X-torsionfree.

(2). We may apply Theorem 1.13. and Lemma 1.11., and use exactness of $- \otimes \tilde{R}/(1-X)\tilde{R}$ on $\tilde{R}$-gr.

(3). Since $\tilde{R}$ is left Noetherian, the complex

$$0 \longrightarrow \tilde{A} \longrightarrow \tilde{B} \longrightarrow \tilde{C} \longrightarrow 0$$

has finitely generated homology hence it is prefinite in view of Lemma 1.11. and clearly it is also X-torsion. By repeated application of Theorem 1.13. we obtain that the complex

$$0 \longrightarrow \tilde{Q}_{\tilde{S}}^\mu(\tilde{A}) \longrightarrow \tilde{Q}_{\tilde{S}}^\mu(\tilde{B}) \longrightarrow \tilde{Q}_{\tilde{S}}^\mu(\tilde{C}) \longrightarrow 0$$

has X-torsion homology. Therefore

$$0 \longrightarrow Q_S^\mu(A) \longrightarrow Q_S^\mu(B) \longrightarrow Q_S^\mu(C) \longrightarrow 0$$

is an exact sequence of filtered $Q_S^\mu(R)$-modules.   □

We present the following result in a rather general form, its corollaries provide short and intrinsically graded proofs of a. o. Proposition 1.7. and Proposition 1.8..

1.15. Proposition. If $\widetilde{M}$ is a prefinite graded $\tilde{R}$-module then

$$\tilde{R}/I^n \otimes_{\tilde{R}} \tilde{Q}_{\tilde{S}}^\mu(\widetilde{M}) \cong Q_{\tilde{S}}^g(\widetilde{M}/I^n\widetilde{M}).$$

Proof. Apply Theorem 1.13. to the exact sequence

$$0 \longrightarrow t_{X^n}(\widetilde{M}) \longrightarrow \widetilde{M} \xrightarrow{X^n} \widetilde{M} \longrightarrow \widetilde{M}/I^n\widetilde{M} \longrightarrow 0$$

□

1.16. Corollaries. (1). If $M \in R$-filt then $G(Q_S^\mu(M)) = \sigma(S)^{-1}G(M)$.

(2). If S and T are multiplicative sets in R with properties as before and if $S \vee T$ is the multiplicative set generated by $S \cup T$ then let us assume $0 \notin \sigma(S) \vee \sigma(T)$ (just to avoid triviality of the statements). Then $Q_S^\mu(Q_T^\mu(M)) \cong Q_{S\vee T}^\mu(M)$ (this follows from Proposition 1.15. nd the fact that $Q_{\overline{S}}^g Q_{\overline{T}}^g = Q_{\overline{S\vee T}}^g$ on graded $\widetilde{R}/I^n$-modules).

1.17. Theorem. Let $\widetilde{R}$ be left Noetherian.

(1). If $\widetilde{M}$ is a graded finitely generated $\widetilde{R}$-module then

$$\widetilde{Q}_{\overline{S}}^\mu(\widetilde{M}) = \widetilde{Q}_{\overline{S}}^\mu(\widetilde{R}) \otimes_{\widetilde{R}} \widetilde{M}$$

(2). $\widetilde{Q}_{\overline{S}}^\mu(\widetilde{R})$ is flat.

Proof. (1). Since $\widetilde{Q}_{\overline{S}}^\mu$ commutes with finite direct sums the claim holds for gr-free graded $\widetilde{R}$-modules of finite rank. For $\widetilde{M}$ consider a presentation

$$\widetilde{G} \longrightarrow \widetilde{F} \longrightarrow \widetilde{M} \longrightarrow 0$$

where $\widetilde{F}$ and $\widetilde{G}$ are gr-free of finite rank. Application of Theorem 1.13. yields the following commutative diagram:

$$
\begin{array}{ccccccc}
\widetilde{Q}_{\overline{S}}^\mu(\widetilde{G}) & \longrightarrow & \widetilde{Q}_{\overline{S}}^\mu(\widetilde{F}) & \longrightarrow & \widetilde{Q}_{\overline{S}}^\mu(\widetilde{M}) & \longrightarrow & 0 \\
\downarrow{\cong} & & \downarrow{\cong} & & \uparrow & & \\
\widetilde{Q}_{\overline{S}}^\mu(\widetilde{R}) \otimes_{\widetilde{R}} \widetilde{G} & \longrightarrow & \widetilde{Q}_{\overline{S}}^\mu(\widetilde{R}) \otimes_{\widetilde{R}} \widetilde{F} & \longrightarrow & \widetilde{Q}_{\overline{S}}^\mu(\widetilde{R}) \otimes_{\widetilde{R}} \widetilde{M} & \longrightarrow & 0
\end{array}
$$

By the five lemma, the vertical map on the right is an isomorphism of graded $\widetilde{Q}_{\overline{S}}^\mu(\widetilde{R})$-modules.

(2). The functor $\widetilde{Q}_{\overline{S}}^\mu(\widetilde{R}) \otimes_{\widetilde{R}} -$ commutes with graded direct limits and it is exact on finitely generated modules by the first part and theorem 1.13.. □

1.18. Corollary. Let $\widetilde{R}$ be left Noetherian, then

(1). The functor $Q_S^\mu(R) \otimes_R -$ preserves strict maps and it is exact on R-modules.

(2). If $M \in R$-filt with good filtration FM, i.e., $\widetilde{M}$ is finitely generated, then $Q_S^\mu(R) \otimes_R M \cong Q_S^\mu(M)$ as filtered R-modules.

Proof. (1). Since $\widetilde{Q}_{\widetilde{S}}^\mu(\widetilde{R})$ is flat as an $\widetilde{R}$-module, flatness of $Q_S^\mu(R)$ follows for example from localization at $\{1, X, X^2, \cdots\}$. A strict morphism f yields a graded morphism $\widetilde{f}$ with X-torsionfree cokernel. Then the flatness of $\widetilde{Q}_{\widetilde{S}}^\mu(\widetilde{R})$ yields that $\widetilde{Q}_{\widetilde{S}}^\mu(\widetilde{R}) \otimes \widetilde{f}$ has X-torsionfree cokernel too.
(2). Trivial from the foregoing.   □

A multiplicative set S is said to be **saturated** if $S = \{r \in R, \ \sigma(r) \in \sigma(S)\}$. Put $\widetilde{S}_{\text{sat}}$ equal to the multiplicative set of homogeneous elements in $\widetilde{S} + \widetilde{R}X$, then it is clear that $\widetilde{S}_{\text{sat}}$ maps to $S_{\text{sat}} = \{r \in R, \ \sigma(r) \in \sigma(S)\}$ (for a multiplicatively closed set S such that $\sigma(S)$ is multiplicatively closed) under the map $\widetilde{R} \to R = \widetilde{R}/(1 - X)\widetilde{R}$. If $\widetilde{S}_{\text{sat}}$ is an Ore set in $\widetilde{R}$ then S_{sat} is an Ore set in R.

1.19. Proposition. If $\widetilde{R}$ is left Noetherian and S is a saturated multiplicatively closed set of R such that $\sigma(S)$ satisfies the second left Ore condition in $G(R)$ then S satisfies the second left Ore condition in R. In case $\sigma(S)$ is a regular left Ore set in $G(R)$, S is a left Ore set in R and $Q_S^\mu(R) = (S^{-1}R)^{\wedge_{F_S}}$.

Proof. Clearly $\widetilde{S} = \widetilde{S}_{\text{sat}}$ since S is saturated. By Lemma 1.1., $\widetilde{S}$ maps to homogeneous sets satisfying the second left Ore condition in $\widetilde{R}/I^n$ for every $n \in I\!N$, (note: first or second refers to the order on the Ore conditions as used in Lemma 1.2.). Now $I = X\widetilde{R}$ is centrally generated and it is an invertible ideal of the Noetherian ring $\widetilde{R}$ so it certainly satisfies the Artin-Rees condition with respect to left ideals L of $\widetilde{R}$, i.e., $L \cap \widetilde{R}X^{h(n)} \subseteq LX^n$ for some $h(n)$ associated to $n \in I\!N$. Combining this Artin-Rees condition with the second left Orecondition for $\overline{S}^{(n)}$, the image of $\widetilde{S}$ in $\widetilde{R}/\widetilde{R}X^n$, with $n \in I\!N$, one easily obtains the second left Ore condition for $\widetilde{S}$ in $\widetilde{R}$. It is obvious that the regularity of $\sigma(S)$ in $G(R)$ allows to lift the first Ore condition as well. In view of Remark 1.9. it follows indeed that microlocalization at S is just a completion of $S^{-1}R$ at the localized filtration.   □

1.20. Corollary. For any S (with $\sigma(S)$ multiplicatively closed as always) we have: $Q_S^\mu(R) = (S_{\text{sat}}^{-1}R)^{\wedge_{F_S}}\text{sat}$, i.e., microlocalization is always just a suitable completion of a localization at some Ore set in R.   .

§2. Quantum Sections and Microstructure Sheaves

In this section we study Zariski filtered rings R such that the associated graded ring $G(R)$ is a **commutative Noetherian domain** and unless otherwise stated we assume that FR is **positive**.

Using the microlocalizations of the foregoing section we may introduce the microstructure sheaf over R over the projective scheme $\text{Proj}(G(R))$. The sections of a coherent sheaf of graded modules over a Zariski open set in $\text{Proj}(G(R))$ correspond to a graded localization but on the level of the microstructure sheaf of a filtered module one has to consider microlocalizations at filters of ideals of $G(R)$ not necessarily stemming from a multiplicatively closed subset of $G(R)$. The microlocalizations may also be used to introduce the quantum-sections of a coherent sheaf of modules over the structure sheaf of $\text{Proj}(G(R))$.

For the theory of localization in terms of Gabriel filters, torsion theories or Serre localizing subcategories, we refer to [Gab], [Gol] and [Ser1]. If $\mathcal{L}(\kappa)$ is a Gabriel filter (of left ideals) of R then for every R-module M, $\kappa(M)$ is the κ-torsion R-submodule of M given by: $\kappa(M) = \{m \in M$, there is an $I \in \mathcal{L}(\kappa)$ such that $Im = 0\}$. Then $M/\kappa(M)$ is $\mathcal{L}(\kappa)$-torsionfree and the localization functor Q_κ associated to $\mathcal{L}(\kappa)$ may be defined by $Q_\kappa(M) = \varprojlim_{I \in \mathcal{L}(\kappa)} \text{Hom}_R(I, M/\kappa(M))$. The canonical R-morphism $j_\kappa \colon M \to M/\kappa(M)$, extends to $j_\kappa \colon M \to Q_\kappa(M)$ and $j_\kappa \colon R \to Q_\kappa(R)$ is an epimorphism in the category of rings (but not surjective in general). When Q_κ is exact and commutes with direct sums we say that $\mathcal{L}(\kappa)$ is **perfect**. Since we only look at Zariski rings every Gabriel filter $\mathcal{L}(\kappa)$ has finite type in the sense that every $I \in \mathcal{L}(\kappa)$ contains a finitely generated left ideal J of R such that $J \in \mathcal{L}(\kappa)$ too. Recall that $Q_\kappa(M)$ is in a canonical way a left $Q_\kappa(R)$-module and this structure is the unique one extending the R-structure of M (via $M/\kappa(M)$).

2.1. Lemma. If $\mathcal{L}(\kappa)$ is a Gabriel filter of R then we obtain a Gabriel filter $\mathcal{L}(\tilde{\kappa})$ on $\tilde{R}$ by putting:

$$\mathcal{L}(\tilde{\kappa}) = \{J \text{ left ideal of } R, \ J \subseteq \tilde{L} \text{ for some } L \in \mathcal{L}(\kappa)\}$$

where $\tilde{L}$ is constructed with respect to the induced filtration $FL = L \cap FR$.

Proof. Let us check the four defining properties for a Gabriel filter.

1. If $J \in \mathcal{L}(\tilde{\kappa})$ and $H \supset$ then $H \in \mathcal{L}(\tilde{\kappa})$, is obvious.

2. If $I, J \in \mathcal{L}(\tilde{\kappa})$ then $I \cap J \supset \tilde{H}_1 \cap \tilde{H}_2 \supset (H_1 \cap H_2)^\sim$ for some H_1, $H_2 \in \mathcal{L}(\kappa)$ and thus $H_1 \cap H_2 \in \mathcal{L}(\kappa)$ as $\mathcal{L}(\kappa)$ is a Gabriel filter.

3. For $L \in \mathcal{L}(\tilde{\kappa})$ and $Ty \in \tilde{R}$ we must check that $[L : \tilde{y}] = \{\tilde{x} \in \tilde{R}, \ \tilde{x}\tilde{y} \in L\}$ is in $\mathcal{L}(\tilde{\kappa})$, hence we have to find a $K \in \mathcal{L}(\tilde{\kappa})$ such that $K\tilde{y} \subseteq L$ and it is clear that it suffices to do this for homogeneous $\tilde{y}$ (in view of 2.), say $\tilde{y} = (\tilde{y}')_{m(y)}$ (see Ch.I. subsection 4.3.) for some

$y \in R$, $m(y) \in I\!N$. Since $L \in \mathcal{L}(\tilde{\kappa})$ there is an $\tilde{I} \subseteq L$ for some $I \in \mathcal{L}(\kappa)$ and there exists an $H \in \mathcal{L}(\kappa)$ such that $Hy' \subseteq I$. Now if $\tilde{h} \in \tilde{H}$ is homogeneous, say $\tilde{h} = (\tilde{h}')_{m(h)}$ for some $h' \in F_{m(h)}H - F_{m(h)-1}H$, then we have $\tilde{h}\tilde{y} = (\widetilde{h'y'})_{m(h)+m(y)}$ (note that we have assumed that $G(R)$ is a domain!). Consequently, $\tilde{H}\tilde{y} \subseteq \widetilde{Hy} \subseteq \tilde{I} \subseteq L$ with $\tilde{H} \in \mathcal{L}(\tilde{\kappa})$.

4. If $H \subseteq L$ with $L \in \mathcal{L}(\tilde{\kappa})$ is such that L/H is $\mathcal{L}(\tilde{\kappa})$-torsion then we have to show that $H \in \mathcal{L}(\tilde{\kappa})$; clearly we do not loose generality if we assume that $L = \tilde{I}$ for some $I \in \mathcal{L}(\kappa)$. Since H is X-torsionfree it follows that $H = \tilde{K}$ for some left ideal K of R. For every $i \in I$ there exists a $J_i \in \mathcal{L}(\tilde{\kappa})$ such that $J_i\tilde{i} \subseteq \tilde{K}$ and it is not restrictive to assume that $J_i = \tilde{E}_i$ for some $E_i \in \mathcal{L}(\kappa)$. Since $\mathcal{L}(\kappa)$ is a Gabriel filter, the property $E_i i \subseteq K$ for every $i \in I \in \mathcal{L}(\tilde{\kappa})$, with $E_i \in \mathcal{L}(\kappa)$, yields $K \in \mathcal{L}(\kappa)$ and therefore we arrive at $\tilde{K} = H \in \mathcal{L}(\tilde{\kappa})$ as desired. $\square$

From the definition of $\mathcal{L}(\tilde{\kappa})$ it is clear that $\mathcal{L}(\tilde{\kappa})$ has a cofinal system of graded left ideals, so we may consider the graded filter $\mathcal{L}^g(\tilde{\kappa}) = \{\tilde{L}, \quad L \in \mathcal{L}(\kappa)\}$, and this defines a rigid torsion theory on $\tilde{R}$-gr. Full detail on graded localization and rigid torsion theory may be found in [NVO1]. We let $\pi: \tilde{R} \to G(R)$ be the canonical epimorphism and define $G\kappa$ on $G(R)$-mod by putting $\mathcal{L}(G\kappa) = \{L \text{ left ideal of } G(R) \text{ such that } L \supset \pi(\tilde{M}) \text{ for some } \tilde{M} \in \mathcal{L}(\tilde{\kappa})\}$. It is easy to check that $\mathcal{L}(G\kappa)$ is a Gabriel filter. Again $\mathcal{L}(G\kappa)$ has a cofinal system of graded left ideals and so we may consider the graded filter $\mathcal{L}^g(G\kappa)$ and the corresponding graded localization in $G(R)$-gr. We let Q_κ^g, resp. $Q_{G\kappa}^g$, denote the graded localization functor at $\mathcal{L}(\tilde{\kappa})$, resp. $\mathcal{L}(G\kappa)$, in $\tilde{R}$-gr, resp. $G(R)$-gr.

When looking at microlocalization it is more natural to start from graded filter $\mathcal{L}^g(\overline{\kappa})$ and then try to lift it to an $\mathcal{L}^g(\tilde{\overline{\kappa}})$ yielding a graded localization on $\tilde{R}$-gr. However if we start from a homogeneous (left) Ore set $\sigma(S)$ in $G(R)$ then S need not be an Ore set but S_{sat} is (cf. Proposition 1.19). When we try this for abstract localization we would define $\mathcal{L}(\tilde{\overline{\kappa}})$ by taking $\beta^g = \{\tilde{L} \subseteq \tilde{R}, \quad \pi(\tilde{L}) \in \mathcal{L}(\overline{\kappa})\}$ for a filterbasis. Were this set a filterbasis then we could define $\mathcal{L}(\kappa)$ by letting $\mathcal{L}(\kappa)$ be generated as a filter by those left ideals L of R such that $L[t, t^{-1}] \cong \tilde{L}_X$ for some $\tilde{L} \in \beta^g$. Much to our regret it seems that β^g need not be a filterbasis for a Gabriel filter. We say that $\overline{\kappa}$ is **saturated** exactly then when β^g as above is a filterbasis for a Gabriel filter we then denote by $\mathcal{L}(\tilde{\overline{\kappa}})$. If $\overline{\kappa}$ is saturated we may define $\mathcal{L}(\kappa)$ as mentioned above and we have $\mathcal{L}(\overline{\kappa}) = \mathcal{L}(G\kappa)$ in this case. In the situation that $\overline{\kappa}$ is saturated we just write $\tilde{\kappa}$ for $\tilde{\overline{\kappa}}$ (and indeed $\mathcal{L}(\tilde{\overline{\kappa}}) = \mathcal{L}(\tilde{\kappa})$ holds when κ is defined and $\tilde{\kappa}$ derived from κ as before) and we will sometimes say that $\mathcal{L}(\kappa)$ **is saturated** instead of $\mathcal{L}(\overline{\kappa})$ is saturated.

2.2. Lemma. Let R be a Zariskian filtered ring with filtration FR and $\mathcal{L}(\kappa)$ a saturated

Gabriel filter on R, then we have:

For $M \in R$-filt: $\tilde{\kappa}(\widetilde{M}) = \kappa\widetilde{(M)}$ and $\widetilde{M}/\tilde{\kappa}(\widetilde{M}) = (M/\kappa(M))^{\sim}$.

Proof. Since $\tilde{\kappa}(\widetilde{M})$ is X-torsionfree it may be written as some $\widetilde{N}$ for $N \subseteq M$ (but FN not necessarily induced by FM). Since we do have that $N = \tilde{\kappa}(\widetilde{M})/(1 - X)\tilde{\kappa}(\widetilde{M})$ it follows that N is κ-torsion and $\widetilde{N} \subseteq \kappa\widetilde{(M)}$. On the other hand, if $\overline{m} \in \kappa\widetilde{(M)}$ is homogeneous then $\overline{m} = (\widetilde{m'})_d$ for some d and $m' \in F_d M - F_{d-1} M$ where $m' \in \kappa(M)$. Thus $LM = 0$ for some $L \in \mathcal{L}(\kappa)$ and from $\widetilde{Lm} = \widetilde{L}\overline{m}$ we conclude that $\overline{m} \in \tilde{\kappa}(\widetilde{M})$, whence $\tilde{\kappa}(\widetilde{M}) = \kappa\widetilde{(M)}$ follows. Next we check that $\widetilde{M}/\tilde{\kappa}(\widetilde{M})$ is X-torsionfree. If for some homogeneous $\tilde{y} \in \widetilde{M}$ we would have $X\tilde{y} \in \tilde{\kappa}(\widetilde{M})$ then $\widetilde{I}X\tilde{y} = 0$ for some $\widetilde{I} \in \mathcal{L}^g(\tilde{\kappa})$. This leads to $I\tilde{y} = 0$ and hence $\tilde{y} \in \tilde{\kappa}(\widetilde{M})$. The sequence of X-torsionfree graded $\widetilde{R}$-modules

$$0 \longrightarrow \kappa\widetilde{(M)} \longrightarrow \widetilde{M} \longrightarrow \widetilde{M}/\tilde{\kappa}(\widetilde{M}) \longrightarrow 0$$

is exact in $\mathcal{F}_X$, so applying the exact functor $\widetilde{R}/(1 - X)\widetilde{R} \otimes_{\widetilde{R}} -: \widetilde{R}$-gr $\rightarrow R$-filt to the forementioned sequence we obtain an exact sequence in R-filt

$$0 \longrightarrow \kappa(M) \longrightarrow M \longrightarrow N_1 \longrightarrow 0$$

which is strict and where N_1 is such that $\widetilde{N_1} = \widetilde{M}/\tilde{\kappa}(\widetilde{M})$. Consequently, $(M/\kappa(M))^{\sim} = \widetilde{M}/\tilde{\kappa}(\widetilde{M})$. □

2.3. Note. Since $G(\kappa(M)) = \pi(\kappa\widetilde{(M)})$ it follows that $G(\kappa(M))$ is $\mathcal{L}(\overline{\kappa})$-torsion, hence for a saturated $\mathcal{L}(\kappa)$ we have that $G(\kappa(M)) \subseteq \overline{\kappa}G(M)$. In the sequel **we shall restrict attention to the case where** $\overline{\kappa}(G(M)) = 0$. Since $\widetilde{M}/\tilde{\kappa}(\widetilde{M})$ is X-torsion free the filtration defined on $\tilde{\kappa}(\widetilde{M})/(1 - X)\tilde{\kappa}(\widetilde{M})$ by dehomogenization is just the filtration $F\kappa(M)$ induced by FM on $\kappa(M)$, consequently if $\overline{\kappa}G(M) = 0$ then $G(\kappa(M)) = 0$ and this in turn yields that $\kappa(M) = 0$ because FM is separated and so it also follows that $\tilde{\kappa}(\widetilde{M}) = 0$ in view of Lemma 2.2..

2.4. Proposition. Let R be a Zariskian filtered ring with filtration FR and $\mathcal{L}(\kappa)$ a saturated Gabriel filter such that $G(R)$ is $\mathcal{L}(\overline{\kappa})$-torsionfree, then $Q_\kappa(R)$ has a (essentially unique) filtration $FQ_\kappa(R)$ extending FR. If M has a good filtration FM and $G(M)$ is $\mathcal{L}(\overline{\kappa})$-torsionfree then $Q_\kappa(M)$ has a (essentially unique) filtration $FQ_\kappa(M)$ extending FM.

Proof. Define $FQ_\kappa(R)$ by putting $F_r Q_\kappa = \{x \in Q_\kappa\}$, there is $I \in \mathcal{L}(\kappa)$ such that $F_R Ix \subset R_{n+r}$ for all $n\}$. It is not hard to check that $FQ_\kappa(R)$ is indeed a filtration. Now if $z \in$

$F_r Q_\kappa(R) \cap R$ then $(F_n I)z \subseteq F_{n+r}R$ for all $n \in \mathbb{Z}$ with $Iz \subseteq R$, some $I \in \mathcal{L}(\kappa)$. If $z \notin F_r R$ say $z \in F_{r+p}R - F_{r+p-1}R$ for $p > 0$, but from $(F_n I)z \subseteq F_{r+n}R$ it then follows that $G(I)_n \overline{z} = 0$, with $\overline{z} \in G(R)_{r+p}$, for all $n \in \mathbb{Z}$, hence $G(I)\overline{z} = 0$ or $\overline{z} \in \overline{\kappa}(G(R)) = 0$, a contradiction. It follows that $F_r Q_\kappa(R) \cap R = F_r R$ and so $FQ_\kappa(R)$ extends FR.

Consider another filtration $F'Q_\kappa(R)$ extending FR to $Q_\kappa(R)$, then for $\omega \in F'_r Q_\kappa(R)$ we have for some $I \in \mathcal{L}(\kappa)$ that $F_n I y \subseteq F_{n+r}R$ for all $n \in \mathbb{Z}$ and consequently $y \in F_{r'}Q_\kappa(R)$ with $r' \leq r$. Exchanging the role of F and F' in the foregoing, we deduce from $y \in F_{r'}Q_\kappa(R)$ that $y \in F'_r Q_\kappa(R)$ with $r \leq r'$ and thus $r = r'$ and $F'Q_\kappa(R) = FQ_\kappa(R)$. The statements concerning M and FM may be proved in a completely similar way. $\square$

The foregoing proposition allows to consider $\widetilde{Q_\kappa(M)}$ with respect to the localized filtration $FQ_\kappa(M)$ stemming from the good filtration FM on M. Note that $Q_\kappa(M)$ is in a natural way a filtered $Q_\kappa(R)$-module when we consider the localized filtrations on both objects, but $FQ_\kappa(M)$ need not be a good filtration even if FM is good! Before comparing $Q^g_{\overline{\kappa}}(\widetilde{M})$ and $\widetilde{Q_\kappa(M)}$ we need a lemma:

2.5. Lemma. With assumptions as in the proposition above we have:

(1). $Q^g_{\overline{\kappa}}(\widetilde{M})$ is X-torsionfree;

(2). $Q^g_{\overline{\kappa}}(\widetilde{M})/\widetilde{M}$ is X-torsionfree, hence $j_\kappa : M \to Q_\kappa(M)$ is strict.

Proof. (1). Suppose $z \in Q^g_{\overline{\kappa}}(\widetilde{M})$ is homogeneous and such that $Xz = 0$, then for some $\widetilde{J} \in \mathcal{L}^g(\widetilde{\kappa})$ we have $\widetilde{J}z \subseteq \widetilde{M}$ and $X\widetilde{J}z = 0$, then $\widetilde{J}z = 0$ because $\widetilde{M}$ is X-torsionfree. Moreover $\widetilde{J}z = 0$ contradicts the fact that $Q^g_{\overline{\kappa}}(\widetilde{M})$ is $\mathcal{L}^g(\widetilde{\kappa})$-torsionfree by definition, hence the claim of (1) must hold.

(2). Suppose that $a \in Q^g_{\overline{\kappa}}(\widetilde{M})$ is such that $Xa \in \widetilde{M}$. For some $\widetilde{TI} \in \mathcal{L}^g(\widetilde{\kappa})$ we have $\widetilde{I}a \subseteq \widetilde{M}$. If $\widetilde{I}a \not\subseteq X\widetilde{M}$ then $0 = \Pi(\widetilde{I}Xa) = \pi(\widetilde{I})\Pi(Xa)$ with $\Pi(Xa) \neq 0$ in $G(M)$, where $\Pi: \widetilde{M} \to G(M)$ is the canonical epimorphism. Since $\overline{\kappa}G(M) = 0$ and $\pi(\widetilde{I}) \in \mathcal{L}(\overline{\kappa})$ we contradict $\Pi(Xa) \neq 0$, hence $Xa \in X\widetilde{M}$. From (1) we know that $Q^g_{\overline{\kappa}}(\widetilde{M})$ is X-torsionfree and it follows that $Xa \in X\widetilde{M}$ entails $a \in \widetilde{M}$. $\square$

2.6. Proposition. With notations and conventions as before,

(1). $Q^g_{\overline{\kappa}}(\widetilde{M}) = \widetilde{Q_\kappa(M)}$;

(2). $Q_\kappa(M) = Q^g_{\overline{\kappa}}(\widetilde{M})/(1 - X)Q^g_{\overline{\kappa}}(\widetilde{M})$;

(3). $G(Q_\kappa(M)) = \widetilde{G}(Q^g_{\overline{\kappa}}(\widetilde{M})) \subseteq Q^g_{\overline{\kappa}}(G(M))$,

Proof. (1). Since $Q_\kappa(M)/M$ is $\mathcal{L}(\kappa)$-torsion it follows that $\widetilde{Q_\kappa(M)}/\widetilde{M}$ is $\mathcal{L}(\tilde{\kappa})$-torsion and so $\widetilde{Q_\kappa(M)} \subseteq Q_{\tilde{\kappa}}^g(\widetilde{M})$. Conversely, since $Q_{\tilde{\kappa}}^g(\widetilde{M})$ is X-torsionfree we may write it as $\widetilde{N}$ for some $N \supset M$ in R-filt and as $\widetilde{M} \hookrightarrow Q_{\tilde{\kappa}}^g(\widetilde{M})$ is exact in $\mathcal{F}_X$ we even have $FN \cap M = FM$. Since $\widetilde{N}/\widetilde{M}$ is $\mathcal{L}^g(\tilde{\kappa})$-torsion it follows that $N/M = (\widetilde{N}/\widetilde{M})/(1-X)(\widetilde{N}/\widetilde{M})$ is κ-torsion and hence $N \subseteq Q_\kappa(M)$. Then $\widetilde{N} \subseteq \widetilde{Q_\kappa(M)}$; indeed since FN induces FM it is clear that $z \in F_r N$ will be in $F_r Q_\kappa(M)$ by definition of the latter (for some $I \in \mathcal{L}(\kappa)$, $F_n I z \subseteq F_{n+r} M$ for all $n \in \mathbb{Z}$) and therefore the inclusion $N \subseteq Q_\kappa(M)$ is a filtered morphism. But $\widetilde{N} = Q_{\tilde{\kappa}}^g(\widetilde{M})$ yields $Q_{\tilde{\kappa}}^g(\widetilde{M}) = \widetilde{Q_\kappa(M)}$.

(2). This is just a restatement of (1).

(3). Exactness of G on the strict exact sequence

$$0 \longrightarrow M \longrightarrow Q_\kappa(M) \longrightarrow Q_\kappa(M)/M \longrightarrow 0$$

yields that $G(Q_\kappa(M))/G(M)$ is $\tilde{\kappa}$-torsion and hence $\bar{\kappa}$-torsion, consequently $G(Q_\kappa(M)) \subseteq Q_{\bar{\kappa}}^g(G(M))$. □

The fact that $G(Q_\kappa(M))$ may be different from $Q_{\bar{\kappa}}^g(G(M))$ is somewhat disturbing, so we look for cases where this flaw can be avoided and we shall show that in practice, that is to say when dealing with the usual good examples of Zariskian filtered rings, we have a sufficient set of nice $\mathcal{L}(\tilde{\kappa})$. Recall that one of the many equivalent properties for $\mathcal{L}(\tilde{\kappa})$ to be perfect is given interms of the localization functor $Q_{\tilde{\kappa}}$ on $\tilde{R}$-mod as follows:
$Q_{\tilde{\kappa}}$ commutes with direct summs and is an exact functor, cf [Gab] or [Gol].
Note that this is equivalent to similar statements for $Q_{\tilde{\kappa}}^g$ on $\tilde{R}$-gr.

2.7. Proposition. If R and M are as before and $\mathcal{L}(\tilde{\kappa})$ is such that it is saturated and perfect, then
(1). $\widetilde{G}(Q_{\tilde{\kappa}}^g(\tilde{R})) = G(Q_\kappa(R)) = Q_{\bar{\kappa}}(G(R))$;
(2). $\widetilde{G}(Q_{\tilde{\kappa}}^g(\widetilde{M})) = G(Q_\kappa(M)) = Q_{\bar{\kappa}}(G(M))$.

Proof. By the exactness of $Q_{\tilde{\kappa}}^g$ we easily obtain:

$$Q_{\tilde{\kappa}}^g(\widetilde{M})/X Q_{\tilde{\kappa}}^g(\widetilde{M}) = Q_{\tilde{\kappa}}^g(\widetilde{M}/X\widetilde{M}) = Q_{\tilde{\kappa}}^g(G(M)).$$

But the action of an $\tilde{I} \in \mathcal{L}^g(\tilde{\kappa})$ on $G(M)$ is exactly the action of $\bar{I} = \pi(\tilde{I})$ on $G(M)$ hence it is easily seen (and well-known) that $Q_{\bar{\kappa}}^g(G(M)) = Q_{\tilde{\kappa}}^g(G(M))$, or (2) follows and (1) follows from (2) by putting $M = R$. □

2.8. Corollary. In the situation of the proposition $\mathcal{L}(\overline{\kappa})$ is perfect too. Moreover if we write $\mathcal{L}(\widetilde{\kappa}(n))$ for the Gabriel filter induced on $\widetilde{R}/X^n\widetilde{R}$ then for every n we have:

$$Q^g_{\widetilde{\kappa}(n)}(\widetilde{M}/X^n\widetilde{M}) = Q^g_{\widetilde{\kappa}}(\widetilde{M}/X^n\widetilde{M})$$

and $\widetilde{\kappa}(n)$ is perfect too.

Proof. The first part follows from the exactness properties of $\widetilde{G}$ and $Q^g_{\widetilde{\kappa}}$; the second part idem up tp replacing the functor $\widetilde{G}$ by the functor $\widetilde{R}/X^n\widetilde{R}\otimes_{\widetilde{R}}-$ that enjoys similar exactness properties.   □

2.9. Example. Let R be a Zariskian filtered ring with filtration FR such that $G(R)$ is a commutative Noetherian domain. For every homogeneous element $f \in G(R)$ let $\langle f \rangle$ be the saturated Ore set obtained by putting $\langle f \rangle = \{r \in R, \quad \sigma(r) \in \{1,\, f,\, f^2,\, \cdots\}\}$ and we write $\langle \widetilde{f} \rangle$ for the homogeneous Ore set defined in $\widetilde{R}$ by homogenizing $\langle f \rangle$ in the usual way. Now $\mathcal{L}^g(\widetilde{\kappa})$ associated to $\langle \widetilde{f} \rangle$ consists of the (graded) left ideals of $\widetilde{R}$ that intersect $\langle \widetilde{f} \rangle$ properly. It is easily verified that $\mathcal{L}^g(\widetilde{\kappa})$ is saturated in the sense of the definition given before and that $\mathcal{L}^g(\overline{\kappa})$ is just consisting of the (graded) left ideals of $G(R)$ intersecting $\{1,\, f,\, f^2,\, \cdots\}$ non-trivially, $\mathcal{L}(\kappa)$ consisting of the left ideals of R intersecting $\langle f \rangle$ non-trivially. Hence $\mathcal{L}(\widetilde{\kappa})$, $\mathcal{L}(\kappa)$, $\mathcal{L}(\overline{\kappa})$ are all perfect.

In the sequel we assume that R is a Zariskian filtered ring with filtration FR and that $\mathcal{L}(\widetilde{\kappa})$ is perfect and saturated. As before we write $\mathcal{L}(\widetilde{\kappa}(n))$ for the Gabriel filter defined on $\widetilde{R}/X^n\widetilde{R}$ and we observe that $\mathcal{L}(\widetilde{\kappa}(n))$ has a filterbasis consisting of left ideals of the form $\widetilde{I}/X^n\widetilde{I}$ with $\widetilde{I} \in \mathcal{L}(\widetilde{\kappa})$, because $\mathcal{L}(\widetilde{\kappa})$ is saturated. We write Q^g_n for the graded localization functor associated to $\mathcal{L}(\widetilde{\kappa}(n))$ and we know that $\mathcal{L}(\widetilde{\kappa}(n))$ is again perfect. The homogenized microlocalization at $\mathcal{L}(\widetilde{\kappa})$ is now defined on the Rees level by:

$$Q^\mu_{\widetilde{\kappa}}(\widetilde{M}) = \varprojlim_n{}^{g} Q^g_n(\widetilde{M}/X^n\widetilde{M}).$$

2.10. Lemma. With notation as above, $Q^\mu_{\widetilde{\kappa}}(\widetilde{M})$ is X-torsionfree.

Proof. Suppose $a \in Q^\mu_{\widetilde{\kappa}}(\widetilde{M})$ is such that $Xa = 0$ and write $a_{(n)} \in Q^g_n(\widetilde{M}/X^n\widetilde{M})$ representing a " at level n " in the inverse limit. For some $n \in I\!N$ we have $Xa_{(n)} = 0$ but $a_{(n)} \neq 0$. For some $\widetilde{I} \in \mathcal{L}(\widetilde{\kappa})$ we have $\widetilde{I}a_{(n)} \neq 0$ in $j_{\widetilde{\kappa}(n)}(\widetilde{M}/X^n\widetilde{M})$ but $X\widetilde{I}a_{(n)} = 0$. For $i \in \widetilde{I}$ let

$b_{(n)}(i) \in \widetilde{M}/X^n\widetilde{M}$ represent $ia_{(n)}$. Since $Xb_{(n)}i$ maps to zero in $j_{\widetilde{\kappa}(n)}(\widetilde{M}/X^n\widetilde{M})$, where $j_{\widetilde{\kappa}(n)}$ stands for the canonical morphism $\widetilde{M}/X^n\widetilde{M} \rightarrow Q_n^g(\widetilde{M}/X^n\widetilde{M})$, it follows that $Xb_{(n)}(i) \in \widetilde{\kappa}(n)(\widetilde{M}/X^n\widetilde{M})$ and therefore we may find $J \in \mathcal{L}^g(\widetilde{\kappa})$ such that $\widetilde{J}Xb_{(n)}(i) = 0$ in $\widetilde{M}/X^n\widetilde{M}$. It follows that $\widetilde{J}b_{(n)}(i) \in X^{n-1}\widetilde{M}/X^n\widetilde{M}$ or $ia_{(n)} \in Q_n^g(X^{n-1}\widetilde{M}/X^n\widetilde{M})$ for all $i \in \widetilde{I}$. Hence we arrive at $\widetilde{I}a_{(n)} \subseteq Q_n^g(X^{n-1}\widetilde{M}/X^n\widetilde{M})$ and also $a_{(n)} \in Q_n^g(X^{n-1}\widetilde{M}/X^n\widetilde{M})$. However, this means that $a_{(n-1)} = 0$. Since we may start the argument at any m larger than n it follows that $a_{(n)} = 0$ for all n and hence $a = 0$ as desired. □

2.11. Lemma. With assumptions as before and assuming that $G(M)$ is $\widetilde{\kappa}$-torsionfree we have that $Q_{\kappa}^{\mu}(\widetilde{M})/\widetilde{M}$ is X-torsionfree.

Proof. Take $a \in Q_{\kappa}^{\mu}(\widetilde{M})$ such that $Xa \in \widetilde{M}$ and represent a by $a_n \in Q_n^g(\widetilde{M}/X^n\widetilde{M})$ then Xa maps to $Xa_n \in \overline{\widetilde{M}} = \widetilde{M}/X^n\widetilde{M}$. For some $\widetilde{I} \in \mathcal{L}^g(\widetilde{\kappa})$ we have $\widetilde{I}a_n \subseteq \overline{\widetilde{M}}$ and $X\widetilde{I}a_n \subseteq X\overline{\widetilde{M}}$. Calculating modulo X yields $\pi(\widetilde{I})\overline{\pi}_n(Xa_n) = 0$ where $\pi_n\colon \overline{\widetilde{M}}/X^n\widetilde{M} \rightarrow \widetilde{M}/X\widetilde{M}$ and $\overline{\pi}_n\colon \overline{\widetilde{M}} \rightarrow G(M)$ are the canonical maps. Since $\pi(\widetilde{I}) \in \mathcal{L}^g(\overline{\kappa})$ it follows that $\overline{\pi}_n(Xa_n) = 0$ or $Xa_n \in X\overline{\widetilde{M}}$. Therefore we either have that $a_n \in \overline{\widetilde{M}}$ or $Xa_n = 0$. The latter case means that $a_n \in X^{n-1}(\overline{\widetilde{M}})$ or $a_{n-1} = 0$. If for some n we do have that $a_n \notin \overline{\widetilde{M}}$ then this is valid for all larger n too and so the chain $(\cdots, a_n, \cdots)$, representing a in the limit, has a tail of zeros on the left or else all $a_n \in \overline{\widetilde{M}} = \widetilde{M}/X^n\widetilde{M}$, that is $a \in (\widetilde{M})^{\wedge x}$ (the $X\widetilde{R}$-adic completion of $\widetilde{M}$). Since $\widetilde{M}$ is obviously $X\widetilde{R}$-closed in the $X\widetilde{R}$-adic completion $(\widetilde{M})^{\wedge x}$ it follows that $a \in \widetilde{M}$. □

In view of Lemma 2.10. we obtain the filtered module $Q_{\kappa}^{\mu}(M) = Q_{\kappa}^{\mu}(\widetilde{M})/(1-X)Q_{\kappa}^{\mu}(\widetilde{M})$ with $FQ_{\kappa}^{\mu}(M)$ being defined by the gradation of $Q_{\kappa}^{\mu}(\widetilde{M})$ in the usual (dehomogenization) way. In view of lemma 2.11. the canonical dehomogenized filtered morphism $j_{\kappa}(M)\colon M \rightarrow Q_{\kappa}^{\mu}(M)$, corresponding to $\widetilde{M} \rightarrow Q_{\kappa}^{\mu}(\widetilde{M})$, is a strict filtered morphism. If FM is separated, i.e., if FM is good, then $FQ_{\kappa}^{\mu}(M)$ is separated (but not good if FM is good in general, although it follows later from our standing assumption that $\mathcal{L}(\widetilde{\kappa})$ is saturated and perfect).

2.12. Lemma. If R and M, $\mathcal{L}(\widetilde{\kappa})$ are as before then $Q_{\kappa}^{\mu}(M)$ is complete with respect to its filtration topology.

Proof. It is not hard to check that $Q_{\kappa}^{\mu}(\widetilde{M})$ is $X\widetilde{R}$-adically complete by using the exactness of the functor Q_n^g. □

For microlocalizations $Q_\kappa^\mu(R)$ we may derive a universal property generalizing the situation of localization at Ore sets.

2.13. Proposition. Let us write $j_\kappa^\mu\colon R \to Q_\kappa^\mu(R)$ for the canonical microlocalization morphism. Given a complete filtered ring B with filtration FB and a filtered ring homomorphism $h\colon R \to B$ such that $Bh(I) = B$ and $G(B)G(I) = G(B)$ for every $I \in \mathcal{L}(\kappa)$ then there exists a unique filtered ring homomorphism $g\colon Q_\kappa^\mu(R) \to B$ such that $h = g \circ j_\kappa^\mu$.

Proof. Since h is a filtered morphism it yields a graded morphism (of degree zero) $\tilde{h}\colon \tilde{R} \to \tilde{B}$. From $h(1_R) = 1_B$ it follows that $\tilde{h}(X) = X_B$ and $\tilde{h}$ is a graded ring homomorphism. From $\tilde{h}(X^n\tilde{R}) = X_B^n\tilde{h}(\tilde{R})$ it follows that $\tilde{h}$ gives rise to the composition of graded morphisms, $\tilde{h}_n\colon \tilde{R}/X^n\tilde{R} \to \tilde{h}(\tilde{R})/X_B^n\tilde{h}(\tilde{R}) \to \tilde{B}/X_B^n\tilde{B}$. Since $Bh(I) = B$ for $I \in \mathcal{L}(\kappa)$ it is clear that $B = Q_\kappa(B)$. Localizing $\tilde{h}_n$ at $\tilde{\kappa}(n)$ we obtain a graded ring homomorphism

$$Q_{\tilde{\kappa}(n)}^g(\tilde{h}_n)\colon \quad Q_{\tilde{\kappa}(n)}^g(\tilde{R}/X^n\tilde{R}) \longrightarrow Q_{\tilde{\kappa}(n)}^g(\tilde{h}(\tilde{R})/X_B^n\tilde{h}(\tilde{R})) \longrightarrow \tilde{B}/X_B^n\tilde{B}$$

where $\tilde{B}/X_B^n\tilde{B} = Q_{\tilde{\kappa}(n)}^g(\tilde{B}/X_B^n\tilde{B})$ by the exactness of $Q_{\tilde{\kappa}(n)}^g$ and the assumptions on B. Note that the condition $G(B)\pi(\tilde{I}) = G(B)$ is necessary here (even in case one inverts an Ore set S it is necessary to specify in the universal property that $h(s)^{-1} \in B$ has the correct order in the sense that if $\deg\sigma(s) = n$ then $\deg\sigma(h(s)^{-1}) = -n$!) it is in fact this condition that allows to conclude that $\tilde{B}\tilde{I} = \tilde{B}$ for $\tilde{I} \in \mathcal{L}(\tilde{\kappa})$. Now taking inverse limits in the graded sense yields a graded morphism $\tilde{g}\colon Q_\kappa^\mu(\tilde{R})\tilde{B} = \varprojlim_n{}^g \tilde{B}/X_B^n\tilde{B}$ (the latter equality holds because B is complete and thus $\tilde{B}$ is $X_B\tilde{B}$-adically complete). From

$$\tilde{R} \xrightarrow{\ j_\kappa^\mu\ } Q_\kappa^\mu(\tilde{R}) \longrightarrow \tilde{B}$$

we derive the desired morphism $g\colon Q_\kappa^\mu(R) \to B$ factorizing h as claimed. $\qquad\square$

2.14. Note. From $Bh(I) = B$ it follows that $N = \tilde{B}/\tilde{B}\widetilde{h(I)}$ is X_B-torsion while $G(B)\pi(\widetilde{h(I)}) = G(B)$ yields that $N/X_B N = 0$ hence $N = X_B N = \cdots = X_B^t N$ for any t but that contradicts the fact that N is X_B-torsion unless $N = 0$.

All properties of microlocalization at Ore sets only depending on the exactness of the functor Q_κ^g (hence also $Q_{\tilde{\kappa}}^g$ is exact) now carry over to the case we are considering without real changes.

2.15. Theorem. With assumptions and notations as before, we have

(1). $Q^\mu_\kappa(\tilde{R})$ is a flat right $\tilde{R}$-module;

(2). If FM is a good filtration on $M \in R$-filt then $Q^\mu_\kappa(\tilde{M}) = Q^\mu_\kappa(\tilde{R}) \otimes_{\tilde{R}} \tilde{M}$.

Proof. Along the lines of Theorem 3.19. in [AVVo]. $\qquad\square$

2.16. Corollary. (1). The functor $Q^\mu_\kappa(R) \otimes_R -$ preserves strict filtered maps and it is exact on R-modules.

(2). If $\tilde{M}$ is finitely generated (i.e., FM is good) then $Q^\mu_\kappa(R) \otimes_R M \cong Q^\mu_\kappa(M)$ as filtered R-modules.

Proof. Modify the proof of Corollary 3.20. in [AVVo]. $\qquad\square$

2.17. Theorem. With assumptions and notations as before, we have $G(Q^\mu_\kappa(M)) = Q^g_{\tilde{\kappa}}(G(M))$ for $M \in R$-filt such that $\overline{\kappa}(G(M)) = 0$.

Proof. We have

$$Q^\mu_\kappa(\tilde{M})/XQ^\mu_\kappa(\tilde{M}) = \varprojlim_n Q^g_{\tilde{\kappa}(n)}(\tilde{M}/X^n\tilde{M}) / \varprojlim_n Q^g_{\tilde{\kappa}(n)}(X\tilde{M}/X^n\tilde{M})$$

The maps in the inverse system $\{Q^g_{\tilde{\kappa}(n)}(X\tilde{M}/X^n\tilde{M}),\ n\}$ are all surjective (exactness of $Q^g_{\tilde{\kappa}(n)}$) and so we may calculate the latter quotient term by term and then take $\varprojlim^g$, hence

$$\begin{aligned}
Q^\mu_\kappa(\tilde{M}/XQ^\mu_\kappa(\tilde{M})) &= \varprojlim_n[Q^g_{\tilde{\kappa}(n)}(\tilde{M}/X^n\tilde{M})/Q^g_{\tilde{\kappa}(n)}(X\tilde{M}/X^n\tilde{M})] \\
&= \varprojlim_n Q^g_{\tilde{\kappa}(n)}(\tilde{M}/X\tilde{M}) \\
&= Q^g_{\tilde{\kappa}(n)}(G(M)) \\
&= Q^g_{\tilde{\kappa}}(G(M))
\end{aligned}$$

where the latter equality follows because $\mathcal{L}^g(\tilde{\kappa}(n))$ induces $\mathcal{L}^g(\overline{\kappa})$ on $G(R)$-gr for every $n \in$ $I\!N$. $\qquad\square$

2.18. Remark. Since $Q^g_{\tilde{\kappa}(n)}(\tilde{M}/X^n\tilde{M})$ is a graded $Q^g_{\tilde{\kappa}(n)}(\tilde{R}/X^n\tilde{R})$-module for every n, it follows that $Q^\mu_\kappa(M)$ is a $Q^\mu_\kappa(R)$-module and so $Q^\mu_\kappa(-)$ is a functor R-filr $\to Q^\mu_\kappa(R)$-filt. Along the lines of Corollary 3.16. (2) of [AVVo] it also follows that equivalent filtrations FM and $F'M$ yield equivalent filtrations $FQ^\mu_\kappa(M)$ and $F'Q^\mu_\kappa(M)$.

2.19. Lemma. With notations and assumptions as before we have: $X \in J^g((Q_\kappa(R)^\wedge)^\sim) = J^g((Q^g_{\tilde{\kappa}}(\tilde{R}))^{\wedge_X})$, where $\wedge_X$ stands for the completion in the $X\tilde{R}$-adic topology.

Proof. We know that $Q^g_{\tilde{\kappa}}(\tilde{R}) = \widetilde{Q_\kappa(R)}$ and by the translation from FR to the $X\tilde{R}$-adic topology on $\tilde{R}$ we know that $(Q^g_{\tilde{\kappa}}(\tilde{R}))^{\wedge_X} = (\widehat{Q_\kappa(R)})^\sim$. So we arrive at the equality $(\widehat{Q_\kappa(R)})^\sim = Q^g_{\tilde{\kappa}}(\tilde{R})^{\wedge_X}$ and therefore the respective graded Jacobson radicals coincide. Put $B = Q^g_{\tilde{\kappa}}(\tilde{R})^{\wedge_X}$ and take $b \in B_{-1}$. Since B is complete in the $X\tilde{R}$-topology there exists an element $1 + bX + b^2X^2 + \cdots$ in B and therefore $1 - bX$ has an inverse $1 + bX + b^2X^2 + \cdots$ for every $b \in B_{-1}$ or $X \in J^g(B)$. $\square$

2.20. Proposition. Let $\bar{\kappa}$ be saturated such that $\tilde{\kappa}$ is perfect as before, then for every $\tilde{I} \in \mathcal{L}(\tilde{\kappa})$ we have $(Q^g_{\tilde{\kappa}}(\tilde{R}))^{\wedge_X} \cdot \tilde{I} = Q^g_{\tilde{\kappa}}(\tilde{R})^{\wedge_X} \cdot \tilde{I} = B$.

Proof. Since $\tilde{\kappa}$ has finite type we may assume that $\tilde{I}$ is finitely generated. We have $B/XB = G(Q^g_{\tilde{\kappa}}(\tilde{R})^{\sim_X}) = G(Q^g_{\tilde{\kappa}}(\tilde{R})) = Q^g_{\bar{\kappa}}(G(R))$ and since $\bar{\kappa}$ is perfect too we have $Q^g_{\bar{\kappa}}(G(R))\pi(\tilde{I}) = Q^g_{\bar{\kappa}}(G(R))$. Consequently $B \cdot \tilde{I} + BX = B$ and the foregoing lemma allows to use the graded Nakayama lemma in order to derive $B\tilde{I} = B$. $\square$

2.21. Theorem. Let $M \in R$-filt be such that $\bar{\kappa}G(M) = 0$ where R and $\bar{\kappa}$ are as before, then $Q^\mu_{\tilde{\kappa}}(M) \cong \widehat{Q_\kappa(M)}$.

Proof. It will be sufficient to establish that $Q^{\widehat{\kappa}}_{\tilde{\kappa}}(M) \cong (\widehat{Q_\kappa(M)})^\sim$, or $Q^\mu_{\tilde{\kappa}}(\tilde{M}) \cong \varprojlim_n Q^g_{\tilde{\kappa}}(\tilde{M})/X^n Q^g_{\tilde{\kappa}}(\tilde{M}) = \varprojlim_n Q^g_{\tilde{\kappa}}(\tilde{M}/X^n\tilde{M})$, where the latter equality follows from the exactness of $Q^g_{\tilde{\kappa}}$. By the saturatedness condition the $\tilde{I}/X^n\tilde{I}$ with $\tilde{I} \in \mathcal{L}(\tilde{\kappa})$ form a filterbasis for $\mathcal{L}(\tilde{\kappa}(n))$ (note that $\tilde{I}/X^n\tilde{I}$ maps to $\pi(\tilde{I}) \in \mathcal{L}(\bar{\kappa})$ under the canonical $\tilde{R}/X^n\tilde{R} \to \tilde{R}/X\tilde{R} = G(R)$), hence $Q^g_{\tilde{\kappa}}$ and $Q^g_{\tilde{\kappa}(n)}$ coincide on $\tilde{R}/X^n\tilde{R}$-modules and we arrive at

$$
\begin{aligned}
Q^\mu_{\tilde{\kappa}}(\tilde{M}) &= \varprojlim_n Q^g_{\tilde{\kappa}(n)}(\tilde{M}/X^n\tilde{M}) \\
&\cong \varprojlim_n Q^g_{\tilde{\kappa}}(\tilde{M}/X^n\tilde{M}) \\
&\cong \varprojlim_n Q^g_{\tilde{\kappa}}(\tilde{M})/(X^n Q^g_{\tilde{\kappa}}(\tilde{M})) \\
&= (Q^g_{\tilde{\kappa}}(\tilde{M}))^{\wedge_X}
\end{aligned}
$$

2.22. Remark. (1). Theorem 2.21. also follows from Proposition 2.20. because $B \cdot \tilde{I} = B$ and clearly $G(B)\pi(\tilde{I}) = G(B)$ allows to apply the universal property and derive the isomorphism from the triangle

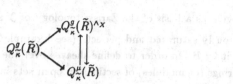

(2). Theorem 2.21. extends the result that $Q_S^\mu(R) = \widehat{S_{\text{sat}}^{-1}} R$ where S is a multiplicative set of R such that $\sigma(S)$ is an Ore set in $G(R)$.

(3). From Lemma 2.19. it is clear that $FQ_\kappa^\mu(R)$ is Zariskian when FR is; indeed, the facts that $Q_\kappa^\mu(R)$ is complete and $G(Q_\kappa^\mu(R)) = Q_\kappa^g(G(R))$ is Noetherian yield that $FQ_\kappa^\mu(R)$ is Zariskian.

In the sequel we consider a Zariskian filtered ring R with filtration FR such that $G(R)$ is a commutative domain that is positively graded (moreover we usually assume that $G(R)$ is generated by $G(R)_1$ over $G(R)_0$). We write $\mathcal{X} = \text{Spect}G(R)$, $\text{Spec}^g(G(R))$ is the set of graded prime ideal of $G(R)$ with the topology induced by the Zariski topology of $\mathcal{X}$, and put $\mathcal{Y} = \text{Proj}G(R) = \{P \in \text{Spec}^g(G(R)), P \not\supset G(R)_+\}$ where $G(R)_+ = \oplus_{n>0}G(R)_n$. It is clear that $\text{Proj}G(R)$ has a basis for its Zariski topology given by the basic open sets $\mathcal{Y}_{\overline{f}}$ for $\overline{f}$ homogeneous in $G(R)$. We write $S(\overline{f}) = \{1, \overline{f}, \overline{f}^2, \cdots\}$ in $G(R)$ and put $\pi^1 S(\overline{f}) = S(\tilde{f})$ in $\tilde{R}$, where we may assume that $\tilde{f}$ represents $\overline{f}$. The image of $S(\tilde{f})$ in R will be denoted by $\langle f \rangle$ where f is the image of $\tilde{f}$, i.e., $\sigma(f) = \overline{f}$. We know, by definition of $\langle f \rangle$ that this is a saturated Ore set of R (because $\sigma(\langle f \rangle) = S(\overline{f})$ is an Ore set of $G(R)$ and $\langle f \rangle$ is obviously saturated) in fact it is the saturated Ore set generated by f in R. Note also that $\langle f \rangle$ is both a left and right Ore set because $S(\overline{f})$ is. We know that $\mathcal{Y}_{\overline{f}} \cap \mathcal{Y}_{\overline{g}} = \mathcal{Y}_{\overline{fg}}$. Of course we have that $\langle fg \rangle$ is contained in the Ore set generated by $\{f, g\}$ in R, say $S(f,g) = \langle \langle f \rangle, \langle g \rangle \rangle$ but of course $\langle fg \rangle \neq S(f,g)$ in general.

2.23. Lemma. With notations as above, for all homogeneous $\overline{f}, \overline{g} \in G(R)$ we have that $Q_{S(f,g)}^\mu(R) = Q_{\langle fg \rangle}^\mu(R)$.

Proof. Since $S(f,g) \supset \langle fg \rangle$ the inclusion $Q_{S(f,g)}^\mu(R) \supset Q_{\langle fg \rangle}^\mu(R)$ (as filtered rings) is obvious. However the elements of $\langle fg \rangle$ are invertible in $Q_{\langle fg \rangle}^\mu(R)$ and therefore f as well as g is invertible in $Q_{\langle fg \rangle}^\mu(R)$. Consequently, the Ore set generated by f and g in R, that is $S(f,g)$, consists of invertible elements in $Q_{\langle fg \rangle}^\mu(R)$. By the universal property of microlocalization (note: the degree assumption on degrees of inverses holds automatically here because $G(R)$ is a domain) it follows that the inclusion first obtained must be an equality.   □

The $\mathcal{Y}_{\bar{f}}$ provide us a basis of the Zariski topology of $\mathcal{Y}$ such that each $\mathcal{L}^g(\bar{f})$ associated to $S(\bar{f})$ is obviously saturated and perfect. (see Example 2.9.); we fix this basis throughout and denote it by $\mathcal{B}$. In order to define sheaves over the topological space $\mathcal{Y}$ it is sufficient to give the rings (or modules) of sections on open sets in some basis of the topology; in the sequel we use $\mathcal{B}$ for this purpose.

The **graded structure sheaf** of $\text{Proj}G(R)$ is obtained by taking for the ring of sections over $\mathcal{Y}_{\bar{f}}$ the graded ring $S(\bar{f})^{-1}G(R)$, the sheaf thus obtained is denoted by $\underline{O}_{\mathcal{Y}}^g$. The **structure sheaf** $\underline{O}_{\mathcal{Y}}$ is defined by associating $(S(\bar{f})^{-1}G(R))_0$ to the basic open $\mathcal{Y}_{\bar{f}}$. By using the corresponding microlocalizations $Q_{(f)}^{\mu}(R)$, resp. $Q_{S(\bar{f})}^{\mu}(\tilde{R})$, we define presheaves (cf. [SVo], [RVo2]) over $\mathcal{Y}$ and we denote these by $\underline{O}_{\mathcal{Y}}^{\mu}$, resp. $\tilde{\underline{O}}_{\mathcal{Y}}^{\mu}$.

The stalk of $\underline{O}_{\mathcal{Y}}^g$ at a given $P \in \text{Proj}(G(R))$ is $Q_P^g(G(R))$ and for $\underline{O}_{\mathcal{Y}}$ we obtain the stalk $(Q_P^g(G(R)))_0$. First let us calculate the stalks of the microstructure sheaves $\underline{O}_{\mathcal{Y}}^{\mu}$ and $\tilde{\underline{O}}_{\mathcal{Y}}^{\mu}$.

2.24. Proposition. Consider $P \in \mathcal{Y}$, then

(1). The stalk $\tilde{\underline{O}}_{\mathcal{Y},P}^{\mu}$ equals the microlocalization at the complement $S(P)$ of P after taking the $X\tilde{R}$-adic completion, that is: $(\tilde{\underline{O}}_{\mathcal{Y},P}^{\mu})^{\wedge x} = Q_{\widetilde{S(P)}}^{\mu}(\tilde{R})$.

(2). The stalk of $\underline{O}_{\mathcal{Y}}^{\mu}$ at P completed in the localized filtration with respect to $S(P)$ yield the microlocalizations at $S(P)$, that is: $(\underline{O}_{\mathcal{Y},P}^{\mu})^{\wedge} = Q_{S(P)}^{\mu}(R)$.

Proof. (1). Since $\mathcal{B}$ is a basis we have to calculate the stalk as follows

$$\tilde{\underline{O}}_{\mathcal{Y},P}^{\mu} = \tilde{A} = \varinjlim_{\bar{f}}\{Q_{S(\bar{f})}^{\mu}(\tilde{R}), \quad \bar{f} \text{ homogeneous in } G(R)\}.$$

Because of our standing assumptions all the maps in this directed system are injective because $G(R)$, R and $\tilde{R}$ are domains; so in fact $\varinjlim_{\bar{f}}$ is just the union of the $Q_{S(\bar{f})}^{\mu}(\tilde{R})$ viewed as subrings of the total microlocalization $Q_{\text{cl}}^{\mu}(\tilde{R})$ obtained by microlocalizing at $G(R)^*$. The latter ring is a gr-skewfield equal to $(Q_{\text{cl}}^g(\tilde{R}))^{\wedge x}$ and the skewfield $Q_{\text{cl}}^{\mu}(R)$ is obtained from it by dehomogenization as usual. It is clear that $\mathcal{L}(S(P)) = \cup\{(\bar{f}), P \notin \mathcal{L}(\bar{f})\}$ and since $\mathcal{L}(S(P))$ as well as all $\mathcal{L}(\bar{f})$, $\mathcal{L}(S(\bar{f}))$ etc$\cdots$ are perfect we know that for every $\tilde{I} \in \mathcal{L}(S(\bar{f}))$, $Q_{\widetilde{S(P)}}^{\mu}(\tilde{R})\tilde{I} = Q_{\widetilde{S(P)}}^{\mu}(\tilde{R})$ as well as $Q_P^g(G(R))\pi(\tilde{I}) = Q_P^g(G(R))$. Therefore we certainly have that $\tilde{A} \subseteq Q^\mu_{\widetilde{S(P)}}(\tilde{R})$ (by the universal properties). On the other hand if we look at $\pi(\tilde{I}) \in \mathcal{L}^g(S(P))$ then $\pi(\tilde{I})$ contains an ideal $\bar{J}$ from $\mathcal{L}^g(S(P))$ of the form $\bar{J} = G(R)\bar{f}^n$ for some homogeneous $\bar{f}$ in $G(R)$. With respect to this $\bar{f}$ we obtain $Q^\mu_{S(\bar{f})}(\tilde{R}) = Q^\mu_{S(\bar{f})}(\tilde{R})$ and hence $\tilde{A}\tilde{I} = \tilde{A}$ holds for all $\tilde{I} \in \mathcal{L}(\widetilde{S(P)})$. Clearly we also have $G(A)\pi(\tilde{I}) = G(A)$ where $G(A) = \tilde{A}/X\tilde{A}$. By the universal property the identity on $\tilde{R}$ extends (as an essentially unique way) to an inclusion $\tilde{A} \hookrightarrow Q^\mu_{\widetilde{S(P)}}(\tilde{R})$ (see before) so that $G(A) = Q^g_P(G(R))$ (because $G(A)\pi(\tilde{I}) = G(A)$ for all $\tilde{I} \in \mathcal{L}(\widetilde{S(P)})$ exactly states that all ideals of $\mathcal{L}(G(1))$ extend to $G(A)$ and the perfectness of $\mathcal{L}(G(P))$ then yield the claimed equality). Since $Q^\mu_{\widetilde{S(P)}}(\tilde{R})$ is $X\tilde{R}$-adically complete it follows easily that $(\tilde{A})^{\wedge x} \subseteq Q^\mu_{\widetilde{S(P)}}(\tilde{R})$ and the equality follows again by the universal property because $\widetilde{S(P)}$ is invertible in $(\tilde{A})^{\wedge x}$ and the latter is now complete by definition.

(2). The statement follows immediately from (1) by dehomogenization (i.e., by the category equivalence $\widetilde{(-)}$: R-filt $\leftrightarrow \mathcal{F}_X$). $\qquad\square$

2.25. Proposition. With notations as before, for every $P \in \mathcal{Y}$ the ring $Q^\mu_{\mathcal{Y},P}$ is a Zariskian filtered ring.

Proof. Write $Q^\mu_{\mathcal{Y},P} = T_P$, $\tilde{Q}^\mu_{\mathcal{Y},P} = \tilde{T}_P$ (note that, indeed, $\tilde{T}_P$ is the Rees ring of T_P). Consider $x \in F_{-1}T_P$, $a \in F_0 T_P$. Then for suitable homogeneous $\bar{f} \notin P$, $\bar{g} \notin P$ we have $x \in F_{-1}Q^\mu_{S(f)}(R)$, $a \in F_0 Q^\mu_{S(g)}(R)$ hence $x \in F_{-1}Q^\mu_{S(fg)}(R)$, $a \in F_0 Q^\mu_{S(fg)}(R)$ because $Q^\mu S(f)(R) \to Q^\mu_{S(fg)}(R)$ is a strict filtered morphism (of degree zero) as is rather early seen. We know that $Q^\mu_{S(fg)}(R)$ is Zariskian filtered by the microlocalized filtration (as it is complete), hence $(1 - ax)^{-1} \in F_0 Q^\mu_{S(fg)}(R)$ and thus $(1 - ax)^{-1} \in F_0 T_P$. This proves that $x \in J(F_0 T_P)$, the Jacobson radical condition part of the Zariskian condition. To verify that $\tilde{T}_P$ is Noetherian let $\tilde{L}$ be a graded left ideal of $\tilde{T}_P$. Then $\tilde{G}(\tilde{L}) = \tilde{L}/X\tilde{L}$ is finitely generated because $\tilde{G}(\tilde{T}_P) = \tilde{T}_P/X\tilde{T}_P = Q^g_{S(P)}(G(R))$ is (left) Noetherian. Say $\tilde{G}(\tilde{L}) = G(R)\bar{\lambda}_1 + G(R)\bar{\lambda}_2 + \cdots + G(R)\bar{\lambda}_t$ and let $\tilde{\lambda}_1, \cdots, \tilde{\lambda}_t \in \tilde{L}$ represent the $\bar{\lambda}_1, \cdots, \bar{\lambda}_t$. In fact since t is finite all the $\bar{\lambda}_1, \cdots, \bar{\lambda}_t$ belong in $Q^g_{S(\bar{f})}(G(R))$ for some invertible homogeneous $\bar{f}$ in $G(R)$ and therefore the $\tilde{\lambda}_1, \cdots, \tilde{\lambda}_t$ may be chosen in $Q^\mu_{S(\bar{f})}(\tilde{R}) \cap \tilde{L}$. Since $Q^\mu_{S(\bar{f})}(\tilde{R})$ is Noetherian, $\tilde{L} \cap Q^\mu_{S(\bar{f})}(\tilde{R})$ is finitely generated and hence we may assume that it is generated by the $\tilde{\lambda}_1, \cdots, \tilde{\lambda}_t$ (up to enlarging the original set of generators if necessary).

Putting $\tilde{L}(\tilde{f}) = \tilde{L} \cap Q^\mu_{S(\tilde{f})}(\tilde{R})$ then it is clear that $\tilde{L} = \cup_{\tilde{f}}\{\tilde{L}(\tilde{f}), \tilde{f} \notin P\}$, since for every $\tilde{f}$,

$\tilde{L}(\tilde{f}) \subseteq \sum_{i=1}^t \tilde{T}_P \tilde{\lambda}_i$ it follows that $\tilde{L} = \sum_{i=1}^t \tilde{T}_P \tilde{\lambda}_i$, hence it is finitely generated. This finishes the proof of the fact that $\tilde{T}_P$ is graded Noetherian but then it is Noetherian in the ungraded sense, as is well known. Consequently T_P is a Zariskian filtered ring as claimed.   □

2.26. Theorem. With notation as before, the presheaves $\tilde{\mathcal{O}}^\mu_{\mathcal{Y}}$ and $\mathcal{O}^\mu_{\mathcal{Y}}$ as defined before are in fact sheaves of rings.

Proof. cf. Theorem 4.2. of [SVo]; the proof given there may be simplified somewhat by restricting by immediately to coverrings by opens in the selected basis $\mathcal{B}$ and it just consists in verifying the sheaf axioms.   □

2.27. Corollary. If $M \in R$-filt has a good filtraion FM such that $G(M)$ is torsionfree then $\tilde{\mathcal{O}}^\mu_M$ and $\mathcal{O}^\mu_M$ may be defined in the obvious way these are sheaves of filtered modules such that $(\tilde{\mathcal{O}}^\mu_{M,P})^{\wedge x} = Q^\mu_{\widetilde{S(P)}}(\widetilde{M})$, $(\mathcal{O}^\mu_{M,P})^{\wedge x} = Q^\mu_{S(P)}(M)$.

2.28. Terminology. We may say that $\mathcal{O}^\mu_{\mathcal{Y}}$ is a sheaf of Zariski rings because the rings of sections $\Gamma(\mathcal{Y}_{\tilde{f}}, \mathcal{O}^\mu_{\mathcal{Y}})$ over basic open sets $\mathcal{Y}_{\tilde{f}}$ (as well as the stalks) are Zariski rings. For some detail on sheaf theory, e.g. concerning coherent sheaves, we refer to [Ser2].

Define a **filtered sheaf of rings** $\underline{\mathcal{R}}$ to be a sheaf of rings endowed with a family of subsheaves of additive groups, $\mathcal{F}_n\underline{\mathcal{R}}$, $n \in \mathbb{Z}$, such that $\mathcal{F}_n\underline{\mathcal{R}} \subseteq \mathcal{F}_{n+1}\underline{\mathcal{R}}$ and $\mathcal{F}_n\underline{\mathcal{R}}\mathcal{F}_m\underline{\mathcal{R}} \subseteq \mathcal{F}_{n+m}\underline{\mathcal{R}}$ for $n, m \in \mathbb{Z}$, the unit section $\underline{I}$ is in $\mathcal{F}_0\underline{\mathcal{R}}$ and $\underline{\mathcal{R}} = \cup_{n\in\mathbb{Z}}\mathcal{F}_n\underline{\mathcal{R}}$ where the right hand term stands for the sheaf associated to the presheaf defined by $\mathcal{Y}_{\tilde{f}} \to \cup_{n\in\mathbb{Z}}\mathcal{R}_n(\mathcal{Y}_{\tilde{f}})$, $\tilde{f} \in \mathcal{B}$. It is clear how to define a sheaf of filtered modules over a sheaf of filtered rings and one can introduce good filtrations on sheaves of filtered modules to obtain sheaf-versions of many of the ring theoretical properties that have been obtained. For example, it is not difficult to prove that a sheaf of Zariski rings $\underline{\mathcal{R}}$ such that the associated graded sheaf $\underline{G(\mathcal{R})}$ is coherent must be coherent, moreover if $\underline{\mathcal{M}}$ is a sheaf of filtered $\underline{\mathcal{R}}$-modules which is locally of finite type then $\underline{\mathcal{M}}$ is coherent if and only if $\underline{G(\mathcal{M})}$ is coherent, etc... . We do not go deeper into the theory of coherent filtered sheaves here.

The sheaf $\mathcal{O}^\mu_{\mathcal{Y}}$ is a sheaf of filtered rings in the above sense; let us write $\underline{\mathcal{R}} = \mathcal{O}^\mu_{\mathcal{Y}}$. Then $\widetilde{\underline{\mathcal{R}}} = \tilde{\mathcal{O}}^\mu_{\mathcal{Y}}$ is a graded sheaf of rings and $\underline{G(\mathcal{R})} = \widetilde{\underline{\mathcal{R}}}/X\widetilde{\underline{\mathcal{R}}}$ is also a graded sheaf. Here $\underline{X}$ stands for the global section of $\widetilde{\underline{\mathcal{R}}}$ determined by $X_{Q^\mu_{S(\tilde{f})}(\tilde{R})}$ for each $\tilde{f} \in \mathcal{B}$.

More explicitly $(\mathcal{F}_n\underline{\mathcal{R}})(\mathcal{Y}_{\tilde{f}}) = F_nQ^\mu_{S(f)}(R)$, $\tilde{f} \in \mathcal{B}$, and $\underline{G(\mathcal{R})}$ is the sheaf associated to the presheaf (only!) $\oplus_{n \in \mathbf{Z}} \mathcal{F}_n \mathcal{R}/\mathcal{F}_{n-1} \mathcal{R}$. In this case, where $\mathcal{R} = Q^\mu_\mathcal{Y}$ we obtain that $\underline{G}(\mathcal{R})$ is nothing but the graded structure sheaf $Q^g_\mathcal{Y}$ of $\text{Proj}(G(R))$ mentioned earlier in this section. Similar micro-structure sheaves may be defined over filtered modules but we do not consider that here (cf. [RVo1], [SVo]). In particular $\mathcal{F}_0 \mathcal{R}$ is a sheaf of rings defined by $(\mathcal{F}_0 \mathcal{R})(\mathcal{Y}_{\bar{f}}) = F_0 Q^\mu_{S(f)}(R)$, $\bar{f} \in \mathcal{B}$. This defines a sheaf of rings isomorphic to the sheaf $(\widetilde{\mathcal{R}})_0$ in $\widetilde{\mathcal{R}}$ and the image of $(\widetilde{\mathcal{R}})_0$ in the graded structure sheaf $Q^g_\mathcal{Y}$ of $\text{Proj}(G(R))$ is nothing but the classical structure sheaf $Q_\mathcal{Y}$ of $\text{Proj}(G(R))$.

2.29. Theorem. The presheaf defined over $\mathcal{Y} = \text{Proj}(G(R))$ by associating $F_0 Q^\mu_{S(f)}(R)$ to $\mathcal{Y}_{\bar{f}}$, $\bar{f} \in \mathcal{B}$, is a Noetherian and coherent sheaf of rings. The sheaf of ideals $\mathcal{F}_{-1} \mathcal{R}$ is coherent and $\mathcal{F}_0 \mathcal{R}/\mathcal{F}_{-1} \mathcal{R} = Q_\mathcal{Y}$ is the structure sheaf of $\text{Proj}(G(R))$.

Proof. All claims will follow from the observations preceding the theorem up to the following argument. When we view $\mathcal{F}_0 \mathcal{R}$ with the filtration $\mathcal{F}_-$ given by $\mathcal{F}_{-n} \mathcal{R}$, $(\mathcal{F}_{-n} \mathcal{R})(\mathcal{Y}_{\bar{f}}) = F_{-n} Q^\mu_{S(f)}(R)$ for positive n, then we obtain $\underline{G}_{\mathcal{F}_-}(\mathcal{F}_0 \mathcal{R}) = (Q^g_\mathcal{Y})_- = \oplus_{n \leq 0}(Q^g_\mathcal{Y})_n$ which is known to be Noetherian and coherent.

Since each $Q^g_{S(\bar{f})}(G(R))_-$ is Noetherian, and $\mathcal{F}_-$ is Zariskian on the sheaf of rings $\mathcal{F}_0 \mathcal{R}$ it follows that each $\mathcal{F}_{-n} \mathcal{R}$ is locally of finite type as an $\mathcal{F}_0 \mathcal{R}$-module. Therefore the coherent as well as the Noetherian property of $\underline{G}_{\mathcal{F}_-}(\mathcal{F}_0 \mathcal{R})$ lift to $\mathcal{F}_0 \mathcal{R}$; in a similar way one establishes that $\mathcal{F}_{-1} \mathcal{R}$ is a Noetherian and coherent sheaf of ideals. $\qquad\qquad\square$

2.30. definition. The sheaf $\mathcal{F}_0 \mathcal{R}$ is called the **sheaf of quantum sections** of $Q_\mathcal{Y}$, in other words the quantum sections of $\mathcal{Y} = \text{Proj}(G(R))$ over the basic open $\mathcal{Y}_{\bar{f}} \in \mathcal{B}$ are given as the ring $F_0 Q^\mu_{S(f)}(R)$ equipped with the negative filtration induced by the mcrolocalized filtration.

2.31. Remark. In view of Proposition 2.24. it follows that the stalk $(\mathcal{F}_0 \mathcal{R})_P$, $P \in \mathcal{Y}$, is such that $((\mathcal{F}_0 \mathcal{R})_P)^{\wedge F_-} = F_0 Q^\mu_{S(P)}(R)$ but this does not mean that the " quantum-stalk " $F_0 Q^\mu_{S(P)}(R)$ is obtained as the stalk of the **sheaf-wise** completion of $\mathcal{F}_0 \mathcal{R}$!

2.32. Corollary. If $M \in R$-filt has good filtration FM then we have a Good filtered sheaf of $\mathcal{R}$-modules $\underline{M} = Q^\mu_M$ with $\underline{G}(\underline{M})$ equal to the graded structure sheaf $Q^g_{G(M)}$ Since the latter is coherent and $\underline{M}$ is locally finite by the assumption that FM is good, it follows that $\underline{M}$ is coherent. Consequently we obtain a coherent $\mathcal{F}_0 \underline{M}$ with a coherent $\mathcal{F}_0 \mathcal{R}$-subModule $\mathcal{F}_{-1} \underline{M}$ such that the sheaf $\mathcal{F}_0 \underline{M}/\mathcal{F}_{-1} \underline{M}$ is nothing but $Q_{G(M)}$ the usual projective structure sheaf of $G(M)$ over $\text{Proj}(G(R))$. Again we call $\mathcal{F}_0 \underline{M}$ the sheaf of **quantum-sections of**

$\underline{Q}_{G(M)}$.

Even though we will restrict attention to quantum sections of the ring in the consequent sections it may be worthwhile to point out the good behaviour of quantum sections with respect to strict morphisms.

2.33. Proposition. If $f: M \to N$ is a strict filtered morphism then f induces a morphism of sheaves $\varphi: \mathcal{F}_0 \underline{M} \to \mathcal{F}_0 \underline{N}$ that is also strict in the sense of filtered sheaves. Moreoverif FM is good then $\mathcal{F}_-$ is Good on $\mathcal{F}_0 \underline{M}$.

Proof. cf. Proposition 4.6. of [SVo]; again this proof is somewhat simplified if one restricts from the start to the special basis $\mathcal{B}$ we have selected here. □

§3. Generalized Gauge Algebras

Historically the n-th Weyl algebra $A_n(\mathbb{C})$ has been introduced as the operator algebra generated by the components $\hat{x}_i$ of the position vector and the components $\hat{p}_j = i\hbar y_j$ of the momentum vector of a quantum particle in n-dimensional space. The uncertainty principle expressing that one cannot have simultaneous knowledge of position and momentum of a particle in the quantum case is equivalent to the non-vanishing of the commutator $[\hat{x}_i, \hat{p}_i] = i\hbar$. The associated graded ring $G(A_n(\mathbb{C}))$, that is a polynomial ring $\mathbb{C}[x_1, \cdots, x_n, y_1, \cdots, y_n]$, is the operator algebra of the classical non-quantum situation. So we see that in a sense the quantum case appears as a (filtered) deformation of the classical (graded) case and in fact the deformation from $G(A_n(\mathbb{C}))$ to $A_n(\mathbb{C})$ may be described in terms of the well known Poisson brackets $\{-,-\}$ that can be defined on $G(A_n(\mathbb{C}))$ with respect to the filtration $FA_n(\mathbb{C})$. On the other hand one may view the relations:

$$G(A_n(\mathbb{C})) \cong \widetilde{A_n(\mathbb{C})}/X\widetilde{A_n(\mathbb{C})},$$
$$A_n(\mathbb{C}) \cong \widetilde{A_n(\mathbb{C})}/(1-X)\widetilde{A_n(\mathbb{C})}$$

may also be viewed as expressing this deformation principle. Here the Rees algebra $\widetilde{A_n(\mathbb{C})}$ is the positively graded algebra generated by the degree one element X and x_iX, y_iX with $1 \leq i \leq n$, satisfying the commutation relations: $[x_i, x_j]X^2 = [y_i, y_j]X^2 = 0$, $[x_i, y_j]X^2 = \delta_{ij}X^2$. we introduce new indeterminates $X = X$, $X_i = x_iX$, $y_iX = Y_i$ and rewrite the relations in their homogenized form as: $[X_i, X_j] = [Y_i, Y_j] = 0$ and $[X_i, Y_j] = \delta_{ij}X^2$. This states that we may consider $\widetilde{A_n(\mathbb{C})}$ as a quadratic extension of the enveloping algebra of the Heisenberg Lie algebra and $A_n(\mathbb{C})$ is the filtered dehomogenization of this quadratic extension. With the philosophy in mind that " quantization " may be obtained as a deformation from the graded to the filtered case, and in fact from the graded commutative to the filtered noncommutative situation.

In general, it is clear that the quantum sections introduced in the foregoing section deserve this name because they are in a rather canonical way deformed from the classical sections of $\text{Proj}(G(R))$, considered in case $G(R)$ is a commutative positively graded Noetherian domain. The theory of Zariskian filtrations established in the first three chapters of this book has the advantage that the assumption that $G(R)$ is commutative is nowhere assumed (in the general results). Since the important examples of enveloping algebras of Lie algebras and rings of differential operators have commutative associated graded rings it may be necessary to motivate the consideration of noncommutative associated graded rings in somewhat more detail.

An **n-dimensional quantum space** (or rather its function ring) is an affine positively graded (noncommutative) $\mathbb{C}$-algebra Q, say $Q = \mathbb{C} \oplus Q_1 \oplus \cdots \oplus Q_m \oplus \cdots$, that is a quadratic algebra in the sense of Manin, [Man], that is: Q_1 generates Q as a $\mathbb{C}$-algebra and the defining relations are quadratic, and Q is regular in the sense of Artin, Schelter, [AS], that is, gl.dim$Q$ = GK.dim$Q = n$ and $\text{Ext}^i_Q(\mathbb{C}, Q) = \delta_{in}\mathbb{C}$. In fact more recently it turned out that we may just as well require Auslander regularity as introduced earlier. For $n = 2$ and $n = 3$ a complete classification of quantum spaces has been obtained ([ATV1], [ATV2]). The case $n = 4$ is largely unfinished and for $n > 4$ almost nothing is known.

A **generalized gauge algebra** is a positively filtered algebra $\mathcal{G}$ (necessarily of finite global dimension) such that its associated graded ring $Q = G(\mathcal{G})$ is a quantum space as above. Of course Weyl algebras, enveloping algebras and the rings of differential operators usually considered are gauge algebras of the " almost commutative" type but there are many others. Later we include Witten's gauge algebras as examples, of course without explaining much about the Chern-Simons gauge theory in which these appear.

Another recently booming ingredient of the theory of quantum groups and quantum spaces is the noncommutative geometry. Obviously when using terminology like " spaces " it is awkward that the objects referred to are really just algebras, albeit special algebras that may be thought of as rings of functions on some undefined " noncommutative geometrical object " nobody has ever seen but everybody is invited to believe in. in case $\mathcal{G}$ is such that $Q = G(\mathcal{G})$ is commutative then $\text{Proj}(Q)$ is a nice geometrical space parametrizing some features of $\mathcal{G}$ but in fact one would like to be able to consider $\text{Proj}(\tilde{\mathcal{G}})$ because that would hold all information about $\text{Proj}(G(\mathcal{G}))$, via the canonical embedding $\text{Proj}(G(\mathcal{G})) \hookrightarrow \text{Proj}(\tilde{\mathcal{G}})$ associated to $\tilde{\mathcal{G}} \to G(\mathcal{G}) \to 0$, and an " affine " part corresponding to $\mathcal{G}$ that can be understood from the relation $\tilde{\mathcal{G}}_{(x)} \cong \mathcal{G}[t, t^{-1}]$ because the graded $\tilde{\mathcal{G}}_{(x)}$-objects then correspond bejectively to $\mathcal{G}$-objects. Now M. Artin has introduced **Proj**(Q) **for a quantum** n-**space** Q, cf. [Art], as the quotient category of the category of graded finitely generated Q-modules with respect to the localizing subcategory of graded modules of finite length. Let us write κ_+ for (the kernel functor determined by the graded Gabriel filter) $\mathcal{L}^g(\kappa_+)$ in Q, i.e., $\mathcal{L}^g(\kappa_+)$ is the filter of graded left ideals of Q containing a power of Q_+. The quotient category (Q, κ_+)-mod consists of the κ_+-localized modules obtained as $\varinjlim_n \text{Hom}_Q(Q^n_+, M/\kappa_+ M)$. (see

the begining of §2.). Hence $\text{Proj}(Q)$ is just the localized category but starting from the finitely generated instead of all graded Q-modules. To see this it suffices to note that having finite length for such a module comes down to being annihilated by some Q^n_+, or to having finite $\mathbb{C}$-dimension. Experience with the Brauer group learns that $\text{Proj}(Q)$ will not be nice in dimension larger than 3, indeed $\text{Br}(\text{Proj}(R))$ for a commutative positively graded affine domain R equals the categorical Brauer group $\text{Br}(R, \kappa_+)$ (cf. [CVo]), but even the relative (i.e., projective) Azumaya algebras appearing when the dimension is larger than 3 are very hard to describe. In fact, in M. Artin's definition $\text{Proj}(Q)$ is defined as above but together with a given " shift " functor (cf. Definition 1, 2, [ArT]); for a commutative ring one may recover the underlying scheme structure from J-P. Serre's global sections theorem. As pointed out above, the scheme structure of $\mathrm{Proj}(Q)$ is not easily understood. A perhaps desperate try to get a grasp on the underlying geometry starts from the consideration of particular modules, i.e., the point-and line-modules, cf. [ATV2], as well as the fat point modules introduced in [ArT]. On the other hand if one restrict attention to generalized gauge algebras deriving from so-called **innocent quantum spaces**, i.e., those possessing a central regular element t of degree one and being Noetherian, then the Proj of such a generalized gauge algebra or rather of its Rees ring may be covered by a finite number of " affine " parts corresponding to certain quantum sections. In this way the study of (fat) point-modules reduces to the study of finite dimensional representations of algebras.

3.1. Theorem. Let $\mathcal{G}$ be a Noetherian gauge algebra then $\mathcal{G}$ and $\widetilde{\mathcal{G}}$ are Noetherian regular algebras. If $G(\mathcal{G})$ is an n-dimensional quantum space then $\mathrm{gl.dim}\widetilde{\mathcal{G}} = 1+n$ and $\mathrm{gl.dim}\mathcal{G} \leq n$; moreover $\widetilde{\mathcal{G}}$ is an $n + 1$-dimensional quantum space.

Proof. See Theorem 5.2.5. of Ch.II., this can be applied because we have assumed the Noetherian condition here. The final statement is easily verified. □

Let us first provide a few concrete examples by calculating explicitly certain rings of quantum sections; the case of enveloping algebras will be treated in some detail in the consequent section. A delicate point made visible by these exercises is that we always try to avoid the completion intrinsically present in the definition of quantum sections by restricting attention to saturated Ore sets.

3.2. Example. Quantum sections for $A_1(\mathbb{C})$.

we have seen before that $\widetilde{A_1(\mathbb{C})} = \mathbb{C}[X,Y,Z]/(XZ - ZX, YZ - ZY, XY - YX - Z^2)$ where we have now changed notation so that Z is the regular central homogeneous element of degree one (before denoted by X) and we have put $X = xZ$, $Y = yZ$ and $G(A_1(\mathbb{C})) \cong \mathbb{C}[\overline{x},\overline{y}]$. Consider $S(\overline{x}) = \{1, \overline{x}, \overline{x}^2, \cdots\}$ in $\mathbb{C}[\overline{x},\overline{y}]$. The saturation of $S(\overline{x})$ in $A_1(\mathbb{C})$, denoted $S(x)$, consists of all $f \in A_1(\mathbb{C})$ such that $\sigma(f) = \overline{x}^n$ for some n. Note that all elements of the form $\lambda + x$, $\lambda \in \mathbb{C}$, are contained in $S(x)_{sat}$. The homogeneous Ore set in $\widetilde{A_1(\mathbb{C})}$ is $\widetilde{S(x)} = \{fZ^n, \ f = x^n + \sum_{k+l<n} \alpha_{kl} x^k y^l\}$ and we may write fZ^n as a homogeneous form of degree n in the new variables X, Y, Z. For example $x^2 + y + 1 \in A_1(\mathbb{C})$ corresponds to $X^2 + YZ + Z^2$. In $((\widetilde{S(x)})^{-1}\widetilde{A_1(\mathbb{C})})_0$ we find the elements $YX^{-1}, X^{-1}Y, ZX^{-1}, X^{-1}Z$ and it is easily verified that these satisfy the following relations: $YX^{-1} - X^{-1}Y = (ZX^{-1})^2$ and $ZX^{-1} - X^{-1}Z = 0$. Write $p = YX^{-1}$, $q = ZX^{-1}$. then

$$
\begin{aligned}
qp &= ZX^{-1}YX^{-1} \\
&= Z(YX^{-1} - Z^2X^{-2})X^{-1}
\end{aligned}
$$

$$= YZX^{-2} - Z^3X^{-3}$$
$$= XY^{-1}ZX^{-1} - Z^3X^{-3}$$
$$= pq - q^3$$

Hence we obtain a defining relation $[p,q] = q^3$. Note that every element of the form $X^n + ZF_{n-1}(X,Y,Z)$ where $F_{n-1}(X,Y,Z)$ is homogeneous of degree $n-1$, has to yield an invertible $1 + ZX^{-1}(X^{-1-n}F_{n-1}(X,Y,Z))$ in the completion of $((\widetilde{S(x)})^{-1}A_1\widetilde{(\mathbb{C}}))_0$ because $ZX^{-1} \in J^g(((\widetilde{S(x)})^{-1}A_1\widetilde{(\mathbb{C}}))_0)$ as $Z \in J^g(A_1\widetilde{(\mathbb{C}}))$. Therefore the algebra $\mathbb{C}\langle p,q \rangle/(pq - qp - q^3) = Q_x$ determines the quantum sections up to completion but the canonical filtration on this ring cannot be read off by looking at the new generators p and q as it is defined in terms of x and y. Nevertheless $\widehat{Q_x} = Q^\mu_{S(x)}(R)$ is enough to guarantee the desirable properties e.g, Auslander regularity, ..., as we will establish in the sequel.

3.3. Example. Consider the two-dimensional Lie algebra $g = \mathbb{C}\,x + \mathbb{C}\,y$ with $[x,y] = x$. Then $\widetilde{U(g)}$ is the quadratic algebra generated by X, Y, Z satisfying: $XZ - ZX = 0$, $YZ - ZY = 0$, $XY - YX = XZ$. Let $\sigma(S)$ be $\{1, y, y^2, \cdots\}$ in $\mathbb{C}[x,y] = G(U(g))$. The saturated Ore set in $U(g)$ is $\overline{S} = \{f = y^n + \sum_{k+l<n} \alpha_{kl}x^ky^l\}$ and this lifts to a homogeneous Ore set in $\widetilde{U(g)}$, $\tilde{S} = \{fZ^n, \quad f \in \overline{S}$ having $\sigma(f) = y^n\}$. Some relations in degree zero of $\tilde{S}^{-1}\widetilde{U(g)}$ derive from the defining relations above: $ZY^{-1} = Y^{-1}Z$, $Y^{-1}X - XY^{-1} = Y^{-1}XZY^{-1}$. Consider the canonical generators $p = XY^{-1}$ and $q = ZY^{-1}$, then we calculate:

$$qp = ZY^{-1}XY^{-1}$$
$$= Z(XY^{-1} + Y^{-1}XZY^{-1})Y^{-1}$$
$$= XZY^{-2} + ZY^{-1}XY^{-1}ZY^{-1}$$
$$= pq + qpq$$

yielding the rather odd relation $[p,q] = -qpq$. However, since $Y - Z \in \tilde{S}$ we must invert $1 - q$ in the quantum sections so we may rewrite the commutation relation as $qp = p\frac{q}{1-q}$ and so we may look at the skew polynomial ring $\mathbb{C}[[q]][p, \gamma]$ where γ is the automorphism defined by $p \mapsto \frac{q}{1-q}$ and see that $\mathbb{C}[[q]][p, \gamma]$ determines the quantum sections of $U(g)$.

3.4. Example. Let g be the Lie algebra sl_2 and change the sl_2-basis such that $[Y,Z] = X$, $[Z,X] = Y$ and $[X,Y] = Z$. Consider the multiplicative set $\sigma(S) = \{1, X, X^2, \cdots\}$. The reader may check that the commutation formulas determining the quantum sections of $U(sl_2)$ at $\sigma(S)$ may be given as:

$$[A, B] = (A^2 + B^2 + 1)C$$

$$[A, C] = AC\frac{C^2}{1+C^2} + B\frac{C^2}{1+C^2}$$
$$[B, C] = BC\frac{C^2}{1+C^2} - A\frac{C^2}{1+C^2}$$

where $A = YX^{-1}$, $B = ZX^{-1}$ and $C = TX^{-1}$.

Now we turn to the " schematic " considerations.

3.5. Example. We reconsider the case $R = A_1(\mathbb{C})$. First let us point out that it suffices to describe a sheaf on the basis of the topology given by the $\mathcal{Y}(f)$, f homogebeous of positive degree in $\mathbb{C}[x, y]$. When checking the proof of the fact that $\underline{Q}_{\mathcal{Y}}^{\mu}$ is a sheaf one sees that the real localizations at the corresponding Ore sets of R, that is the $\overline{S(f)}$, define a subsheaf $\underline{Q}_{\mathcal{Y}}^{\lambda} \subseteq \underline{Q}_{\mathcal{Y}}^{\mu}$ such that $\underline{Q}_{\mathcal{Y}}^{\mu}$ is the sheaf-wise completion (this is always calculated sectionwise, not stalkwise!) of $\underline{Q}_{\mathcal{Y}}^{\lambda}$ at the filtration $F\underline{Q}_{\mathcal{Y}}^{\lambda}$ sectionwise defined by the localized filtrations in $\overline{S(f)}^{-1}R$. We will describe the quantum sections by describing $\underline{Q}_{\mathcal{Y}}^{\lambda}$ i.e., up to a completion that does not interfere with the commutation relations. Now $\mathrm{Proj}(G(R)) = P^1$ in this case and we cover P^1 by $\mathcal{Y}(x) \cong \mathrm{Spec}\mathbb{C}[\frac{y}{x}]$ and $\mathcal{Y}(y) \cong \mathrm{Spec}\mathbb{C}[\frac{x}{y}]$ corresponding to $S(x)$ and $S(y)$ respectively. Of course $\mathcal{Y}(x) \cap \mathcal{Y}(y) \cong \mathrm{Spec}\mathbb{C}[\frac{x}{y}, \frac{y}{x}]$ corresponding to $S(xy)$. Over $S(x)$, $S(y)$ and $S(xy)$ we have to calculate the localizations at the corresponding Ore sets in R, this yields the following cases:

$S(x)$: We have seen before that we obtain a defining relation $[p, q] = q^3$ with $p = YX^{-1}$ and $q = ZX^{-1}$.

$S(y)$: Using the anti-symmetrity between x and y it is easy to obtain the defining relation $[p^{-1}, v] = -v^3$, where $p_{-1} = XY^{-1}$, $v = ZX^{-1}$. Note that $v = qp^{-1}$.

$S(xy)$: We can glue the above rings together their embedding in the ring generated by p, p^{-1}, q (or p, p^{-1}, v).

This may be schematically picturized as follows:

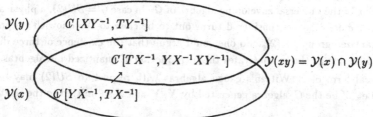

3.6. Example. We now let R be the enveloping algebra $U(g)$ where g is again the two-dimensional non-abelian Lie algebra considered before. Again $\mathrm{Proj}(G(R)) = P^1$ and we consider the same covering as in foregoing example.

$S(x)$: Since x is a normalizing element in $U(g)$ it suffices to localize classically at $\{1, x, x^2, \cdots\}$ in order to describe the quantum sections at $S(x)$ up to completion. The canonical generators are $p = YX^{-1}$ and $q = ZX^{-1}$ and one calculates: $pq = YX^{-1}ZX^{-1} = ZYX^{-2} = Z(X^{-1}Y + X^{-1}) = qp + q^2$, hence $[p, q] = q^2$, which defines the exceptional 2-dimensional quantum plane.

$S(y)$: It is easily seen that we obtain the quantum sections from $\mathbb{C}[[\omega]][p^{-1}, \gamma]$ where $\omega = ZX^{-1}$ and γ is the automorphism determined by $\omega \mapsto \frac{\omega}{1-\omega}$ (see Example 5).

$S(xy)$: Note that $\omega = qp^{-1}$. the schematic picture obtained is:

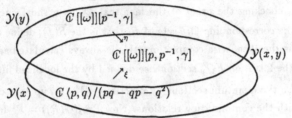

where ξ maps q to ωp and η is clear.

So in constructing the quantum " scheme " of $U(g)$ we glue a 2-dimensional quantum space with a very " localized " algebra.

3.7. Example. E. Witten's gauge algebras.

Quantum groups appear in Statistical Mechanics as the Hopf algebras associated to solutions of the Yang-Baxter equations. The latter equations in turn are connected to (representations of) the braid group of a suitable knot. These theories from Mathematical Physics and Knot Theory have provided classes of nice examples of Hopf algebras and other filtered rings that have raised considerable interest with the noncommutative algebraists. Another partial explaination for the existence of quantum groups has been given by E. Witten in [Wit]; Chern-Simons gauge theory with gauge group $\mathcal{G}$ can be phrased in terms of a Hopf algebra deformation of the universal enveloping algebra of $\mathcal{G}$. In case $\mathcal{G} = SU(2)$, a physically important case, the deformation considered turns out to be close to Woronowich's representation of the quantized group $SU(2)_q$. So one could argue that the existence of three dimensional Chern-Simons gauge theory motivates the introduction of quantized Lie algebras.

The general form of E. Witten's gauge algebras with respect to $SU(2)$ may be given as follows. Let W be the $\mathbb{C}$-algebra generated by X, Y and Z submitted to the relations listed below:

$$W \quad \begin{cases} XY + \alpha XY + \beta Y = 0 \\ YZ + \gamma ZY + \delta X^2 + \varepsilon X = 0 \\ ZX + \xi XZ + \eta Z = 0 \end{cases}$$

$$G(W) \quad \begin{cases} XY + \alpha XY = 0 \\ YZ + \gamma ZY + \delta X^2 = 0 \\ ZX + \xi XZ = 0 \end{cases}$$

where $G(W)$ is the associated graded ring associated to the (standard) filtration of W obtained by putting X, Y and Z (modulo the relations) in filtration degree one.

We see that $G(W)$ is a three-dimensional quantum space oftype S_1 in the classification of [ATV1], that means a line plus conic situation. in general three dimensional quantum spaces have a canonical normalizing element of degree 3; in this particular case this element is the product of two normalizing elements, one of degree 1 (being the X) and one of degree 2 that may be understood as a deformation of the Casimir operator, lending it a physical meaning. This situation is typical for all three-dimensional gauge algebras, that is to say that they have normalizing elements of degree one and two in the associated graded ring and therefore one may consider the microlocalizations at these elements because the fact that they are normalizing elements that they generate Ore sets.

Let us look at the particular case of $W_q(sl_2)$, the $\mathcal{C}$-algebra generated by X, Y, Z satisfying the relations:

$$\begin{cases} \sqrt{q} XZ - \sqrt{q}^{-1}ZX = \sqrt{q+q'}Z \\ \sqrt{q}^{-1}XY - \sqrt{q}YX = -\sqrt{q+q'}Y \\ YZ - ZY = (\sqrt{q} - \sqrt{q}^{-1})X^2 - \sqrt{q+q^{-1}}X \end{cases}$$

where classically $q = \exp(\frac{2\pi i}{k+2})$, k being the so-called coupling constant of Chern-Simons gauge theory. It has been observed by E. Witten that $W_q(sl_2)$ has finite dimensional representations that are deformations of the simple finite dimensional representations of $U(sl_2)$ and the tensor-products of these representations are associative. Moreover other finite dimensional representations appear " from infinity " when q distances itself from 1 and these are called bad representations by E. Witten. The central quadratic element resembling the Casimir operator is here obtained as $c = \sqrt{q}^{-1}ZY + \sqrt{q}YX + X^2$. If we put $A = 1 - c(\sqrt{q} - \sqrt{q}^{-1})(\sqrt{q+q^{-1}})^{-1}$ and use this to perform the following coordinate change:

$$\begin{aligned} x &= (X - (\sqrt{q} + \sqrt{q}^{-1})(\sqrt{q+q^{-1}})^{-1}c)(\sqrt{q+q^{-1}})^{-1}A^{-1} \\ y &= Y(\sqrt{q+q^{-1}})^{-1}\sqrt{A}^{-1} \\ z &= Z(\sqrt{q+q^{-1}})^{-1}\sqrt{A}^{-1} \end{aligned}$$

The defining equations of $W_q(sl_2)$ can be rewritten as

$$\begin{cases} \sqrt{q}\,xz - \sqrt{q}^{-1}zx = z \\ \sqrt{q}^{-1}xy - \sqrt{q}yx = -y \\ q^{-1}zy - qyz = x \end{cases}$$

where we now have to realize that we are working in the localization obtained by inverting the element A. The $\mathbb{C}$-algebra spanned by x, y, z satisfying the above relations, $U_q^\omega(sl_2)$ say, is a Hopf algebra having only the good finite dimensional representations and $U_q^\omega(sl_2)$ localized at $\sqrt{A}^{-1}$ equals the original $W_q(sl_2)$ localized at $\sqrt{A}$. We may homogenize $W_q(sl_2)$ and obtain $\widetilde{W_q}(sl_2)$ determined by the homogenized relations:

$$\begin{cases} \sqrt{q}\,xz - \sqrt{q}^{-1}zx = zt \\ \sqrt{q}^{-1}xy - \sqrt{q}\,yx = -yt \\ q^{-1}zy - qyz = xt \end{cases}$$

and apply the theory explained in Ch.III. in order to obtain that $\widetilde{W_q}(sl_2)$ is a Noetherian Auslander regular domain of global dimension 4; a result of lifting maximal orders also yields that $\widetilde{W_q}(sl_2)$ is a maximal order and it satisfies the Cohen-Macaulay property. It is possible to study $\mathrm{Proj}(\widetilde{W_q}(sl_2))$ in some detail and clarify properties of its point- and line-modules and reduce the study of fat points to representation theory. We do not go into this here, but it is clear that the use of localizations at special homogeneous elements prompts the schematic approach to the study of Proj.

3.8. Example. Weyl algebra again (see 3.5. too).

Reconsider the first Weyl algebra $A_1(\mathbb{C})$. We use the calculation made above to extend the schematic picture in Example 3.5. by glueing to the open set corresponding to $\mathcal{Y}(Z)$, i.e., the affine piece corresponding to $A_1(\mathbb{C})$. We obtain the following diagram of glueing data:

$$\mathbb{C}\langle XZ^{-1}, YZ^{-1}, ZX^{-1}\rangle \hookleftarrow A_1(\mathbb{C}) = \mathbb{C}\langle XZ^{-1}, YZ^{-1}\rangle \hookrightarrow \mathbb{C}\langle XZ^{-1}, YZ^{-1}, ZY^{-1}\rangle$$
$$[XZ^{-1}, YZ^{-1}] = 1$$

$$\mathbb{C}\langle YX^{-1}, ZX^{-1}\rangle \qquad\qquad \mathbb{C}\langle XY^{-1}, ZY^{-1}\rangle$$
$$[YX^{-1}, ZX^{-1}] = (ZX^{-1})^3 \qquad\qquad [XY^{-1}, ZY^{-1}] = (ZY^{-1})^3$$

$$\mathbb{C}\langle ZX^{-1}, YX^{-1}, XY^{-1}\rangle$$

To find its point modules we have to study the one-dimensional representations and their glueing data

$$\phi \quad \hookrightarrow \quad \phi \quad \hookleftarrow \quad \phi$$

$$V((ZX^{-1})^3) \qquad\qquad V((ZY^{-1})^3)$$

$$V((ZX^{-1})^3) \cap V((ZY^{-1}])^3)$$

corresponding to the fact that the associated degree 3 divisor of the quantum 3-space $A_1(\widetilde{\mathbb{C}})$ is Z^3. This may be pictured in the usual P^2

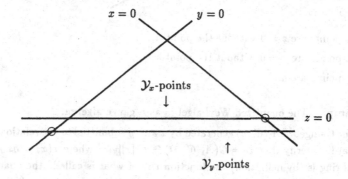

where the circle means that the intersection point is missing and we have drawn a double copy of the same $z = 0$ locus.

3.9. Example. Reconsider the situation of Example 3.6.. Using the computations and notations of that Example 8, we obtain the following diagram of glueing data:

$$U(g)$$

$$\mathbb{C}\langle XZ^{-1}, ZX^{-1}, YZ^{-1}\rangle \hookleftarrow \mathbb{C}\langle XZ^{-1}, YZ^{-1}\rangle \hookrightarrow \mathbb{C}[[ZY^{-1}, YZ^{-1}]][XY^{-1}, \gamma]$$

$$[XZ^{-1}, YZ^{-1}] = XZ^{-1}$$

$$\mathbb{C}\langle YX^{-1}, ZX^{-1}\rangle \qquad\qquad \mathbb{C}[[ZY^{-1}]][XY^{-1}, \gamma]$$

$$[YX^{-1}, ZX^{-1}] = (ZX^{-1})^2$$

$$\mathbb{C}[[ZY^{-1}]][XY^{-1}, YX^{-1}, \gamma]$$

Only the top left corner (glueing the enveloping algebra to the exceptional quantum space) resembles the commutative case. The other two corners have shrunk in dimension. It may be helpful in understanding these phenomena to look at the picture of point-modules in

$\mathrm{Proj}(\widetilde{U(g)})$ pictured in the usual P^2:

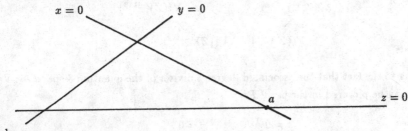

where:

  the $\mathcal{Y}_x$-points are $z = 0$ without the point a,

  the $\mathcal{Y}_z$-points are $x = 0$ without the point a,

  the $\mathcal{Y}_y$-points are a.

3.10. Example. The quantized Weyl algebra as a gauge algebra.

Consider the $\mathbb{C}$-algebra $A_1(\mathbb{C}, q)$ generated by x and y submitted to the relation $xy - qyx = 1$. It is not hard to verify that $B = G(A_1(\mathbb{C}, q)) \cong \mathbb{C}[x][y, \tau]$ where $\tau(x) = qx$ and this skew polynomial ring is (by definition) the function ringof what is called " the quantum plane ". K. Goodeal, cf. [good], called $A_1(\mathbb{C}, q)$ the quantized algebra (what is in a name). In B the set $S(x) = \{1, x, x^2, \cdots\}$ is an Ore set yielding a homogeneous Ore set $\tilde{S} = \{\tilde{f} = x^n + \sum_{k+l<n} \alpha_{kl} x^k y^l T^{n-k-l}\}$ in the Rees ring $A_1(\widetilde{\mathbb{C}}, q)$, where we now use T for the canonical central regular element of degree one. This Rees ring is the $\mathbb{C}$-algebra generated by the elements X, Y, T, submitted to the following relations:

$$ W \quad \begin{cases} [X, T] = [Y, T] = 0 \\ XY - qYX = T^2 \end{cases} $$

The quantum sections of $A_1(\mathbb{C}, q)$ at $S(x)$ is the ring obtained by completing the $\mathbb{C}$-algebra generated by $A = YX^{-1}$, $B = TX^{-1}$ submitted the relation: $AB - qBA = B^3$. If we perform a further coordinate change given by $A \mapsto A' = (q - 1)A + B^2$ we obtain the latter algebra as the quantum plane $\mathbb{C}_q[A', B]$! For notational convenience let us write $R = A_1(\mathbb{C}, q)$. Both X and Y are normalizing elements in the quantum plane $\mathbb{C}_q[X, Y]$, so we may define a schematic structure on $\mathrm{Proj}(\tilde{R})$ by giving the following " glueing " rules, where we just write the diagram of ring morphism corresponding to what should be the restriction morphisms on the open sets covering Proj as a space

$$\mathbb{C}\langle XT^{-1}, TX^{-1}, YT^{-1}\rangle \hookleftarrow \mathbb{C}\langle XT^{-1}, YT^{-1}\rangle \hookrightarrow \mathbb{C}\langle XT^{-1}, YT^{-1}, TY^{-1}\rangle$$
$$XT^{-1}YT^{-1} - qYT^{-1}XT^{-1} = 1$$

$$\mathbb{C}\langle YX^{-1}, TX^{-1}\rangle \qquad\qquad \mathbb{C}\langle XY^{-1}, TY^{-1}\rangle$$
$$YX^{-1}TX^{-1} - qTX^{-1}YX^{-1} = (TX^{-1})^3 \qquad XY^{-1}TY^{-1} - \tfrac{1}{q}TY^{-1}XY^{-1} = -\tfrac{1}{q}(TY^{-1})^3$$

$$\mathbb{C}\langle YX^{-1}, XY^{-1}, TX^{-1}\rangle$$

and again one can visualize the point modules as points in P^2. The picture corresponds to the fact that the associated degree 3 divisor is $T(T^2 + (q-1)XY)$:

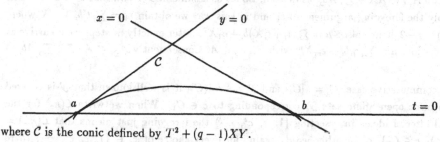

where $\mathcal{C}$ is the conic defined by $T^2 + (q-1)XY$.

$Y(T)$ points are $\mathcal{C} - \{a, b\}$,

$Y(X)$ points are $\mathcal{C} - \{b\} \cup (T = 0) - \{b\}$,

$Y(Y)$ points are $\mathcal{C} - \{a\} \cup (T = 0) - \{b\}$.

The scheme structure of $A_1(\mathbb{C}, q)\tilde{}$ described above is critical in defining the scheme structure on $\operatorname{Proj}(\widetilde{W})$ where $\widetilde{W}$ is the 4-dimensional quantum space of the Rees ring of the Witten gauge algebras. For, it is possible to change the polarization on $\operatorname{Proj}(G(W))$ as in [Art] to obtain $\operatorname{Proj}(A_1(\mathbb{C}, q)\tilde{})$ and use the foregoing in order to define a scheme structure on $\operatorname{Proj}(G(W))$ that is then lifted via quantum sections to $\operatorname{Proj}(\widetilde{W})$ in the way described in the sequel (this technique has been claimed in [LeVo] but it turned out the proof was lacking so we include that here). In a similar way one can study the fat-point modules of multiplicity n (as in [Art]) in an innocent projective quantum space by glueing together the n-dimensional representation of the scheme components.

Now we return to establish the affine covering principle for a large class of Rees rings, including iterated Ore extension, gauge algebras iterated from the quantum-plane, enveloping algebras of Lie algebras etc... .

3.11. Proposition. Let A be a positively graded ring such that $A_0 = k$ is a field and

$A = k\langle A_1 \rangle$; assume that A has a central homogeneous element of degree 1, X say. Denote $-$: $A \to A/AX = \overline{A}$ for the canonical epimorphism and write $A_1 = kX \oplus \sum_{i \in I} ka_i$, I some index set. If there is an $n \in I\!N$ for which we have $\overline{A}_{\geq n} = \oplus_{p \geq n} \overline{A}_p \subset \sum_j \overline{Ab_j}$, for some homogeneous $\overline{b}_j \in \overline{A}$, then for every $m \in I\!N$, there exists a $k \in I\!N$ such that: $A_{\geq k} \subset AX^m + \sum_j Ab_j$, where b_j is a homogeneous representative in A for $\overline{b}_j$ in $\overline{A}$.

Proof. Put $s = n + m$, for the given m, and consider anay homogeneous monomial in the a_i, μ say, of degree $l \geq s$. Since $l \geq n$ we have $\overline{\mu} \in \sum_j \overline{Ab_j}$ and thus $\mu = \sum_i t_i b_i + c_1 X$ wheredeg$(t_i) = l - $ deg(b_i) and c_1 of degree $l - 1$. If $m = 1$ then the proof is finished because we obtain $A_{\geq s} \subset AX + \sum_j Ab_j$, as desired. So let us assume $m > 1$ and thus $l - 1 > n$. Then we apply the foregoing argument to c_1 and in this way we obtain $c_1 = \sum_i t_i' b_i + c_2 X$ where deg$(c_2) = l - 2$, hence also: $\mu = \sum_i (t_i + t_i' X) b_i + c_2 X^2$. After exactly m steps we do arrive at an expression: $\mu = \sum_i \gamma_i b_i + c_m X^m$ with c_m, $\gamma_i \in A$. Consequently $A_{\geq s} \subset AX^m + \sum_j Ab_j$. $\square$

In the commutative case, $C = k[C_1]$ and $\mathcal{Y} = \text{Proj}(C)$, it is well known that $\mathcal{Y}$ is covered by the basic open affine sets $\mathcal{Y}_c$ corresponding to $c \in C_1$. When we write $\mathcal{L}(\kappa_c)$ for the Gabriel filter of ideals intersecting $\{1, c, c^2, \cdots\}$ the foregoing just means that $\mathcal{L}(\kappa_+) = \cap\{\mathcal{L}(\kappa_c), c \in C_1\}$, or in other words again: for every choice of $x_c \in \{1, c, c^2, \cdots\}$ there is an $n \in I\!N$ such that $(C_+)^n \subset \sum_{c \in C_1} Cx_c$. This just expresses that the global sections of Proj (as a scheme) are determined by the sections over the $\mathcal{Y}_c$ for all $c \in C_1$. Of courese if C is Noetherian, C_1 is finite dimensional over k, and there is a finite covering $\mathcal{Y}_{c_i}$, c_1, $\cdots$, c_n being a k-basis of C_1.

In the noncommutative case $A = k\langle A_1 \rangle$ we have to obtain enough Ore setes generated by certain $f \in A_1$ such that we have:

$$(*) \qquad\qquad \mathcal{L}(\kappa_+) = \cap\{\mathcal{L}(\kappa_f), \text{ for certain } f \in A_1\}$$

where $\mathcal{L}(\kappa_+)$ is the Gabriel filter generated by the A_+^n and each $\mathcal{L}(\kappa_f)$ is the Gabriel filter consisting of left ideals of A having a nontrivial intersection with the Ore set $\langle f \rangle$. Whenever a set of such $\langle f \rangle$ exists such that $(*)$ holds, we say that A is **schematic**. If moreover finitely many such $\langle f \rangle$ exist such that $(*)$ holds for them then we say that the **finite affine covering property** (finite-AC) holds.

The schematic property phrased in a categorical way just states that we have a pull-back situation of quotient categories and corresponding (reflectors) localization functors:

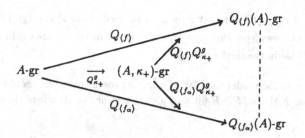

3.12. Theorem. Let R be a positively filtered ring with $k = F_0 R$ being a field. If $G(R)$ is schematic then $\tilde{R}$ is schematic and if $G(R)$ has finite-AC then $\tilde{R}$ has finite-AC. In case FR is Zariskian the finite-AC case reduces to the schematic case.

Proof. We have a regular homogeneous $X \in \tilde{R}_1$ such that $\tilde{R}/X\tilde{R} = G(R)$, so the first claim follows from the proposition. In case FR is Zariskian, both $G(R)$ and $\tilde{R}$ are Noetherian and each $F_i R$ is a finite dimensional k-vectorspace. $\qquad\square$

3.13. Example. The quantum-plane, or in fact any skew polynomial ring, has finite-AC.

Proof. Any skew polynomial ring $k[x_1, \cdots, x_n, \varphi_1, \cdots, \varphi_n] = \Lambda$ with $\varphi_i \in \mathrm{Aut} k$, has Ore sets $\langle x_i \rangle$, $i = 1, \cdots, n$, where $\langle x_i \rangle$ is just $\{1, x_i, x_i^2, \cdots\}$ here because these are normalizing elements. It is clear that any $L = \sum_{i=1}^n \Lambda x_i^{(n_i)}$ contains $(\Lambda_+)^m$ whenever $m \geq \max\{n(i), i = 1, \cdots n\}$. $\qquad\square$

3.14. Corollary. Any filtered iterated Ore extension and also every gauge algebra iterated from the quantum plane has finite-AC.

Proof. A filtered Ore extension has associated graded ring a skew polynomial ring. The rest follows by iteration and Example 3.13.. $\qquad\square$

For commutative rings $C = k[C_1]$ that are domains, we know that $Q_{\kappa_+}(C)$ is the normal closure of C, so in the normal case (that is $C = \overline{C}$ is integrally closed) we have $Q_{\kappa_+}(C) = C$.

3.15. Corollary. Let R be as in Theorem 3.12. such that $G(R)$ is a schematic domain, FR is Zariskian and $G(R)$ is a maximal order, then

$$R = \cap\{Q_{S(f)}(Q), \quad f \text{ varying such that } \mathcal{Y}_{\bar{f}} \text{ is a finite cover for } \mathcal{Y}\}.$$

Proof. It will suffices to show that $A = \tilde{R} = \cap\{Q_{S(\tilde{f})}(A), \tilde{f} \text{ as above }\}$ and therefore we only have establish that $A = Q_{\kappa_+}(A)$ or that A is $\mathcal{L}(\kappa_+)$-closed. Now $\tilde{A}$ is a (gr-) maximal order and so $A = Q_{\kappa_+}(A)$ follows from this (one may also use a variant of the explicite proof given for enveloping algebras in the next section).   □

3.17. Remark. We leave to the reader the task of phrasing some consequences of foregoing results for good filtrations FM on $M \in R$-filt and global sections of structure sheaves over Proj.

§4. Quantum Sections of Enveloping Algebras

We consider a finite dimenisonal Lie algebra g over a field k; in this section shall write $k = \mathbb{C}$ but that is not essential. We write $U(g)$ for the enveloping algebra of g and $\widetilde{U(g)}$ for its Rees ring with respect to the usual filtration. If $g = \mathbb{C}\, x_1 \oplus \cdots \oplus \mathbb{C}\, x_n$, $[x_i, x_j] = \sum_{k=1}^n \gamma_{ij,k} x_k$, then $\widetilde{U(g)}$ is generated by X_0, X_1, $\cdots$, X_n over $\mathbb{C}$ with relations $[X_i, X_j] = \sum_{k=1}^n \beta_{ij,k} X_k X_0$. The results of Ch. III. imply that $\widetilde{U(g)}$ is a quadratic Auslander regular algebra of global dimension $n + 1$ and it satisfies the Cohen-Macaulay property too. Following M. Artin [Art] we define the quantum space $\mathcal{G}_q(g)$ of g to be $\mathrm{Proj}(\widetilde{U(g)})$. In [LeVb] the point-modules of $\widetilde{U(g)}$, i.e., the so-called point-variety of $\mathcal{G}_q(g)$, have been characterized, here we are able to rederive this result from the schematic approach.

For $x \in g$ let $\mathrm{ad}x\colon g \to g$ be the adjoint representation $[x, -]\colon g \to g$. Let E be the set of all eigenvalues of the adjoint representation of x and let $\mathbb{Z}E$ be the additive subgroup of $\mathbb{C}$ generated by E. Put $f = X + \gamma X_0 \in \widetilde{U(g)}_1$ then the minimal Ore set in $\widetilde{U(g)}$ generated by $X \in \widetilde{U(g)}_1$ is $S_f = \langle\{X + (\gamma - e)X_0, \ e \in \mathbb{Z}E\}\rangle$. For $f \bmod X_0 = X \bmod X_0$ we write x again (it should not be ambiguous to view x in g or in $G(U(g)) = \mathbb{C}\,[x_1, \cdots, x_n]$). The quantum sections may be obtained from $(S_f^{-1}\widetilde{U(g)})_0$ up to completion. In this section we do not have to study the completion because the latter localized ring is easily described, hence we call $(S_f^{-1}\widetilde{U(g)})_0$ the ring of **algebraic quantum sections** over $S(x)$ and study this in its own right. we put $\Gamma(f) = \Gamma(f, \mathcal{O}_q(g)) = (S_f^{-1}\widetilde{U(g)})_0$.

Since S_f contains elements of degree 1, its localization $S_f^{-1}\widetilde{U(g)}$ must be strongly graded and hence by the well known category equivalence of strongly graded rings we obtain: $S_f^{-1}\widetilde{U(g)}$-gr $\approx \Gamma(f, \mathcal{O}_q(g))$-mod. Hence it follows that $\Gamma(f)$ is Auslander regular. Since $G(U(g))$ is a polynomial ring over $\mathbb{C}$ it is clear that $\widetilde{U(g)}$ is schematic and it has the finite-AC property. Let us now verify directly that $U(g)$ may be obtained as an intersection of (in fact: exactly two suffice) certain localizations of Ore sets generated by degree one elements. We let σ be the principal symbol map (see Ch. I.) but note that for $x_i \in g$ we now write $\sigma(x_i) = x_i$ in $\mathbb{C}\,[x_1, \cdots, x_n]$.

4.1. Proposition. If $x_i \neq x_j$ in g then $U(g) = S(x_i)^{-1}U(g) \cap S(x_j)^{-1}U(g) = T$.

Proof. Suppose $z = s_i^{-1}a_i = s_j^{-1}a_j$ with $a_i, a_j \in U(g)$ and $s_i \in S(x_i)$, $s_j \in S(x_j)$, hence $\sigma(s_i) = x_i^{n_i}$, $\sigma(s_j) = x_j^{n_j}$ by definition of the minimal Ore sets as above. Since $G(U(g))$ is a domain, σ is multiplicative, hence we arrive at $\sigma(z) = x_i^{-n_i}\sigma(a_i) = x_j^{-n_j}\sigma(a_j)$, or $x_j^{n_j}\sigma(a_i) =$

$x_i^{n_i}\sigma(a_j) \in (x_i^{n_i})$. Since $j \neq i$ we must have $\sigma(a_i) \in (x_i^{n_i})$ and thus $\deg\sigma(a_i) \geq n_i$. Therefore $\deg\sigma(z) \geq 0$ holds for all $z \in S(x_i)^{-1}U(g) \cap S(x_j)^{-1}U(g)$. Now write $\sigma(a_i) = x_i^{n_i}\sigma(b_1)$ for some $b_1 \in U(g)$. Suppose $z \notin U(g)$. Then $\sigma(z) = \sigma(b_1)$ and $z - b_1 \in T$. But $\deg\sigma(z - b_1) <$ $\deg\sigma(z)$ and $\deg\sigma(z - b_1) \geq 0$ because $z - b_1 \in T$, hence we may start the argument again with $z - b_1$, etc..., until after a finite number of steps we have $z - b_1 - \cdots - b_r \in T$ and $\deg\sigma(z - b_1 - \cdots - b_r) < 0$ but that is impossible hence $z \in U(g)$. $\qquad\square$

In order to explicitly write down the equations defining $\Gamma(f)$, where $f = X + \gamma X_0$ and $X \in \widetilde{U(g)}_1$ corresponds to $x \in g$, we choose a basis $\{y_1, \cdots, y_n = x\}$ of g such that the matrix of $\mathrm{ad}x$ will be in Jordan canonical form. More precisely, there exists a subset I of $\{1, \cdots, n\}$ and a function $\alpha\colon I \to E$ such that $\mathrm{ad}x(y_i) = \alpha(i)y_i$ when $i \in I$, but $\mathrm{ad}x(y_i) = \alpha(j)y_i + y_{i+1}$ if $i \notin I$, where $j = \min(I \cap \{i+1, \cdots, n\})$. It is easier to extend α to the whole $\{1, \cdots, n\}$ by putting $\alpha(i) = \alpha(j)$ for $j = \min(\{i, \cdots, n\} \cap I)$. Let Y_i be the element in $\widetilde{U(g)}_1$ corresponding to y_i and put $Z_i = Y_i(X + \gamma X_0)^{-1}$, $i = 1, \cdots, n$, and $T = X_0(X + \gamma X_0)^{-1} = (X + \gamma X_0)^{-1}X_0$. Thus we have $Z_n = 1 - \gamma T$. With this notation we obtain:

4.2. Theorem. $\Gamma(f)$ is the algebra generated by $Z_1, \cdots, Z_{n-1}$ over $\mathbb{C}[T]_{\langle\langle 1+\beta T, \beta\in \mathbf{z}E\rangle\rangle}$ with the following defining relations:

a) for $i, j \in I - \{n\}$:
$$Z_iZ_j = Z_jZ_i(1 + \alpha(i)T)(1 + \alpha(j)T)^{-1} + \sum_{k=1}^n \beta_{ij,k}Z_kT(1 + \alpha(j)T)^{-1}.$$

b) for $i \in I - \{n\}$, $j \in \{1, \cdots, n\} - I$ we have:
$$Z_iZ_j = Z_jZ_i(1 + \alpha(i)T)(1 + \alpha(j)T)^{-1} + \sum_{k=1}^n \beta_{ij,k}Z_kT(1 + \alpha(j)T)^{-1}$$
$$-Z_iZ_{j+1}T(1 + \alpha(j)T)^{-1}$$

c) for $i, j \in \{1, \cdots, n\} - I$ we have:
$$Z_iZ_j = Z_jZ_i(1 + \alpha(i)T)(1 + \alpha(j)T)^{-1} + \sum_{k=1}^n \beta_{ij,k}Z_kT(1 + \alpha(j)T)^{-1}$$
$$-Z_iZ_{j+1}T(1 + \alpha(j)T)^{-1} + Z_jZ_{i+1}T(1 + \alpha(j)T)^{-1}$$

d) for $i \in I - \{n\}$:
$$TZ_i = Z_iT(1 + \alpha(i)T)^{-1}$$

e) for all $i \in \{1, \cdots, n\} - I$, we have:
$$TZ_i = Z_iT(1 + \alpha(i)T)^{-1} - TZ_{i+1}T(1 + \alpha(i)T)^{-1}.$$

Proof. The relations a) may be written in $U(g)$ as

$$(x - \alpha(i))y_i = y_ix, \quad i \in I$$
$$(x - \alpha(i))y_i = y_ix + y_{i+1}, \quad i \notin I.$$

In $\widetilde{U(g)}$, these relations become:

$$Y_i(X + \gamma X_0) = (X + \gamma X_0)Y_i - \alpha(i)Y_iX_0, \quad i \in I$$
$$Y_i(X + \gamma X_0) = (X + \gamma X_0)Y_i - \alpha(i)Y_iX_0 - Y_{i+1}X_0, \quad i \notin I.$$

By multiplying these on both sides by $(X + \gamma X_0)^{-1}$, we get:

$$(X + \gamma X_0)^{-1}Y_i = (Y_i - \alpha(i)(X + \gamma X_0)^{-1}Y_iX_0)(X + \gamma X_0)^{-1} \quad i \in I$$
$$(X + \gamma X_0)^{-1}Y_i = (Y_i - \alpha(i)(X + \gamma X_0)^{-1}Y_iX_0)(X + \gamma X_0)^{-1}$$
$$-(X + \gamma X_0)^{-1}Y_{i+1}X_0(X + \gamma X_0)^{-1} \quad i \notin I.$$

Consequently, for every $i, j \in I - \{n\}$:

$$\begin{aligned}
[Z_i, Z_j] &= Y_i(X + \gamma X_0)^{-1}Y_j(X + \gamma X_0)^{-1} - Y_j(X + \gamma X_0)^{-1}Y_i(X + \gamma X_0)^{-1} \\
&= Y_i(Y_j - \alpha(j)(X + \gamma X_0)^{-1}Y_jX_0)(X + \gamma X_0)^{-2} \\
&\quad Y_j(Y_i - \alpha(i)(X + \gamma X_0)^{-1}Y_iX_0)(X + \gamma X_0)^{-2} \\
&= \sum_{k=1}^{n} \beta_{ij,k}Z_kT + \alpha(i)Z_jZ_iT - \alpha(j)Z_iZ_jT.
\end{aligned}$$

Since for every $k \in \{1, \cdots, n\}$,

$$1 + \alpha(k)X_0(X + \gamma X_0)^{-1} = (X + \gamma X_0 + \alpha(k)X_0)(X + \gamma X_0)^{-1}$$

is invertible in $(S_f^{-1}\widetilde{U(g)})_0$, we may rewrite foregoing relation as follows: for all $i, j \in I - \{n\}$ we have:

$$Z_iZ_j = Z_jZ_i(1 + \alpha(i)T)(1 + \alpha(j)T)^{-1} + \sum_{k=1}^{n} \beta_{ij,k}Z_kT(1 + \alpha(j)T)^{-1}.$$

The other equations may be derived in a similar way.

Note that, in case x is a semisimple element of g the only relations appearing are those of type a) and d). All monomials of degree 3 vanish if x is a semi-invariant, i.e., if $[x, y] = \lambda(y)x$ for all $y \in g$, λ being a functional in g^*. □

4.3. Question. Is $\Gamma(f)$ a Hopf algebra in a natural way?

4.4. Example. $g = \mathbb{C}\,x \oplus \mathbb{C}\,y$, $[x, y] = x$.
On the complement of the " line " X_0 we put as sections:

$$(\widetilde{U(g)}_{\{1, X_0, X_0^2, \cdots\}})_0 \cong (U(g)[X_0, X_0^{-1}])_0 = U(g)$$

With respect to the ordered basis $\{y, x\}$ the matrix of $\text{ad}x$ is $\begin{pmatrix} 0 & 1 \\ 0 & 0 \end{pmatrix}$, so it is in Jordan form. We have $I = \{z\}$, $\alpha = 0$ and $S_x = \{x^n, n \in I\!N\}$. Thus $\Gamma(X)$ is generated by Z, and T satisfying $TZ_1 = Z_1T - T^2$.

In order to invert the element y we have to localize at $\{Y - nX^0, n \in Z\!\!\!Z\}$. In the ordered basis $\{x, y\}$ the matrix of $\text{ad}y$ has Jordan form $\begin{pmatrix} -1 & 0 \\ 0 & 0 \end{pmatrix}$, so on the complement of the lines $Y - nX_0$ we put the algebra $\Gamma(Y)$, generated by Z_1 over $\mathbb{C}[T]_{\langle\langle 1+nT \rangle\rangle}$ with relation given by $TZ_1 = Z_1T(1 - T)^{-1}$.

4.5. Example. Put $g = sl_2(\mathbb{C}) = \mathbb{C}\,e \oplus \mathbb{C}\,f \oplus \mathbb{C}\,h$ with $[e, f] = h$, $[h, e] = 2e$, $[h, f] = -2f$. For $X = X_0$ we obtain $\Gamma(X_0) = U(g)$.

For $X = e$, put $y_1 = -\frac{1}{2}f$, $y_2 = -\frac{1}{2}h$, $y_3 = e$, then $\text{ad}x$ has Jordan form

$$\begin{pmatrix} 0 & 1 & 0 \\ 0 & 0 & 1 \\ 0 & 0 & 0 \end{pmatrix}$$

Using the relation c) and e) established in the theorem, we obtain:

$$\begin{aligned} Z_1Z_2 &= Z_2Z_1 - 2Z_1T + Z_2^2T \\ TZ_1 &= Z_1T - Z_2T^2 + T^3 \\ TZ_2 &= Z_2T - T^2 \end{aligned}$$

This can therefore be written as an iterated Ore extension

$$\Gamma(E) = \mathbb{C}[T][Z_2, \delta][Z_1, \sigma, \rho]$$

For $X = f$ we find the same as for $X = e$.

For $X = h$, take $\{e, f, h\}$ for the ordered basis.

We see that $I = \{1, 2, 3\}$, $\alpha(1) = 2$, $\alpha(2) = -2$, $\alpha(3) = 0$ and so we arrive at the following relations:

$$\begin{aligned} Z_1Z_2 &= Z_2Z_1(1 + 2T)(1 - 2T)^{-1} + T(1 - 2T)^{-1} \\ TZ_1 &= Z_1T(1 + 2T)^{-1} \\ TZ_2 &= Z_2T(1 - 2T)^{-1} \end{aligned}$$

So the obtained ring is again an iterated Ore extension:

$$\Gamma(H) = \mathbb{C}[T]_{\langle\langle 1+2nT, n \in \mathbb{Z} \rangle\rangle}[Z_1, \sigma]$$

where σ is self-evident (as are σ, ρ in the case $X = e$).

The variety of one-dimensional representations of $\Gamma(H)$ after homogenization and correct embedding in $P^4 = P(t, e, f, h)$ is $V(t(2te, -2tf, h^2 + 4ef))$. Similarly we find that , the corresponding variety for both $\Gamma(E)$ and $\Gamma(F)$ is $V(t(t, h^2 + 4ef))$ and for $\Gamma(T)$ we get $V(e, f, h)$. Glued together schematically this defines $V(t((h^2 + 4ef)(e, f, h), t(e, f, h)))$ which is exactly the point-variety of $\widetilde{U(g)}$ as obtained in [LeS].

This final section summarizes some recent work of L. Le Bruyn, F. Van Oystaeyen, L. Willaert. Similar problems for rings of differential operators remain largely unstudied but this is the topic of work in progress.

REFERENCES

[Art] M. Artin, *Geometry of quantum planes*, in Azumaya Algebras, Actions and Groups (Eds. D. Haile and J. Osterberg), Contemp. Math. Vol. 124, Amer. Math. Soc., Providence, 1992.

[AS] M. Artin, W. Schelter, *Graded algebras of global dimension 3*, Adv. Math., 66(1987), 171–216.

[ATV1] M. Artin, J. Tate, M. Van den Bergh, *Some algebras associated to automorphisms of curves*, in The Grothendieck Festschrift (Eds. P. Cartier, et al), Birkhauser, Basel, 1990.

[ATV2] M. Artin, J. Tate, Van den Bergh, Modules over regular algebras of dimension 3, *Inventiones Math., 106(1991), 335–389.*

[Aus] M. Auslander, *On the dimension of modules and algebras III*, Nagoya Math. J. 9, 1955, 67–77.

[AVo] M. Awami, F. Van Oystaeyen, *On filtered rings with Noetherian associated graded rings*, in Proceedings of Ring Theory Meeting, Granada 1986, Springer Verlag, LNM.

[AVVo] M.J. Asensio, M. Van den Bergh, F. Van Oystaeyen, *A new algebraic approach to microlocalization of filtered rings*, Trans. Amer. Math. Soc. Vol. 316, No. 2, 1989, 537–555.

[Ba1] H. Bass, *Algebraic K-theory*, W. Benjamin, 1968, New York.

[Ba2] H. Bass, *On the ubiquity of Gorenstein rings*, Math. Z. 82, 1963, 8-28.

[B-B] W. Borho, J-L. Brylinski, *Differential operators on homogeneous spaces III*, Preprint IHES, 1984.

[Be1] I.N. Bernstein, *On the dimension of modules and algebras III, Direct limits*, Nagoya math. J. 13, 1958, 83-84.

[Be2] I.N. Bernstein, *Algebraic theory of D-modules*, Preprint, 1982.

[Bj1] J-E. Björk, *Rings of differential operators*, Math. Library 21, North Holland, Amsterdam, 1979.

[Bj2] J-E. Björk, *Filtered Noetherian rings*, Math. Surveys and monographs, No. 24, 1987, 59–97.

[Bj3] J-E. Björk, *Unpublished notes*, 1987.

[Bj4] J-E. Björk, *The Auslander condition on Noetherian rings*, Proc. Sem. Malliavin, LNM, 1404, Springer Verlag, 1988, 137–173.

[Bj5] J-E. Björk, *Notes distributed at the Autumn school at Hamberg University*, Oct. 1-7, 1984.

[Bo] M. Boratinsky, *A change of rings theorem and the Artin Rees property*, Proc. Amer, Math. Soc. 53, 1975, 307–320.

[Bor] A. Borel, *Algebraic D-modules*, Academic press, London-New York, 1987.

[B-V] W. Bruns, U. Vetter, *Determinantal rings*, LNM, 1327, Springer-Verlag, 1988.

[C-E] H. Cartan, S. Eilenberg, *Homological dimension*, Princeton Univ. press, Princeton, N.J., 1956.

[C-G] A.W. Chatters, S.M. Ginn, *Localization in hereditary rings*, J. Alg. 22, 82–88, 1972.

[Ch] M. Chamarie, *Sur les ordres maximaux au sens d'Asano*, Vorlesungen aus dem Fachbereich Mathematik, Universität Essen, Heft 3, 1979.

[Coh] P.M. Cohn, *On the embedding of rings in skew fields*, Proc. London Math. Soc. (3) 11, 1961, 511–530.

[CVo] S. Caenepeel, F. Van Oystaeyen, *Brauer groups and the cohomology of graded rings*, Marcel Dekker, INC. 1989.

[Dix] J. Dixmier, *Enveloping algebras*, Math. Library, North Holland, Vol. 14, Amsterdam, 1977.

[EF] E. Formanek, *Noetherian P.I. rings*, Comm. Alg. 1(1), 1974, 79–86.

[E-H] E.K. Ekström, Ho Dinh Duan, *Purity and equidimensionality of modules for a commutative Noetherian ring with finite dimension*, Reports Stockholms No. 18, 1986.

[FGR] R.M. Fossum, .A. Griffith, I. Reiten, *Trivial extensions of abelian categories, homological algebra of trivial extensions of abelian categories with applications to ring theory*, LNM. 456, Springer-Verlag, 1975.

[Ga] O. Gabber, *The integrability of characteristic varieties*, Amer. J. Math. 103, 1981, 445–468.

[Gab] P. Gabriel, *Des catégories abéliennes*, Bull. Soc. Math. France 90, 1962, 323–448.

[Gi] V. Ginsburg, *Characteristic varieties and vanishing cycles*, Invent. Math. 84, 1986, 327-402.

[Go] A. Goldie, *The structure of Noetherian rings*, LNM. 246, Springer-Verlag, 1972, 213–321.

[Gol] J. Golan, *Localization of noncommutative rings*, M. Dekker, New York, 1975.

G-R] P. Gabriel, R. Rentschler, *Sur la dimension des anneaux et ensembles ordonnées*, C. R. Acad. Sci. Paris, 265(1967), 712–715.

[Ka1] M. Kashiwara, *Algebraic study of systems of partial differential equations*, Master's thesis, University of Tokyo, 1971.

[Ka2] M. Kashiwara, *b-function and holonomic systems*, Invent. Math. 38, 1976, 33–53.

[Kap] I. Kaplansky, *Fields and rings*, The University of Chicago Press, Chicago and London.

[K-L] G.R. Krause, T.H. Lenagan, *Growth of algebras and Gelfand-Kirillov dimension*, Research Notes in Mathematics, 116, Pitman Advanced Publishing Program, 1985.

[Krau] G.R. Krause, *On the Krull dimension of left Noetherian left Matlis rings*, Math. Z. 118, 1970, 207–214.

[K-T] M. Kashiwara, T. Tanisaki, *The characteristic cycles of holonomic systems on a flag manifold*, Invent. Math. 77, 1984, 185–198.

[LB]   L. Le Bruyn, *Gauge algebras*, UIA Preprint, Antwerp 1991.

[LeS] L. Le Bruyn, P. Smith, *Homogenized sl*(2), UIA Preprint, Antwerp 1991.

[Lev] T. Levasseur, *Thesis*, Paris VI, 1983.

[LeVb] L. Le Bruyn, F. Van den Bergh, *On quantum spaces of Lie algebras*, UIA Preprint, Antwerp 1991.

[LeVo] L. Le Bruyn, F. Van Oystaeyen, *Quantum sections and gauge algebras*, UIA Preprint, Antwerp 1991.

[LeVV] L. Le Bruyn, M. Van den Bergh, F. Van Oystaeyen, *Graded orders*, Birkhauser Verlag, 1988.

[Li1] Huishi Li, *Note on pure module theory over Zariskian filtered ring and the generalized Roos theorem*, Comm. Alg. Vol. 19, 843–862, 1991.

[Li2] Huishi Li, *Note on microlocalization of filtered rings and the embedding of rings in skew fields*, Bull. Math. Soc. Belgique, (serie A), Vol. XLIII, 1991, 49–57.

[Li3] Huishi Li, *Noetherian gr-regular rings are regular*, To appear.

[Li4] Huishi Li, *Rees rings of grading filtration and an application to Weyl algebras*, To appear.

[Li5] Huishi Li, *On the stability of graded rings*, To appear.

[LVo1] Huishi Li, F. Van Oystaeyen, *Strongly filtered rings, applied to Gabber's integrability theorem and modules with regular singularities*, Proc. Sem. Malliavin, LNM. 1404, Springer-Verlag, 1988.

[LVo2] Huishi Li, F. Van Oystaeyen, *Filtrations on simple Artinian rings*, J. Alg. Vol. 132, 1990, 361–376.

[LVo3] Huishi Li, F. Van Oystaeyen, *Zariskian filtrations*, Comm. Alg. Vol. 17, 1989, 2945–2970.

[LVO4] Huishi Li, F. Van Oystaeyen, *Global dimension and Auslander regularity of graded rings*, Bull. Math. Soc. Belgique, (serie A), Vol. XLIII, 1991, 59–87.

[LVo5] Huishi Li, F. Van Oystaeyen, *Dehomogenization of gradings to Zariskian filtrations and applications to invertible ideals*, Proc. Amer. Math Soc. vol. 115, No. 1, 1992, 1–11.

[LVo6] Huishi Li, F. Van Oystaeyen, *Sign gradation on group ring extensions of graded rings*, To appear in J. of pure and applied algebra.

[LVoW] Huishi Li, F. Van Oystaeyen, E. Wexler-Kreindler, *Zariski rings and flatness of completion*, J. Alg. Vol. 138, 1991, 327–339.

[LVV1] Huishi Li, M. Van den Bergh, F. Van Oystaeyen, *Note on the K_0 of rings with Zariskian filtrations*, K-Theory, Vol. 3, 1990, 603-606.

[LVV2] Huishi Li, M. Van den Bergh, F. Van Oystaeyen, *The global dimension and regularity of Rees rings for non-Zariskian filtration*, Comm. Alg. Vol. 18, 1990, 3195–3208.

[Ma] G. Maury, *Nouveaux exemples d'ordres maximaux*, Comm. Alg. Vol. 14, 1986, 1515–1517.

[Man] Yu I Manin, *Quantum groups and noncommutative geometry*, Publ. centre Prech. Math. Montreal, 1988.

[Mats] H. Matsumura, *Commutative algebra*, W.A. Benjamin INC. New York, 1970.

[Mc] J.C. McConnell, *The K-theory of filtered rings and skew Laurent extensions*, LNM. 1146, Springer-Verlag, 1987.

[MNVo] H. Marubayashi, E. Nauwelaerts, F. Van Oystaeyen, *Graded rings over arithmetical orders*, Comm. Alg. 12(6), 1984, 745–775.

[M-R] J.C. McConnell, J.C. Robson, *Noncommutative Noetherian rings*, J. Wiley, London, 1987.

[Naga] M. Nagata, *On the structure of complete local rings*, Nagoya math. J. 1(1950), 63–70.

[NNVo] C. Năstăsescu, E. Nauwelaerts, F. Van Oystaeyen, *Arithmetically graded rings revisited*, Comm. Alg. 14(10), 1986, 1991–2017.

[No1] D.G. Northcott, *An introduction to homological algebra*, Cambridge Univ. Press, London and New York, 1960.

[No2] D.G. Northcott, *Lessons on rings, modules and multiplicities*, Cambridge Univ. Press, London and New York, 1968.

[NVo1] C. Năstăsescu, F. Van Oystaeyen, *Graded ring theory*, Math. Library, 28, North Holland, Amsterdam, 1982.

[NVo2] C. Năstăsescu, F. Van Oystaeyen, *The dimension of ring theory*, D. Reidel Publ. Co., 1987.

[P] C. Procesi, *Rings with polynomial identities*, Marcel Dekker INC., New York, 1977.

[Ps1] P.F. Smith, *The Artin-Rees property*, LNM. in Proc. Sem. Malliavin, Springer-Verlag.

[Ps2] P.F. Smith, *New examples of maximal orders*, Comm. Alg. 2(2), 1989.

[Roo1] J-E. Roos, *Compléments ál étude des quotients primitifs des algébres enveloppantes des algébres de Lie sémi-simple*, C. R. Acad. Sci. Paris. Sér. A-B.

[Roo2] J-E. Roos, *Bidualité et structure des foncteurs dérivés un anneau régulier*, C. R. Acad. Sci. Paris, 254, 1962.

[Rot] J.J. Rotman, *An introduction to homological algebra*, New York, 1979.

[Row1] L. Rowen, *Ring theory Vol. 2*, Academic Press, 1988.

[Row2] L. Rowen, *On rings with central polynomials*, J. Alg. Vol. 31, 1974, 393–426.

[RVo1] A. Radwan, F. Van Oystaeyen, *Coherent sheaves over microstructure sheaves*, UIA Preprint, Antwerp 1991.

[RVo2] A. Radwan, F. Van Oystaeyen, *Microstructure sheaves and quantum sections over formal schemes*, UIA Preprint, Antwerp 1991.

[S] O. Schilling, *The theory of valuations*, AMS. Math. Suur. IV, 1950.

[Sch] P. Schapira, *Microdifferential systems in the complex domain*, Springer Verlag, Berlin, 1985.

[Ser1] J.P. Serre, *Algébre locale, multiplicités*, LNM. 11, Springer-Verlag, 1975.

[Ser2] J.P. Serre, *Faisceaux algébriques Cohérents*, Ann. Math. 61, 1955, 197-278.

[Sj] G. Sjödin, *On filtered modules and their associated graded modules*, Math. Scand. 33, 1973, 229–240.

[Spr] T. Springer, *Microlocalization algébrique*, Sem. Algebra Dubreil-Malliavin, LNM, Springer Verlag, Berlin, 1984.

[Sta] J.T. Stafford, *Non-holonomic modules over Weyl algebras and enveloping algebras*, Invent. Math. 79, 1984, 619–638.

[SVo] R. Sallam, F. Van Oystaeyen, *A microstructure sheaf and quantum section over a projective scheme*, To appear in J. Alg..

[VDB1] M. Van den Bergh, *On a theorem of Cohen-Montgomery*, Proc. Amer. Math. Soc. 94, 1985, 562–564.

[VDB2] M. Van den Bergh, *A note on graded K-theory*, Comm. Alg. 14(8), 1986, 1561–1564.

[Ve1] A. Van den Essen, *Algebraic microlocalization*, Comm. Alg. 1985.

[Ve2] A. Van den Essen, *Modules with regular singularities over filtered rings*, P.R.I. for Mathematical Science, Kyoto Univ., Vol. 22, No. 5, 1986.

[VG1] J. Van Geel, *Places and valuations in noncommutative ring theory*, Lecture notes, Vol. 71, Marcel Dekker, New York, 1981.

[VG2] J. Van Geel, *Primes and value functions*, Ph.D. Thesis, Antwerp Univ., 1980.

[Vo1] F. Van Oystaeyen, *Prime spectra in noncommutative algebra*, LNM. 444, Springer-Verlag, 1978.

[V02] F. Van Oystaeyen, *Graded P.I. rings and generalized crossed product Azumaya algebras*, Chin. Ann. Math. 8B, 1, 1987, 13–21.

[Vo3] F. Van Oystaeyen, *A note on graded P.I. rings*, Bull. Soc. Math. Belgique, 32(1980), 22–28.

[VVo] M. Van den Bergh, F. Van Oystaeyen, *Lifting maximal orders*, Comm. Alg. 2(2), 1989.

[Wu] Quanshui Wu, $gr(\mathcal{D}_n)$ and $gr(\mathcal{E}_p)$ are not Noetherian rings with pure dimension, To appear.

[Wu1] Quanshui Wu, *On the Formula $d(M) + j(M) = 2n$ over the rings $\mathcal{D}_n(\widehat{calD}_n)$ and $\mathcal{E}_p$*, To appear in China Science, 1993.

[Z-S] O. Zariski, P. Samuel, *Commutative algebra*, Vol. I and II, D. Van Nostrand, 1958, 1960, New Printing, Springer, New York.

SUBJECT INDEX

K-Monographs in Mathematics

1. A.A. Ranicki (ed.), A.J. Casson, D.P. Sullivan, M.A. Armstrong, C.P. Rourke and G.E. Cooke: *The Hauptvermutung Book*. A Collection of Papers on the Topology of Manifolds. 1996   ISBN 0-7923-4174-0
2. L. Huishi and F. van Oystaeyen: *Zariskian Filtrations*. 1996
    ISBN 0-7923-4184-8

KLUWER ACADEMIC PUBLISHERS – DORDRECHT / BOSTON / LONDON